OXFORD SERIES ON ADVANCED MANUFACTURING

SERIES EDITORS

J. R. CROOKALL

MILTON C. SHAW

OXFORD SERIES ON ADVANCED MANUFACTURING

---

1. William T. Harris: *Chemical Milling: The Technology of Cutting Materials by Etching* (1976)
2. Bernard Crossland: *Explosive Welding of Metals and its Applications* (1982)
3. Milton C. Shaw: *Metal Cutting Principles* (1984)
4. Shiro Kobayashi, Soo-Ik Oh, and Taylan Altan: *Metal Forming and the Finite Element Method* (1989)
5. Norio Taniguchi: *Energy Beam Processing of Materials* (1989)
6. Nam P. Suh: *The Principles of Design* (1990)
7. N. Logothetis and H. P. Wynn: *Quality through Design* (1990)
8. John L. Burbidge: *Production Flow Analysis* (1990)
9. J. Francis Reintjes: *Numerical Control: Making a New Technology* (1991)
10. John Benbow and John Bridgwater: *Paste Flow and Extrusion* (1993)
11. Andrew J. Yule and John J. Dunkley: *Atomization of Melts for Powder Production and Spray Deposition* (1994)

QUALITY THROUGH DESIGN

Experimental Design, Off-line Quality Control and Taguchi's Contributions

N. Logothetis
TQM Hellas Ltd, Athens

and

H. P. Wynn
City University, London

CLARENDON PRESS · OXFORD
1994

Oxford University Press, Walton Street, Oxford OX2 6DP

Oxford New York
Athens Auckland Bangkok Bombay
Calcutta Cape Town Dar es Salaam Delhi
Florence Hong Kong Istanbul Karachi
Kuala Lumpur Madras Madrid Melbourne
Mexico City Nairobi Paris Singapore
Taipei Tokyo Toronto

and associated companies in
Berlin Ibadan

Oxford is a trade mark of Oxford University Press

Published in the United States
by Oxford University Press Inc., New York

First published, 1989
First published (with corrections) in paperback, 1994

A catalogue record for this book is available from the British Library

Library of Congress Cataloging in Publication Data

Logothetis, N.
Quality through design : experimental design, off-line quality control, and Taguchi's contributions / N. Logothetis and H. P. Wynn.
p. cm. — (Oxford series on advanced manufacturing; 7)
Bibliography: p. Includes index.
1. Taguchi methods (Quality control) I. Wynn, H. P. (Henry P.)
II. Title. III. Series.
TS156.L64 1989 658.5'62—dc20
89–32048 CIP

ISBN 0 19 851993 1 (Hbk)
ISBN 0 19 859395 3 (Pbk)

Printed in Great Britain on acid-free paper by
The Universities Press (Belfast) Ltd.

To Elena and Jan

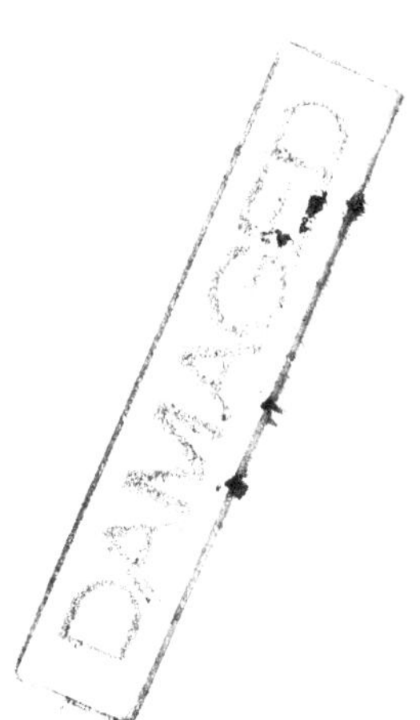

PREFACE

This book is written for engineers but we realize that it will also be read by statisticians. We hope that it will be read, too, by managers, students, quality improvement specialists, and other professional people who have heard about 'Taguchi methodologies' and want to find out what they are all about. 'Taguchi methodologies' have been criticized for being statistically unsound and obscure in places. We share some of these reservations. They have also been described as reinventing the wheel; for example, orthogonal arrays. Such critiscism should not detract from the enormous sociological, philosophical, and economic contribution they are making. Sociological, because of the breaking of barriers between engineers and statisticians and between design engineering and production engineering. Philosophical, because of their concentrated effort to extend experimentation into noisy environments. Economic, because of their dramatic and proven application to quality improvement.

Of course, there is a danger in claiming them as a panacea for industry eager for quick solutions to the problems of international competition. They must be used in careful conjunction with advances in engineering design and product development.

Our approach has been twofold. First, we hope to have laid bare the methods, 'warts and all', with enough real examples to give a feel for applications. Second, we have drawn on methods available in statistics which, together with the special Taguchi methodology and philosophy, should form the backbone of a post-Taguchi methodology in off-line quality improvement. Foremost amongst these are the response surface methods of G. E. P. Box and the optimal design methods of J. Kiefer and, of course, their co-workers and followers.

To write a book like this in the middle of a revolutionary period is fraught with the risk of attack from competing paradigms, problem-hungry theoreticians and data-hungry statisticians. Our only salvation is to plead for mercy from our friends in statistics while hoping to make friends in engineering.

Unless otherwise indicated, every practical example or case study included in this book is the result of an actual case study that was recently undertaken within the General Electric Company (GEC), UK. The authors are deeply grateful to the members of all the different GEC sites who allowed the experiments to take place and to be completed. They would also like to acknowledge the American Supplier Institute (ASI) Inc., Dearborn, Michigan, USA, for granting permission to reproduce the orthogonal arrays and triangular matrices from Taguchi methods: Orthogonal arrays and linear graphs; tools for quality engineering, by G. Taguchi and S. Konishi (1987).

The first author would especially like to thank: Mr Derek H. Roberts, deputy Managing Director of GEC, for his continuous moral and active support for the ideas of total Quality Control and for his help in the establishment of 'Quality by Design' techniques within GEC; Dr Cyril Hilsum, Director of Research of GEC, for making the publication of this book possible; Mr F. M. Clayton, Chief Mathematician of GEC–Hirst Research Centre (HRC), for numerous contributions during the assessment period of the off-line Quality Control techniques; and Mr J. P. Salmon, assistant statistician of HRC, for his efficient analyses of the experimental data.

The second author would especially like to thank Dr Alan Jebb of City University who co-directs City University Quality Unit (CUQU) and who provided much intellectual and moral support in engineering design. Professors Jerry Sacks, Don Ylvisaker, Toby Mitchell, and Will Welch and many others were inspirational during visits to workshops at the University of Illinois and earlier in Berkeley and UCLA. Dr Jeff Hooper of AT&T Bell Laboratories supplied a vision of a post-Taguchi world probably as accurate as anyone's.

Hirst Research Centre—GEC N. L.

City University H. P. W.
1988

CONTENTS

1
INTRODUCTION

In this introduction, we try to identify some main themes underlying the new quality improvement revolution. We shall examine how Taguchi's philosophy fits in both with the management principles of Dr W. E. Deming and with some prevailing philosophies and methodologies in Western industry. This is to encourage proper implementation of the techniques that have contributed greatly to Japan's industrial success.

1.1 Japanese quality improvement

As part of the United Kingdom National Quality Campaign, a group of people from management, banking, purchasing, the trade unions, and the media went to Hong Kong, Japan, and the United States in the summer of 1984. The purpose of this *Pacific Basin Study Mission* was to assess the importance of quality management to the business success of the three economies. All of the representatives became alarmed at the strength of the competitive prices challenge faced by UK enterprises in world markets and were convinced of the need for urgent and radical changes of attitudes. Upon their return the participants contributed to a booklet entitled · · · *you won't do it.*

The words of the title were spoken by a leading Japanese economist. When asked why Japanese companies had allowed access to their factories and talked so openly about their methods, he replied, 'It would take you ten years to get where we are now and by that time we shall be even further ahead. Besides, *we know you won't do it*'.

1.1.1 Costs

W. A. Shewhart, who can be considered the founder of modern quality control, introduced the idea that 'the better the quality the lower the cost', in his 1931 book. This idea, which to some seems at first contradictory, is the foundation of the Japanese approach. Here is a quotation from Hagime Karatsu given in the booklet:

> As inferior products are eliminated through innovation in the manufacturing process, materials and labour as well as energy can be saved while producing

> the same volume of products. In addition, when a smaller volume of inferior goods is produced, machines less frequently have to be stopped for adjustment, and materials less often have to be replaced in order to produce satisfactory products; this reduces the operation rate, and so can lead to higher productivity.

As the manufacturing system itself improves in quality, the less is the need for mass inspection to achieve product quality. Products, then, should not have to be withdrawn from the market or returned by the customer for replacement or repair. This reduces warranty costs, increases the company's reputation and customer goodwill, and leads to increased sales. There is, then, no conflict between quality and price. Japan has succeeded in producing high-quality goods at competitive prices.

1.1.2 The 'Deming effect'

For thirty years, Dr W. E. Deming worked to improve quality in Japan. His quality programme is claimed to work because of the constant involvement of senior management in the use of statistical techniques. According to W. E. Conway, Chairman and President of Nashua Corporation:

> Dr Deming is the founder of the third wave of the Industrial Revolution. The Japanese manufacturers utilizing the statistical control of quality (as taught by Deming), are sweeping the world in the second half of the twentieth century, just as American manufacturers utilizing mass production swept the world in the first half . . . . The Japanese principle is based upon the statistical control of quality, introduced to them by an American!

In Japan, statistical methods and charts have for years helped managers to measure, evaluate, and *continually improve* operations. Dr Deming points out that approximately 85% of the problems in any operation can be directly traced to the system, 'to common causes that affect the whole group of workers, and that can only be fixed by management.' Statistics can help to teach managers to focus on processes, rather than blaming individuals.

Until fairly recently, managers in most companies in the West did not see the connection between quality and a competitive position. Their attitude of going for short-term profits and blaming business failures on fiscal policy and worker attitudes was deeply ingrained. Professor Kaoru Ishikawa, the father of Total Quality Control in Japan, described how the Quality Control Research Group of JUSE (Union of Japanese Scientists and Engineers) interpreted statistical quality control so that every worker could understand and use it. The success of quality in Japan he says, 'is due not to government-led activity, but to dedicated people who have come voluntarily from public, private and academic bodies to promote quality.' Changing attitudes needs co-operation at all levels. There is a 'Deming prize' awarded annually to Japanese companies for improved use of statistics in organization, consumer research, product design, and production.

Following the success of his methods in Japan, Deming has encapsulated his philosophy in fourteen cardinal 'points for management' (see Appendix A, Deming 1982, Mann 1985, and Scherkenbach 1986). His methods are in contrast with the more mechanistic management-by-objective approaches fostered, for example, in some business schools. He stresses bringing personal and corporate goals into closer harmony, and using quality as a foundation for this. Management and labour unite in a common aim of improving the system, rather than relying on personalized performance-led assessment. Statistics becomes a common language which can be used at all levels in the organization and provide the information to anticipate, identify, and correct mistakes. Above all, the purpose is to reduce wasteful *variability* in the system by 'doing it right the first time'.

1.2 From Deming to Taguchi

The technical issue which concerns us most in this book is neatly summarized in Deming's point No. 3.

> Cease dependence on inspection to achieve quality. Eliminate the need for mass inspection by building quality into the product in the first place.

The stages of a product's life can be summarized as follows:

$$\text{design} \rightarrow \text{production} \rightarrow \text{use}.$$

If one of Deming's main achievements has been to shift quality improvement backwards from inspection to statistical process control, Taguchi's has been to make the further step back from production to design. The design stage is the off-line stage. Using the methods described in this book, a design can be made robust against variability 'downstream' in both production and the use environment. This can avoid re-design during production, inspection, and, at the extreme, recall of a product after distribution. The idea is the extension 'upstream' of 'the better the quality the lower the cost'.

Professor Genichi Taguchi, director of the Japanese Academy of Quality and four times recipient of the Deming award, has built a philosophical framework for carrying out off-line quality improvement. It is this philosophy which is his lasting contribution. It is worth listing at this early stage the main components.

(a) *Social philosophy.* Taguchi defines quality in broad terms which include an idea of the loss imparted to society as a whole from poor quality. This is closely related to his notions of quality-loss, which we shall address in detail in the next section.

(b) *Experimental design.* The use of experimental design in off-line quality improvements is critical. Taguchi castigates textbooks on experimental design in the West as being too passive and academic. He advocates the wide use of orthogonal arrays (Chapter 3, Appendix D), and demands a more active and creative approach. His point is reinforced by the almost complete lack of application of experimental design to engineering design in the West until the last few years. His comments are somewhat unfair when applied to production where response surface methods have been used for years. But, in general, the active rather than passive use of the scientific method, and statistics in particular, is something which we wholeheartedly support.

(c) *Minimization of variability.* Concentration on the minimization of variability distinguishes Taguchi's methods from traditional tolerance-based quality control. Thus we find achievement of target with minimum variability as a main criterion. This is considered to be a better way of yielding the highest proportion of products within the desired specification than strict adherence to sometimes artificial tolerance bands.

(d) *Classification of factors.* Factors are divided into those over which the designer has some control (design or control factors), and those over which he has none (noise factors). Amongst the former one can, hopefully, isolate those whose levels can be changed to decrease variability. (Variability here is, precisely, variability over the noise factors.) The basic method, then, uses factors, sometimes called signal factors, which have no effect on variability to correct the output onto the target. In summary:

> minimize the variability in the response and then achieve the target for the response.

It should be said that carrying out the procedure in two stages is not strictly necessary although it has considerable engineering appeal.

(e) *Inner versus outer noise.* The classification in (d) uses noise as outer noise, the downstream variation caused by manufacture and use. Taguchi is also keen to distinguish this from material wear, tolerances on bought-in components, and so on. This is an important distinction. Classical reliability, for example, can be considered as the study of internal noise. The distinction can blur when material variability interacts with manufacture, but it is none the less useful.

A more detailed assessment of the above methods will be given in Chapter 6 but a few comments are needed here. We have said that the philosophy is original and has had a profound effect in practice. However, the statistical methods fall short of the kind of rigour which is used in mainstream statistics and some areas of applied statistics, such as medical statistics. This is a pity, and has led to demarcation disputes in some companies in the United States and consternation at a few open meetings. Our main aim in this book is to run together the philosophy and standard statistical methods available in the West. If the marriage remains a little unhappy, then that reflects both our own limitations and the fluid state of the art. Short sections at the end of each chapter may help to anchor the ship.

Taguchi's ideas about 'quality loss' and also his attitude about 'On-line Quality Control' are described in detail in Sections 1.3 and 1.4. We include them in this first chapter because we believe that 'quality-loss-function analyses' can be very important for setting up objectives.

1.3 Taguchi's 'quality-loss'

1.3.1 Definition of quality

'Quality is the loss a product causes to society after being shipped, other than any losses caused by its intrinsic function.'

The above can very well be taken as a definition of 'non-quality' rather than of 'quality'. But the essence of these words lies in the fact that Taguchi is opposed to treating quality questions as 'value questions'. He believes that the desirability of a product is determined by the societal loss it generates from the time it is shipped to the customer; the smaller the loss, the higher the desirability. For example, losses due to poor and varied performance of a product, failure to meet the customer's requirements of fitness for use or for prompt delivery, and harmful side-effects caused by the product, are all societal losses. However, the loss that a product may inflict on society through its intrinsic function is not a quality loss. For example, despite the many losses that people sustain from fights or accidents while under the influence of alcohol, it would be nonsense to manufacture non-intoxicating liquor. The resulting product would not have the intrinsic function of liquor. The question of what functions society should allow products to have is a cultural and legal problem, not an engineering one. On the other hand, the harmful hangover caused by the contents of the liquor, the side-effects of its preservatives, or the spoil and variability caused by the absence of preservatives would be quality questions.

So, in the context of the definition of quality, 'loss' should be restricted to loss caused by variability of function and loss caused by harmful side-effects.

1.3.2 Loss-function

Variations from desired functional specifications cause loss of quality. There is a direct loss due to warranty and increased service costs and to dissatisfied customers. There is also an indirect loss due to market share loss and to increased marketing efforts to overcome uncompetitiveness.

An appropriate quality improvement programme should have as a main objective the minimization of the variation of product performances about their target values. The smaller the performance variation the better the quality. The larger the deviation from the target, the larger the customer's loss. This loss can be approximately evaluated by Taguchi's 'loss-function' which unites the financial loss with the function specification through a quadratic relationship obtained as shown below.

1.3.2.1 When the target value is finite

If Y is the value of the response for a particular performance characteristic for which there is a finite target value of m, then the loss $L(Y)$ due to the deviation of Y from m, can be expanded by a Taylor series around m as:

$$L(Y) = L(m + Y - m) = L(m) + \frac{L'(m)}{1!}(Y - m) + \frac{L''(m)}{2!}(Y - m)^2 + \cdots.$$

Since by assumption $L(m) = 0$, and since $L(Y)$ is minimal at $Y = m$, $L'(m)$ is zero. So the loss can be approximated by only the third term in the above expansion as

$$L(Y) \approx k(Y - m)^2.$$

The 'loss-coefficient' k can be determined on the basis of information about the losses (in monetary terms) caused by exceeding the customer's tolerances.

If the customer's tolerance for a particular product with target value m is

$$(m + D),$$

and the cost to the customer when this tolerance is exceeded is M, then

$$k = \frac{M}{D^2}, \tag{1.1}$$

and so

$$L(Y) = \frac{M}{D^2}(Y - m)^2. \tag{1.2}$$

If the tolerance interval is defined as

$$[m - D_1, m + D_2],$$

with cost M_2 if $Y < m - D_1$ and M_2 if $Y > m + D_2$, then $k_i = M_i/D_i^2$, $i = 1$, 2, and

$$L(Y) = \begin{cases} \dfrac{M_1}{D_1^2}(Y - m)^2 & \text{if} \quad Y < m. \\ \dfrac{M_2}{D_2^2}(Y - m)^2 & \text{if} \quad Y > m. \end{cases} \tag{1.3}$$

Note that the loss-coefficients could be determined because the loss $L(Y)$ was known for a particular value of Y; for example, it was known that $M = kD^2$, from which (1.1) followed.

Example: Shoe sizes are provided at 1 cm intervals. If a tight shoe is less bearable than a loose one, then the tolerances can be set asymmetrically as

$$D_1 = 0.4 \quad \text{and} \quad D_2 = 0.6.$$

If these tolerances are exceeded, the shoe must be tailor-made and the losses include the additional tailoring or adjustment, the delay, transport, and other social costs. Assuming that M_1 and M_2 are both £50, then

$$\text{if } Y < m, \quad k_1 = \frac{M_1}{D_1^2} = \frac{50}{(0.4)^2} = 312.5,$$

and

$$\text{if } Y > m, \quad k_2 = \frac{M_2}{D_2^2} = \frac{50}{(0.6)^2} = 138.9.$$

A person with 37.3 size foot (i.e. $m = 37.3$) must choose a shoe of either 37.0 or 38.0 cm. Using (1.3) we have

$$L(Y) = \begin{cases} 312.5 \times (37.0 - 37.3)^2 = £28.1. \\ 138.9 \times (38.0 - 37.3)^2 = £68.1. \end{cases}$$

This person will accordingly prefer the slightly tighter 37 cm size.

1.3.2.2 When the target value is infinite

If Y is the value of the response for which the target value is infinite (the 'larger-the-better' characteristic, $0 \leq Y \leq \infty$), then the loss $L(Y)$ in the neighbourhood of $Y=\infty$ has the following Laurent expansion:

$$L(Y)=L(\infty)+\frac{L'(\infty)}{1!}\times\frac{1}{Y}+\frac{L''(\infty)}{2!}\times\frac{1}{Y^2}+\cdots.$$

Since by assumption $L(\infty)=0$, and since $L(Y)$ is minimal near $Y=\infty$, it is reasonable to stipulate that $L'(\infty)=0$. It follows that $L(Y)$ can be approximated by

$$L(Y)\approx K_{\mathrm{I}}\times\frac{1}{Y^2}.$$

As in the 'finite-target' case, the loss coefficient can be estimated if the loss is known at a particular point $Y=Y_0$, in which case

$$L(Y_0)=K_{\mathrm{I}}\times\frac{1}{Y_0^2}\rightarrow K_{\mathrm{I}}=Y_0^2\times L(Y_0). \tag{1.1'}$$

Indeed, if the customer's lower value for a characteristic is D_{I} and the loss at that point and lower is M_{I} then

$$K=D_{\mathrm{I}}^2\times M_{\mathrm{I}},$$

and so

$$L(Y)=\frac{D_{\mathrm{I}}^2\times M_{\mathrm{I}}}{Y^2}. \tag{1.2'}$$

1.3.3 Producer's tolerance

The loss to the producer, which occurs when a purchased product is found to be faulty at the installation site, includes costs of transporting the faulty product to the factory, and back to the retail store, when it is fixed, plus the rework costs. So, the producer's tolerance cannot be the same as the consumer's tolerance.

Suppose $m\pm D$ is the consumer's tolerance (finite target) and M is the societal loss when this tolerance is exceeded, then, if M^0 is the producer's loss when the product is rejected (at the installation site), from (1.2)

$$M^0=\frac{M}{D^2}(Y-m)^2. \tag{1.4}$$

Then, the producer's tolerance $m\pm D^0$ can be found from (1.4) by solving for $D^0=|Y-m|$ which yields

$$D^0=D\times\sqrt{M^0/M}. \tag{1.5}$$

If there is an infinite target, the producer's tolerance D_{I}^0 can similarly be obtained from (1.2′) as $D_{\mathrm{I}}^0=D_{\mathrm{I}}\sqrt{M_{\mathrm{I}}/M_{\mathrm{I}}^0}$ where M_{I}^0 is the producer's loss when the product is rejected. So (1.5) provides the means of setting the tolerances in product specifications assuming that the average cost M^0 of reworking a rejected article into a satisfactory article is known under an adequate process capability. However, when there is an inadequate process capability, sometimes articles not satisfying the tolerance D^0 have to be scrapped, and when a unit is rejected a replacement might also be rejected. In general, if p is the proportion of products expected not to meet the specifications at the production site (and having to be scrapped), then the producer's costs increase to $M^0/(1-P)$ and the producer's tolerance is then given by

$$D^0 = D \times \sqrt{\frac{M^0/(1-P)}{M}}. \tag{1.6}$$

Of course, the proportion of rejected products, and therefore the proportion of accepted ones $q = 1 - p$, depends on the producer's tolerance D^0; so, if

$$Q(D^0) = q = 1 - p,$$

from (1.6), D^0 is the solution of the following (non-linear) equation

$$D^0 = D \times \sqrt{\frac{M^0}{M \times Q(D^0)}}. \tag{1.7}$$

Since D^0 is unknown, $Q(D^0)$ is also unknown. However, if the distribution of the characteristic value Y is known, the tolerance D^0 can be obtained by successive approximations, starting with (1.5) to obtain a first approximation for D^0 [i.e. by first assuming that $Q(D^0) = 1$] as the following example shows.

Example: Suppose that the characteristic value Y is normally distributed with mean m at the target value and standard deviation σ. Then using the formula for the normal distribution function (see Appendix F), $Q(D^0)$ is given by

$$Q(D^0) = \int_{m-D^0}^{m+D^0} \frac{1}{\sigma\sqrt{2\pi}} \exp\left\{-\frac{1}{2}\left(\frac{Y-m}{\sigma}\right)^2\right\} \mathrm{d}Y,$$

or

$$Q(D^0) = \int_{-D^0/\sigma}^{+D^0/\sigma} \frac{1}{\sqrt{2\pi}} \exp(-t^2/2)\,\mathrm{d}t. \tag{1.8}$$

Now, knowing D, M, and M^0, eqn (1.5) can provide a first approximation for D^0; then (using normal statistical tables) eqn (1.8) can be used for a first approximation of $Q(D^0)$ which can be substituted in (1.7) for a second approximation of D^0. Continuing sequentially, (1.8) and (1.7) can alternatively be used for further approximations of $Q(D^0)$ and D^0.

Taguchi's recommendation: 'It is *fundamentally incorrect* to set wide tolerances (D^0) on the basis of process capability considerations [e.g. on the basis of σ, (1.8) and (1.7)]. If $p < 0{\cdot}2$ (i.e. if the expected defect rate is less than 20%), (1.5) is adequate for setting the producer's tolerance; otherwise something *has to be done* to improve the process capability (to decrease the variability represented by σ).' Indeed, high variability is the cause of a high number of defects and rejects, not only at the production line but most importantly at the user's environment. To avoid betraying the people's trust, a company must develop and produce products of low functional variability. This variability is apparent when, for example, the product deteriorates fast or when it is unsuitable for its environment. The obvious reason for the unexpected deterioration and unsuitability is that insufficient effort was put into making the product insensitive to the effects of uncontrollable factors. This robustness cannot be achieved at the production stage—and setting wide production tolerances is certainly not the answer. It is wrong to compromise on the tolerance limits because of lack of process capability. It would be far better to estimate the cost of improving process capability and request a price adjustment. This cost does not have to be high, especially if the improvement efforts are concentrated at the product design stage (off-line stage). In fact, if robustness is achieved at this stage, it can actually lead to a reduction in costs and therefore to more competitive prices. Taguchi advocates design optimization for lower cost as *the* primary requirement in improving competitive position in world markets. The result—not the objective—is greatly improved quality through the achievement of product and process uniformity.

1.3.4 Why 100% inspection and 'zero defects' is not enough

By definition, the loss incurred when a product's characteristic deviates from its ideal value, is directly proportional to the square of this deviation from the target. In general, if σ^2 represents the mean squared deviation from target, then because of (1.2)

$$\text{loss} = k\sigma^2, \tag{1.9}$$

where, as before

$$k = \frac{\text{loss from not meeting tolerance}}{(\text{tolerance})^2}.$$

This 'quadratic loss' approach demonstrates the need for variability minimization and for achieving the target. It also advocates that merely 'meeting specifications' is a poor measure of quality.

The normal distribution of Fig. 1.1 corresponds to the 'concentration near the target with minimum variability' principle. On the other hand, the uniform distribution corresponds to the '100% inspection resulting in

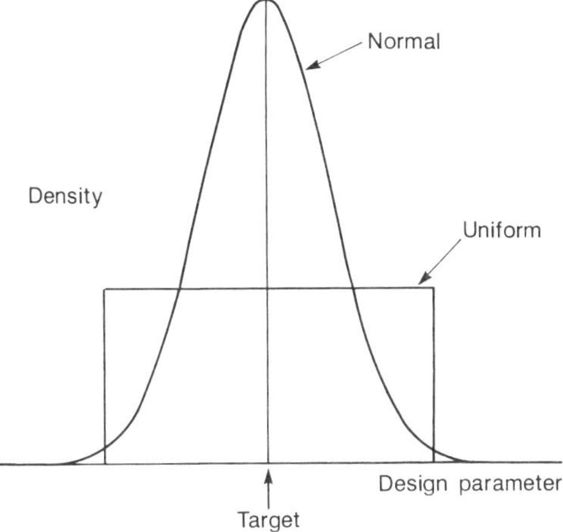

FIG. 1.1. Normal versus uniform distribution.

zero defects' principle. The sharp edge of this distribution is accounted for by the fact that inferior grade sets are produced and are 'inspected out'. In this case, although all the supplied items have met the spec. during inspection, there is no guarantee that they will go on meeting the spec. in the user's environment. Because, when there is a large variability around the target value, as in the case of the uniform distribution of Fig. 1.1, there is little difference between an item just inside the spec. and an item just outside it. They are both of inferior quality and result in a societal loss. The mass inspection and consequential scrap/rework or service under warranty costs add even more loss than the customer dissatisfaction and market loss caused by poor quality.

Example: Suppose that the normal and uniform distribution of Fig. 1.1 represent the distributions of products received by suppliers N and U respectively. Supplier U's products all fall within the specification $m \pm D$, while N supplies a proportion of products, 5% say, that is just outside the spec.

If δ is the standard deviation of the normal distribution then (from Normal Tables T3, Appendix E)

$$D = 1.96\delta,$$

and the societal loss incurred by supplier N is [from (1.9)].

$$L_{\mathrm{N}} = K\sigma_{\mathrm{N}}^2 = k\delta^2.$$

The distribution of supplier U's product could represent what one typically receives from a (well-working) 100% inspection scheme: a product meeting the specification of $m \pm 1.96\delta$. However, this distribution is not centred on the design target, m, and is considerably more variable than N's. In fact, the variance of this uniform distribution is given by (see Appendix F)

$$\sigma_{\mathrm{U}}^2 = \frac{\{(m+D)-(m-D)\}^2}{12} = \frac{(2 \times 1.96 \times \delta)^2}{12} = 1.28\delta^2.$$

So the societal loss due to supplier U is

$$L_{\mathrm{U}} = k\sigma_{\mathrm{U}}^2 = 1.28 \times k\delta^2.$$

In other words, the loss due to supplier U is 28% greater than the loss due to N, despite the fact that U supplies zero defects while N supplies 5% defectives, and U may have to bear the additional cost of 100% inspection, scrap or rework, whereas supplier N does not.

1.4 Taguchi's 'on-line quality control' systems

The actions Taguchi recommends to be taken on a regular basis during normal production are not based on the conventional control chart methodology (which attempts to find the cause of variability and remove it). In fact, he strongly believes that the main objective of an on-line QC system should be prevention and not rectification. The out-of-control region in the control charts should never be allowed to be reached. This means that variability in the production process should be diagnosed and dealt with *before* it becomes a cause for concern.

Sources of variability on a production line could be the following:

> Process drift, machine and tool failure due to wear and tear, variability in materials and purchased components, variability in execution due to human error, and so on.

General reworking and 100% inspection of all product units do not guarantee long-run elimination of this variability (see Section 1.3.4). On-line QC methods should be targeted *at the process* in an effort to predict or diagnose the cause of trouble, and then to adjust or correct before inferior quality products are produced.

Techniques for 'preventive' on-line quality control are dealt with in Sections 1.4.1 and 1.4.2.

1.4.1 Diagnosis and adjustment/Preventive maintenance

The production process is diagnosed at regular intervals. During a 'diagnosis period' a product unit is inspected. If the unit is not faulty, production is continued. If the unit is defective the production is stopped and the cause is found and dealt with by the appropriate process adjustment. Another unit is then produced under the adjusted process, inspected, and if it is good, normal production restarts.

Two characteristics are of interest here:

(a) the 'optimum diagnostic interval', i.e. how many items will be allowed to be produced before a unit is inspected;

(b) the cost (loss) per unit associated with this type of on-line quality control procedure.

If we define the following parameters:

$$m = \text{mean interval between failures} = \frac{(\text{total number of products})}{\text{number of failures}};$$

A = the cost of scrapping a defective item;
B = the inspection cost;
C = the process adjustment costs (i.e. costs associated with stopping, adjusting, and restarting the process);
T = the 'time-lag' interrelated by the diagnostic interval, i.e. the number of units produced before the process is stopped after a product unit is found abnormal on diagnosis;

then (see 'Process adjustment theory', Taguchi 1981) the optimum diagnostic interval D_I is given by

$$D_I = \sqrt{\frac{2(m+T)B}{A - C/m}}. \tag{1.10}$$

The quality-control loss $L(QC)$ per unit associated with the diagnosis and adjustment procedure is given by

$$L(QC) = \frac{B}{D_I} + \frac{(D_I + 1) \times A}{2 \times m} + \frac{(C + T \times A)}{m}. \tag{1.11}$$

Note that the mean interval between failures m significantly affects the loss: the lower the m (i.e. the lower the product quality) the higher the quality-loss.

Example 1: In the manufacturing process of stator bars for turbine generators, there is an annual production of 20 000 bars. In the preceding year, 20 bars were diagnosed as faulty. The mean interval between failures m is accordingly:

$$m = \frac{20\,000}{20} = 1000 \text{ units}.$$

The inspection cost of determining whether a stator bar is faulty or not, is £500 ($=B$). A defective bar is scrapped at a loss of £10 000 ($=A$). The tool cost, labour cost, and other costs associated with stopping the process, replacing the faulty part of the process which caused the trouble, and then restarting production, is £25 000 ($=C$). From past experience it can be assumed that $T=1$.

From (1.10) we can estimate D_{I} as

$$D_{\mathrm{I}}=\sqrt{\frac{2(1000+1)500}{10\,000-25\,000/1000}}\simeq 10 \text{ units.}$$

If the diagnostic interval is set at 10 units, the quality-control cost $L(QC)$ will be [using (1.11)]

$$L(QC)=\frac{500}{10}+\frac{(10+1)\times 10\,000}{2\times 1000}+\frac{(25\,000+1\times 10\,000)}{1000}=140.$$

So, there is a £140 cost per unit, and on an annual basis there is a cost of £140 × 20 000 = £2.8 million.

A pass/fail quality-control system can be rationalized by reducing the cost $L(QC)$ given by (1.11). Of course, $L(QC)$ can be reduced by finding ways to reduce inspection (B), and diagnostic and adjustment costs (C). But, if change in the current inspection, diagnostic and adjustment methods is to be avoided, quality-control savings can be achieved by maximizing the mean time between failures m. Preventive maintenance by regular replacement of process tools can increase m.

One type of preventive maintenance is the periodic replacement without diagnosis of a process part which may cause failure. For instance, if the mean life of a process tool is the equivalent to the production of 10 000 units, the tool must be replaced after machining the 7000th unit without checking whether it is still serviceable.

If C_{R} is the replacement cost of a part which is replaced periodically every n units (usually $C_{\mathrm{R}}\leq C$), then the additional cost of preventive maintenance (PV) is given by

$$PV=C_{\mathrm{R}}/n. \tag{1.12}$$

Then, combining (1.11) and (1.12) the quality-control costs per unit, including the PV cost is given by

$$L(QC,PV)=PV+L(QC). \tag{1.13}$$

Example 2: Suppose that a process part which usually causes a problem in the stator-bar production process (Example 1) is replaced periodically every $n=600$ units, well before the mean interval between failures of 1000 units. Assume that the failure rate before the 600th unit is 0.03

which includes failures due to other causes as well. Finally, assume that $C_R = C = £25\,000$. The predicted mean interval between failures can now be calculated to be

$$m = \frac{600}{0.03} = 20\,000 \text{ units.}$$

Then the new optimum diagnostic interval is [using (1.10)]

$$D_I = \sqrt{\frac{2 \times (20\,000 + 1) \times 500}{(10\,000 - 25\,000/20\,000)}} \approx 45 \text{ units.}$$

Using (1.13), the loss is now

$$\begin{aligned} L(QC, PV) &= PV + L(QC) \\ &= \frac{C_R}{n} + \frac{B}{D_I} + \frac{(D_I + 1) \times A}{2 \times m} + \frac{(C + T \times A)}{m} \\ &= \frac{25\,000}{600} + \frac{500}{45} + \frac{46 \times 10\,000}{2 \times 20\,000} + \frac{(25\,000 + 1 \times 10\,000)}{20\,000} = 66. \end{aligned}$$

So, with preventive maintenance, there is a loss of £66 per unit and so, on an annual basis a loss of £66 × 20 000 = £1.32 million.

This represents a saving of almost £1.5 million in annual quality control costs (original annual cost £2.8 m). This significant saving was due to the increase of the mean time to failure m by preventive maintenance. Preventive maintenance is thus equivalent to a process or specific method that cuts the failure rate without increasing the cost.

1.4.2 Prediction and correction/feedback and feedforward control

A quantitative characteristic to be controlled is measured at regular intervals; the measure value is used to predict the (mean) characteristic product value on the assumption that production is continued without adjustment. If the predicted product value differs from the target product value by a significant amount, then the level of a 'target-control factor' (see Chapter 6) is modified to reduce the difference.

The simplest method of predicting the (mean) characteristic product value Y_p for all the products in the interval up to the next measurement, is to use the measured value itself. If the mean squared deviation from target σ^2 is known, this can help determine the optimum measurement interval and the optimum amount of correction for differences between the predicted value Y_p and the target value Y_t.

Indeed, the optimum measurement interval will depend on the loss incurred due to the variability, i.e.

$$L = k\sigma^2,$$

where

$$k = \frac{\text{loss from not meeting tolerance}}{(\text{tolerance})^2}.$$

The optimum amount of correction is given by

$$-\beta(Y_p - Y_t), \tag{1.14}$$

where

$$\beta = \begin{cases} 1 - \dfrac{1}{F_0} & \text{when} \quad F_0 = \dfrac{(Y_p - Y_t)^2}{\sigma^2} > 1. \\ 0 & \text{when} \quad F_0 \leq 1. \end{cases}$$

1.4.2.1 Feedback control

If a measure of the process variability σ^2 is not readily available, but it is known that the mean squared drift is proportional to the production quantity, the following feedback control system procedure can be followed.

Define:

m = the mean interval between adjustments,
$= \dfrac{\text{total number of products}}{\text{number of adjustments}}$;
A = the loss caused by a defective unit;
B = the cost of measurement;
C = the process of adjustment costs;
T = the time lag interrelated by the measurement interval, i.e. the number of units produced before the process is stopped after a product unit is found abnormal on measurement;
D = the current control limit, i.e. the amount of deviation from the target value upon which the decision is taken for a process adjustment to the target;
Δ = the tolerance of the objective characteristic.

As in Section 1.4.1, two measures of interest are: the 'optimum measurement interval', and the loss per unit when this feedback QC system is followed.

The optimum measurement interval M_I can be estimated by

$$M_I = \Delta \times \sqrt{\frac{2mB}{AD^2}}. \tag{1.15}$$

The quality-control loss per unit is given by

$$L(QC) = \frac{B}{M_I} + \frac{C}{m} + \frac{A}{\Delta^2}\left\{\frac{D^2}{3} + \left(\frac{M_I + 1}{2} + T\right) \times \frac{D^2}{m}\right\}. \tag{1.16}$$

Example 3: The daily production volume of a component is 10 000 units. The current average adjustment rate of the production process is 4 times a day. Consequently, the mean interval between adjustments is

$$m = \frac{10\,000}{4} = 2500 \text{ units}$$

The loss from a defective unit is $A = £20$, the measurement cost is $B = £50$ and the adjustment cost $C = £10$. The component's tolerance for its dimension is $\pm 10\ \mu\text{m}$ ($\Delta = 10$) and the dimension is adjusted to target whenever it is found to be outside the control limit of $\pm 4\ \mu\text{m}$ ($D = 4$). There is usually a time lag of $T = 3$. According to (1.15), the optimum interval between measurements is

$$M_I = 10 \times \sqrt{\frac{2 \times 2500 \times 50}{20 \times 4^2}} \simeq 280.$$

So, assuming that measurements will take place every 280 units, the quality control loss per unit can be found from (1.16) as

$$L(QC) = \frac{50}{280} + \frac{10}{2500} + \frac{20}{10^2}\left\{\frac{4^2}{3} + \left(\frac{280 + 1}{2} + 3\right) \times \frac{4^2}{2500}\right\} = £1.43.$$

1.4.2.2 Feedforward control

On the basis of the current value for D, an optimum control limit D_0 can be estimated by

$$D_0 = \left\{\frac{3C}{A} \times \frac{D^2}{m} \times \Delta^2\right\}^{\frac{1}{4}}. \tag{1.17}$$

On the basis of the new D_0 and the old D, a forward prediction for the long-term mean interval between adjustments m_0 can be approximately made by

$$m_0 = m\left(\frac{D_0}{D}\right)^2.$$

Of course, if actual data can be obtained, the actual value m should be used for the estimation of M_I and $L(QC)$.

Nevertheless, using the optimum M_I, D_0 and the predicted m_0, and applying (1.16) one can made a forward prediction L_F(QC) for the future loss per unit (i.e. the loss under the 'optimum solution') which can be compared with the current loss under the current values of D and m, and under the current (perhaps non-optimum) interval between measurements (i.e. the current M_I).

Example 4: Considering the process described in Example 3, an optimum limit can be found by (1.17) as

$$D_0 = \left(\frac{3 \times 10}{20} \times \frac{4^2}{2500} \times 10^2\right)^{\frac{1}{4}} \approx 1.$$

Therefore, optimally, the process should be adjusted to target whenever a unit's dimension is found to be outside the control limit of $\pm 1\ \mu$m.

Under the procedure, the quality control cost per unit can be predicted to be (using (1.16)):

$$L_F(QC) = \frac{50}{280} + \frac{10}{2500} + \frac{20}{10^2}\left\{\frac{1^2}{3} + \left(\frac{280 + 1}{2} + 3\right) \times \frac{1^2}{2500}\right\} = £0.26.$$

This indicates an (over) fivefold reduction in predicted quality costs (from £1.43 per unit) under this optimal control limit.

1.5 History of experimental design

The idea that science needs careful experimentation goes back at least as far as Francis Bacon with his 'crucial experiment' (*experimentum crucis*): if an experiment in which all the conditions are the same except A produces an effect, then A is the cause.

The method is a corner-stone of a branch of the philosophy of science identified in this century with the falsification theory of Karl Popper. John Stuart Mill developed a theory of experimentation with prescriptions which sound modern. Here is the 'second canon' from his book 'A system of logic': 'If an instance in which the phenomenon under investigation occurs and an instance in which it does not occur have every circumstance in common save one, that one occurring only in the former, the circumstance in which alone the two instances differ is the effect, or the cause, or an indispensable part of the cause of the phenomenon.' His method, brilliantly explained, is essentially the 'change one factor at a time' method of experimentation. Mill draws an important distinction between controlled experimentation and passive observation for which he uses, respectively, the words 'artificial' and 'spontaneous' experiments. This distinction is one which will concern us, particularly in Chapter 6. Indeed, the methods described can be seen as a way of trying to tame the haphazardness of passive observation to learn something about variation in less controlled situations.

Having laid out some simple prescriptions for the one variable at a time method the philosophers have largely ignored a more detailed theory of experimentation. The breakthrough into scientific theory started with R. A. Fisher in the 1920s working at the Agricultural Field Station at Rothamsted in the United Kingdom. He was concerned in arranging trials of fertilizers (treatments) on plots to guard against underlying effects of moisture, gradient or whatever else might contribute to the effect of fertilizers. The idea was to muddle up or randomize the allocation or arrange the allocation in a pseudo-random complex layout such as a 'Latin square'.

Later a full theory of factorial experimentation was superimposed on these early ideas, the rows, columns, and treatments themselves being thought of as factors but allowing combinatorial experiments in more general 'factorial settings'. In the 1930s and 1940s a range of combinatorial structures was discovered having interesting links into algebra: orthogonal Latin squares, balance incomplete blocks, Youden squares, orthogonal arrays, factorial fractions, screening designs, and so on. Three main ideas were used in this theory:

(1) balance, for each factor the levels occur equally often;

(2) estimability, every parameter (effect) must be capable of being estimated;

(3) orthogonality, a technical term which says roughly that it is easy to extract and separate out the effect of different factors.

As early as 1934, L. H. C. Tippett, Fisher's co-worker, used a highly fractional factorial design in an industrial experiment in textile manufacturing (a $\frac{1}{125}$th fraction using only 25 trial runs instead of the full factorial requiring 3125 runs). The first systematic account of how to choose appropriate fractions was given by D. J. Finney (1945), a co-worker of Frank Yates, Fisher's successor at Rothamsted Experimental Station. C. R. Rao (1947) provided an extension to these ideas with his fractional orthogonal arrays. Plackett and Burman (1946) have also greatly contributed to this field, with their operations research work in Britain during World War II.

In the 1950s and 1960s response surface methodology led to more versatile modelling but still using the basic linear model theory of factorial experimentation.

The idea was that there is a space, or region, of allowable sites (combinations of factor levels) at which we can experiment. For each observation site x_i we observe a response

$$Y(x_i) = f(x_i, \theta) + \varepsilon_i,$$

where f is known but θ unknown and ε_i is error.

Experimental design was liberated from its agricultural roots and a variety of special designs (sets of x_i's) were suggested. G. E. P. Box, N. R. Draper, and others developed a range of response surface designs which allowed, for example, in addition to factorial designs, centre points, star points, and so on. 'Evolutionary operation' is a technique for sequentially searching the design space for the optimum.

The definition of a good design was sharpened up both by Box and co-workers and, notably, by J. C. Kiefer and J. L. Wolfovitz. The latter were key figures in the growth of the decision theoretic school of statistics dating back to Abraham Wald. They managed to set up the experimental design problem as a rather special optimization problem and solve a number of cases.

Algorithmic methods were developed by Federov, Wynn, Pazman, Mitchell and others, and East and West Europe has made many contributions. The success of this work is demonstrated by the incorporation of the ideas into computer aided experimental design packages. It should be said, however, that the best packages use a mixture of classical combinatorial and optimization methods. Most of the examples used in this book use the former which are to be recommended in many situations but the future also lies with being able to experiment well in complex situations.

One way of thinking of the great advances in the science of experimentation during this century is as the final demise of the simple 'one factor at a time' method, although it should be said that there are still organizations which have never heard of factorial experimentation and use up many man-hours wandering a crooked path.

1.6 Robust engineering design

The present and future impact of Taguchi's methods is difficult to assess. One way is to look at their effects on three different areas, each with their own prevailing methodologies. These are

(a) engineering design,

(b) quality management, and

(c) product development.

We concentrate on engineering design in this section. Quality management is a general term for the administration of all aspects of quality improvement within a company. It covers, in addition to the methods of this work, statistical process control, inspection, accounting methods, training, and the implementation of the total quality referred to above.

Product development is a term used to describe the process of the introduction of a product and its management through production to sales. Progressive companies will put together interdisciplinary teams to control this activity, which may cut across more traditional management boundaries. The important birth-to-death idea contained in this has also been expressed using phases such as 'design to use', 'product life history', and so on.

The first lesson that the design-led philosophy of Taguchi gives is to push as much information back to the designer from the downstream life of the product.

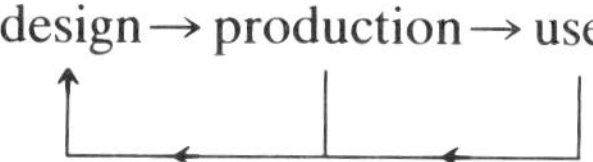

At the very least, it requires breaking down the barriers between the design and manufacturing functions which are still very separate in some companies. Another sociological improvement can be made by bringing sales and market research expertise into the product development team.

Taguchi summarizes his engineering design methodology into three phases: system design, parameter design, and tolerance design (see Chapter 6). In terms of existing engineering design methodology this is an innovation.

We summarize here some current methods.

Conceptual design. This acknowledges the fact that the creative aspects of engineering design are particularly its dynamic and adaptive aspects. The designer is considered to be at the centre of a circle, impinging on which are all the things to be considered: manufacture, assembly, sales aesthetics, and so on. The creative engineer has to sort through these sometimes conflicting requirements. Rabins (1986) summarizes these ideas.

Systematic design. This is a West German invention in which a 'systems' approach is adopted. The main idea is to systematize the design process, breaking it down into manageable boxes which then interact in suitable ways. The scheme can be different, obviously, for different branches, such as electrical, mechanical, software, and so on. A central component is a large list of different configurations through which the designer must search, assessing each one and homing in on the most feasible and suitable (Pahl and Beitz 1984).

Design for assembly, design for manufacture. For companies using mass-production techniques, ease of assembly and ease of manufacture generally became dominant themes well before the current developments. The notion of feeding manufacturing information back to the designer was already incorporated into these methodologies. Ease can be converted into a cost so that design can be indexed in a similar way to the costs associated with manufacture. In should be said, however, that quality has been largely absent as a component of this cost.

Re-design. The pressure of competition from the Far East, and the realization that some products in the West are below standard, forces re-design on many companies. Also, suppliers to large companies have come under stricter quality specifications, requiring them to speed up and improve the re-design process. One result of this has been the build up of in-house data-bases such that previous testing data and manufacturing costs can be quickly assessed. The systematization of the re-design process has taken over from the haphazard operation of correcting a design due to major faults downstream. Some attempt can then be made to re-design ahead of time to capture a market lead, rather than waiting until complaints from customers build up.

CAD and knowledge engineering. Virtually all of engineering design has been affected by the computer, and computer-aided-design packages have now been available for some twenty years. In fields such as aircraft, automotive, and electrical circuit design, they are highly sophisticated and 'intelligent'. They may incorporate optimizers, differential-equation solvers, and extensive graphics capabilities. At the other extreme the packages may not be much more than drafting facilities to replace or augment more traditional methods.

Good CAD packages should contain or 'bolt onto' data-bases containing details of components' manufacturing, costs, materials, and engineering rules. All this is loosely referred to as *knowledge engineering.*

As product development, quality management, and Taguchi methodologies develop alongside the steady computerization of industry, one can begin to see the kind of facilities that a designer will need to carry out his function properly. Methodology becomes the use, within a product-development framework, of the facilities available in a creative fashion, with costs kept low and quality high. Here is a rough list of the kind of modules which are needed. They are in no special order.

1. CAD software and graphics.
2. Rule and component data-bases.
3. Manufacturing data-bases with costs.
4. Manufacturing data-bases with machine capabilities, tolerances, and so on.
5. Model building and simulation packages (unless part of CAD).
6. A statistical analysis package.
7. An optimization package.
8. A tolerance design package.
9. Market research data.

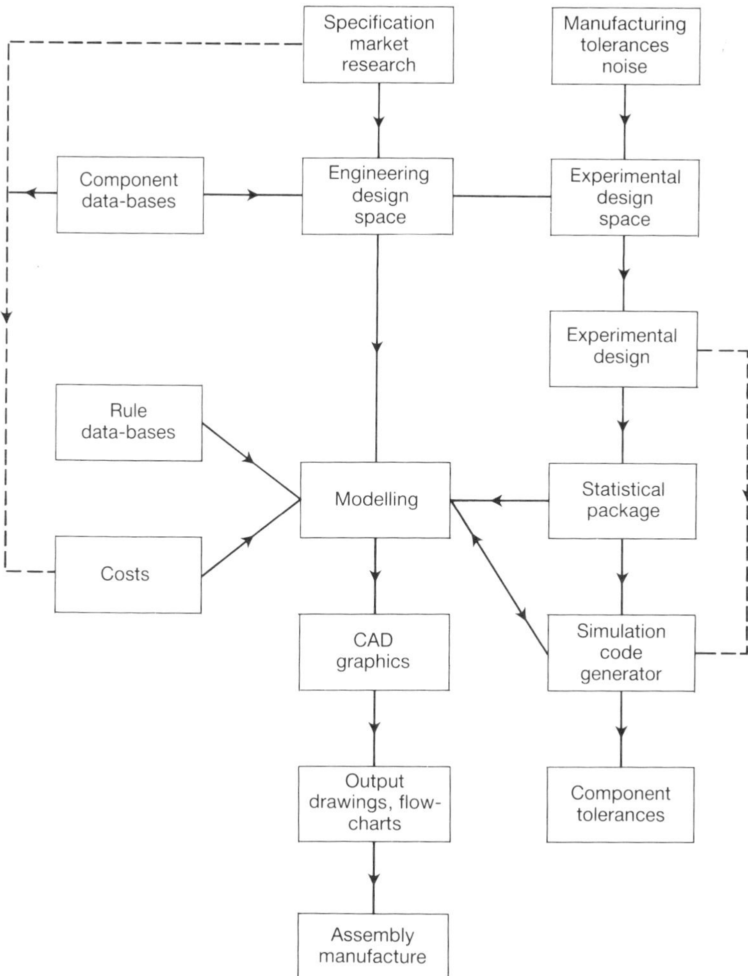

FIG. 1.2. Methodology, use of modules.

Figure 1.2 summarizes how such modules might be used. It contains within it some ideas from engineering design and some of those from Taguchi. The conceptual/system design stage builds up a design space of possible configurations. This is then enlarged to include downstream variables to provide an experimental space. Through off-line testing and statistical analysis, a computer model can be built up which can then be used for simulation, optimization, tolerance design, and so on. Data-bases can be used at any stage to supplement modelling and testing costs and engineering know-how. Many companies now have some flow diagram of this kind to aid product design or re-design. Figure 1.2 runs together some of the thinking from engineering design with Taguchi's three stages. This book is concerned with the right-hand side of the diagram.

The fundamental idea that one purpose of engineering design is to design for robustness to manufacturing and use needs a terminology. *Robust engineering design* serves this purpose and may be considered more general in scope than Taguchi methodology which may in time be reserved for his specific recommendations. In the article (Jebb and Wynn 1989) on which this section is based, the word *immersion* was used to bring home the idea that the time behaviour of a product could only be ascertained by capturing in some way the possible environments into which the product will be placed during its lifetime. To immerse himself in these environments, conceptually or in testing, the designer needs information and imagination, the two ingredients of good science.

2

FUNDAMENTALS OF DATA ANALYSIS

In this chapter we outline the basic statistical techniques needed for the analysis of many types of experimental data. The methods should require no prior statistical expertise and should make understandable the case-studies in subsequent chapters. Some non-standard methods due to Taguchi are included in Chapter 4. The theoretical background of the methods in this and subsequent chapters is outlined in more detail in Appendix B.

2.1 Basic statistical inference

2.1.1 Descriptive statistics

We are often constrained to examine only a small set of observations called a sample, usually written

$$Y_1, \ldots, Y_n.$$

Here, the total number of observations n is called the sample size. From this sample we want to make inferences or decisions about the background 'population'. This could be all the employees in a company, all the batches in a production line, all the items in a batch, and so on. This background population is sometimes called the 'parent population'.

Two assumptions are usually made which are idealizations:

1. The parent population is so large that it can be approximated by a continuous distribution.
2. The sample is chosen randomly.

A main purpose of statistical methods is to try to estimate or make statements about this parent population which is considered to be stable. The sample itself, of course, is not stable in the sense that if we sampled again (say later in the day) we could obtain a different set of values. We tend to use Greek letters μ, σ^2, etc., for aspects of the parent population and seek estimates of them using the sample values.

The most important characteristics of a population under investigation are those describing the population's central tendency and dispersion (variability). Using simple statistical techniques, inferences can be made about these population characteristics by examining the equivalent characteristics of only a small sample taken from the parent population. There are many measures of centrality and dispersion, and these are described in detail in Appendix B. At present, we will concentrate only on the most important ones.

The most commonly used measure of central tendency is the arithmetic average of all the observations, known as the 'mean'. With respect to a sample, the 'sample mean' is given by

$$\bar{Y} = \frac{\sum_{j=1}^{n} Y_j}{n}, \tag{2.1}$$

where $\sum$ is the summation symbol and $\sum Y_j$ represents the sum of all data values in the sample with n being the sample size. The sample mean provides an excellent estimate of the population mean (usually denoted by the symbol μ) except when the population is highly 'skewed' (highly unsymmetrical).

Degrees of freedom (df): The number of observations that can be varied independently of each other.

Variance: The sum of squares of the deviations from the mean, divided by the degrees of freedom. For the case of a sample:

$$\text{sample variance} = s^2 = \frac{\sum_{j=1}^{n} (Y_j - \bar{Y})^2}{n-1}. \tag{2.2}$$

In the estimate s^2, the value $n - 1$ describes the degrees of freedom for estimation of variance. More generally, this is the number of observations minus the number of parameters estimated. In this case, the '-1' comes from the estimation of μ by $\bar{Y}$.

The numerator of (2.2) can be easily calculated considering the relationship

$$\sum_{j=1}^{n} (Y_j - \bar{Y})^2 = \sum_{j=1}^{n} Y_j^2 - n(\bar{Y})^2. \tag{2.3}$$

The sample variance s^2 provides an excellent estimate of the population variance usually denoted by σ^2.

The above lead to the definition of the most useful measure of variability:

Standard deviation: The square root of the variance.

So

$$\text{sample standard deviation} = s = \sqrt{\frac{\sum_j (Y_j - \bar{Y})^2}{n-1}}. \tag{2.4}$$

Equation (2.3) can be used in the calculation of s.

The sample standard deviation is a measure of the average deviation of the observations from their sample mean. Its main attraction is its usefulness in conjunction with the sample mean for making inferences about the population mean and variance. On the basis of $\bar{Y}$ and s, a $(100-\alpha)\%$ 'confidence interval' for the unknown population mean μ can be constructed using the formula

$$\bar{Y} \pm t(\mathrm{df};\, \alpha) \cdot s\sqrt{\frac{1}{n}}, \tag{2.5}$$

where $t = t(\mathrm{df};\, \alpha)$ is a value depending on the degress of freedom $(\mathrm{df} = n-1)$ and on the 'level of significance' α; these t values can be found from the t-tables (see Table T1 in Appendix E). For example, if $n = 10$ and one is interested in a 95 per cent confidence interval (i.e. $\alpha = 5$ or 5 per cent or simply 0.05) then (2.5) can be applied with $n = 10$ and

$$t(\mathrm{df};\, \alpha) = t(9;\, 5\%) = 2.26,$$

found from Table T1 with reference to a 'two-sided' (interval) with $\mathrm{df} = 9$ and $\alpha = 5\%$ (or 0.05).

Note that the entries of the columns headed 'one-sided' in Table T1 are the t-values applicable when only a 'one-sided' confidence limit is required, in which case the following should be used

$$(100-\alpha)\%\ \text{upper confidence limit: } \bar{Y} + t(\mathrm{df};\, \alpha) \cdot s\sqrt{\frac{1}{n}}, \tag{2.6}$$

$$(100-\alpha)\%\ \text{lower confidence limit: } \bar{Y} - t(\mathrm{df};\, \alpha) \cdot s\sqrt{\frac{1}{n}}. \tag{2.7}$$

Remark: In some statistical books the t-tables are presented without a distinction being made between a 'one-sided' and 'two-sided' interval (or 'test'—see Section 2.1.2). In such cases the entries are headed either by the significance level α, or by $P = 100 - \alpha$, known as the 'percentile point' (the percentile point clearly represents the confidence level for the interval and $\alpha = 100 - P$). In either case, if the t-value for a two-sided confidence interval (or a 'two-sided test' see Section 2.1.2) is required [e.g. to be used in (2.5)], one should look for

$$t(\mathrm{df};\, \alpha/2)$$

rather than $t(\mathrm{df};\, \alpha)$ which is now applicable only in (2.6) and (2.7) for one-sided limits (or 'one-sided tests'—see Section 2.1.2).

2.1.2 Hypothesis testing

Making and testing hypotheses is a basic scientific procedure. Statistics tells us how to perform it in the presence of random fluctuations. If we do not take these into account we may erroneously reject a hypothesis when it is true or accept it when it is false.

Statistical hypothesis testing is divided into three steps. These are as follows:

Step 1: State the hypotheses to be tested. There are two hypotheses of interest:

The null hypothesis (H_0): a simple statement about a population characteristic usually having a specific value, for example

$$H_0\colon\ \mu = k\ (k \text{ being any real number}), \text{ or}$$
$$H_0\colon\ \sigma^2 = m\ (m > 0) \text{ etc.}$$

or a statement about two or more populations usually specifying 'no change' or 'no difference', for example in the case of two populations

$$H_0\colon\ \mu_1 = \mu_2 \text{ (i.e. } \mu_1 - \mu_2 = 0) \text{ or even } \mu_1 - \mu_2 = k\ (k \neq 0) \text{ or}$$
$$H_0\colon\ \sigma_1^2 = \sigma_2^2 \text{ etc.}$$

The alternative hypothesis (H_1): a simple statement about a population characteristic usually being larger (or smaller) than a specific value —'one sided test'—or generally different from a specific value—'two sided test', e.g.

$$H_1\colon\ \mu > k \text{ or } \mu < k \text{ (one sided) or}$$
$$H_1\colon\ \mu \neq k \text{ or } \sigma^2 \neq m \text{ (two sided) etc.}$$

or a statement about two or more populations usually specifying that a change or a difference has taken place in the populations, for example

$$H_1\colon\ \mu_1 > \mu_2, \quad \text{or} \quad \mu_1 < \mu_2 \text{ (one-sided test), or}$$
$$H_1\colon\ \mu_1 \neq \mu_2, \quad \text{or} \quad \sigma_1^2 \neq \sigma_2^2 \text{ (two-sided test), etc.}$$

Note that when determining whether there is an 'increase' or 'decrease' we are dealing with a 'one-sided test' rather than a 'two-sided test' which is associated with a test for 'difference'.

Step 2: Calculate a measure called 'test statistic' (*TS*) on the basis of sample characteristics such as the sample mean, sample standard deviation and the sample size, *assuming that* H_0 is true. The choice for the appropriate *TS* depends on the hypothesis under consideration and its value will indicate whether to accept or reject the null hypothesis: the larger (the more 'significant') the value of *TS*, the less the chance of being able to accept H_0.

Step 3: Make a decision: compare the value of *TS* with a critical value (*cv*) found from appropriate statistical tables (from Appendix E). The statistical table depends on the hypothesis under consideration; for example, if the hypothesis is concerned with 'means', usually the t-table (Table T1) is consulted. Whatever the case, the cv's depend upon a pre-specified level of significance α, already mentioned in Section 2.1.1, i.e. $cv = cv(\alpha)$.

The value of TS is rendered 'significant' if it is larger than the critical value $cv(\alpha)$. In such a case we say that 'we cannot accept the null hypothesis at the α-level of significance', running an $\alpha\%$ risk we are wrong by saying so. Alternatively we can say that 'there is evidence to suggest that the null hypothesis has to be rejected in favour of the alternative hypothesis at the $\alpha\%$ level' (running an $\alpha\%$ risk by doing so). Conversely, if $TS < cv(\alpha)$ we say that 'we are not able to reject the null hypothesis at the $\alpha\%$ level'.

Note that the level of significance α, when expressed in decimals, is the probability of being wrong when we reject a null hypothesis. This is sometimes called the 'probability of Type I error' as opposed to the 'probability of Type II error' (β) which is the probability of being wrong when we do not reject a null hypothesis.

The above 3-step procedure will be demonstrated below in testing various hypotheses concerning means and variances.

2.1.2.1 The two-sample t-test

A useful and popular significance test, and one of the most powerful of statistical tools, for comparing two population means is the 'two-sample t-test', which will be examined through a case-study.

A case-study*:* In order to assess the effect of the type of insulation used in overhead power cables, the breakdown voltage was measured for 10 cables using plastic insulation and for 9 cables using rubber insulation. The cables were chosen at random, one from each of 10 batches of plastic insulated cables and one from each of 9 batches of rubber insulated cables. The results are shown in Table 2.1.

The breakdown voltage is required to be as high as possible. At first glance the rubber insulation seems to result in better performance cables since the sample mean of the rubber insulated cables $\bar{Y}_r$ is larger than the sample mean of the plastic insulated cables $\bar{Y}_p$. However, comparisons

TABLE 2.1

	Breakdown voltage (kV)										$\bar{Y}$	s
Plastic insulation	2.8	2.9	2.5	2.6	3.2	3.0	2.8	3.5	2.9	2.7	2.890	0.292
Rubber insulation	2.9	3.0	3.1	3.2	2.9	3.3	2.8	3.2	2.7		3.011	0.203

should not be based only on sample means. A far better approach is to use a significance test, which uses as a test statistic a measure based on the sample means $\bar{Y}_r$ and $\bar{Y}_p$, as well as on the sample variances s_r^2 and s_p^2. Note that $n_r = 9$, $n_p = 10$.

The steps for such a significance test are as follows:

Step 1:
Null hypothesis (H_0): $\mu_p = \mu_r$ or $\mu_r - \mu_p = 0$, i.e. the population mean breakdown voltages are the same for both types of insulation; in the long run there is no difference in the performance (breakdown voltage) of plastic or rubber insulated cables.

Alternative hypothesis (H_1): $\mu_p < \mu_r$ or $\mu_r - \mu_p > 0$, i.e. the population mean breakdown voltage for the rubber insulated cables is larger than the mean breakdown voltage for the plastic insulated cables (one-sided test).

Step 2: Calculate the following test statistic *assuming that* H_0 *is true*

$$TS = \frac{|(\bar{Y}_r - \bar{Y}_p) - (\mu_r - \mu_p)|}{s\sqrt{\left(\dfrac{1}{n_r} + \dfrac{1}{n_p}\right)}},$$

where s, the 'pooled standard deviation', is some kind of an average of the two sample standard deviations s_r and s_p and is given by

$$s = \sqrt{\frac{\Sigma\,(\mathrm{df})s^2}{\Sigma\,(\mathrm{df})}} = \sqrt{\frac{(n_r - 1)s_r^2 + (n_p - 1)s_p^2}{(n_r - 1) + (n_p - 1)}}.$$

If H_0 is true then $\mu_r - \mu_p = 0$ and TS is expressed only through the sample characteristics

$$\bar{Y}_r = 3.011, \quad \bar{Y}_p = 2.89, \quad s_r = 0.203, \quad s_p = 0.292, \quad n_r = 9, \quad n_p = 10,$$

$$s = \sqrt{\frac{(9-1)(0.203)^2 + (10-1)(0.292)^2}{9 + 10 - 2}} = 0.254,$$

and

$$TS = \frac{|3.011 - 2.89|}{0.254\sqrt{\frac{1}{9} + \frac{1}{10}}} = 1.04.$$

Step 3: To make a decision on whether to accept or reject H_0, we need to assess whether TS is small enough or large enough (significant) in comparison with a critical value. The critical value is a t-value to be found in Table T1 (Appendix E) depending on $n_r + n_p - 2 = 17$ degrees of freedom and on a level of significance α which is the risk we are prepared to undergo for being wrong if we eventually decide to reject H_0. Choosing $\alpha = 5\%$, the one-sided-test t-value corresponding to 17 degrees of freedom is

$$cv = t(17; 5\%) = 1.74.$$

Since $TS < cv$ we cannot reject the null hypothesis of equal population means at the 5% level of significance. So, there is no evidence to suggest that the type of insulation used has an effect on the cable performance as far as breakdown voltage is concerned.

Remark: The two-sample t-test can be useful for testing the general null hypothesis

$$H_0: \mu_A - \mu_B = k$$

where k is any real number. Therefore, the general form of the test statistic assuming H_0 is true, is given by

$$TS = \frac{|(\bar{Y}_A - \bar{Y}_B) - k|}{s\sqrt{\left(\frac{1}{n_A} + \frac{1}{n_B}\right)}}.$$

However, the two-sample t-test, as any other t-test or confidence interval using t-values, is applicable only when certain assumptions underlying the calculations are reasonably upheld. These assumptions are:

(A1) The samples have been selected at random. We have already mentioned this earlier.

(A2) The parent populations form a normal distribution. We will revert to the subject of normality in Section 2.1.3.

(A3) The variability of items is the same in both populations. The easiest way of checking this is to carry out a significance test for homogeneity of variance, called the F-test, described below.

2.1.2.2 Comparing variances

In the case-study of Section 2.1.2.1, prior to comparing mean performances through the two-sample t-test, one might be interested in assessing the validity of assumption (A3), i.e. whether $\sigma_p^2 = \sigma_r^2$. The following procedure might be followed:

$$H_0: \sigma_p^2 = \sigma_r^2$$

i.e. the variability in breakdown voltages is, in the long run, the same for both types of insulation.

$$H_1: \sigma_p^2 \neq \sigma_r^2 \text{ (two-sided test).}$$

Test statistic:

$$TS = \frac{\text{larger sample variance}}{\text{smaller sample variance}} = \frac{s_p^2}{s_r^2} = \frac{(0.292)^2}{(0.203)^2} = 2.07.$$

Critical value: using Tables T2 in Appendix E (the F-tables), two-sided with $n_p - 1 = 9$ degrees of freedom for the larger sample variance (numerator) and $n_r - 1 = 8$ degrees of freedom for the smaller sample variance (denominator), at 5% level, we have $cv = 4.36$.

Since $TS < cv$, one cannot reject the null hypothesis of equal population variances at the 5% level. Therefore, the assumption (A3) of 'homogeneity of variance' between the considered populations is satisfied.

2.1.3 *The normal distribution*

In order to assess the quality of individual items, it is necessary to take larger samples than those used to make deductions about the population mean. Such a 'large sample' is the one depicted in Table 2.2 concerning the breakdown voltages of 50 plastic insulated cables, randomly chosen, five from each of 10 batches.

In this section we will examine a simple method of summarizing the shape of the distribution of such single values and then compare this shape with that of a commonly occurring shape for measurements—referred to as the 'normal distribution'.

2.1.3.1 Histograms and frequency distributions

The breakdown voltages of Table 2.2 can be presented in compact form if they are summarized as shown in Table 2.2a.

TABLE 2.2

Batch	Breakdown voltage (kV)				
1	2.70	2.80	2.45	2.62	2.90
2	3.01	2.90	3.20	3.13	2.77
3	2.72	3.05	2.68	3.11	2.78
4	2.50	2.68	3.13	2.40	2.92
5	2.30	2.80	2.93	2.69	3.16
6	2.94	2.75	3.15	3.22	3.30
7	3.32	3.25	2.95	2.85	3.02
8	2.86	2.65	3.09	2.94	2.96
9	2.55	2.73	2.96	2.75	2.87
10	2.98	3.28	3.17	3.40	3.50

TABLE 2.2a

Breakdown voltage	2.30 to 2.499	2.50 to 2.699	2.70 to 2.899	2.90 to 3.099	3.10 to 3.299	3.30 to 3.499	3.50 to 3.699
No. of cables (frequency)	3	7	12	14	10	3	1

Table 2.2a represents the 'frequency distribution' of the 50 breakdown voltages. Each of the 50 values has been allocated to one of the ten groups listed in the top row of the table. The number of cables in each group (interval) known as 'frequency', is indicated immediately below the group. It is clear that there is a concentration of cables in the three central groups. This can be seen even more clearly in a graphic presentation of this frequency distribution in Fig. 2.1, known as a 'histogram'. The height of each rectangle in a histogram is proportional to the frequency. If the frequency tabulation used unequal intervals, it would be necessary to make the height of each rectangle proportional to the ratio

$$\frac{\text{frequency in the interval}}{\text{length of the interval}}.$$

The histogram of Fig. 2.1 indicates that the distribution of the breakdown voltages is nearly symmetrical with a peak around 3.0 kV.

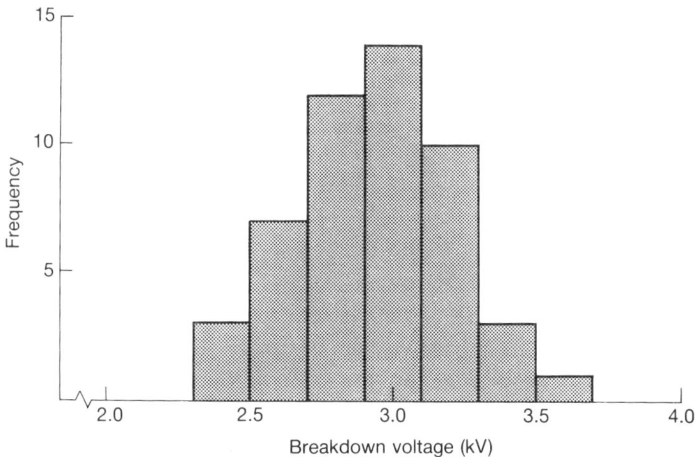

FIG. 2.1. Histogram.

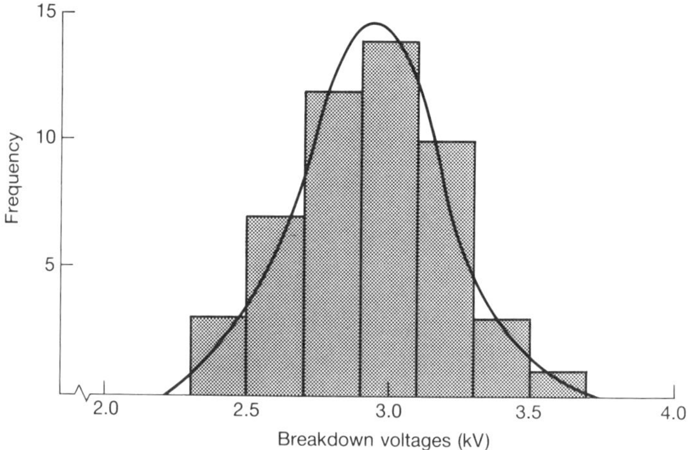

FIG. 2.2. Histogram and normal curve: Breakdown voltages.

However, there is also an obvious spread around the central point, ranging from 2.3 to 3.7 kV. On the basis of the available data, numerical measures of the central tendency and the spread can easily be obtained using simple formulae such as the ones described in Section 2.1.1 and Appendix B.

2.1.3.2 Probability distributions

As Fig. 2.2 shows, a smooth curve has been superimposed on the original histogram of the data of Table 2.2. Now, if the histogram represents the distribution of the breakdown voltages for the 50 cables in the sample, it could be suggested that the curve might represent the distribution of the breakdown voltages of the population of all plastic-insulated cables produced. Clearly, the curve and the histogram are similar but not identical. But if one considers smaller and smaller interval lengths by using larger and larger samples sizes, the resulting histograms will tend to even more nearly approach the smooth curve. The curve can be considered as the limiting value of a histogram when the lengths of the intervals approach zero and when the sample size approaches infinity. This curve is then called the 'probability distribution' describing the population of the items from which the samples originate.

There are many different families of probability distributions. If a variable is subject to errors and these errors are the accumulation of many small deviations then the behaviour of the variable is often well described by the Normal or Gaussian distribution. On the basis of Fig. 2.2 it is reasonable to expect that the breakdown-voltage distribution of the cables will peak near a single value of 2.9 kV, but perturbations in the process will lead to symmetrical departures on each side of the peak, indicating that the population of breakdown voltages could be described by a normal distribution.

The best estimate for the population mean of this normal distribution is the sample mean, calculated using eqn (2.1) to be

$$\bar{Y} = 2.9154,$$

and the best esimate for the population variance is the sample variance given by eqns (2.2) to (2.4)

$$s^2 = 0.07,$$

and so sample standard deviation $s = 0.2646$.

The smooth curve of Fig. 2.2 represents the normal distribution with a mean equal to the sample mean and standard deviation equal to the sample standard deviation. Note that, if we imagine a vertical line on the point 2.9154 ($\bar{Y}$), then 0.2646 (s) is the distance of this line from the points of inflexion of the curve (i.e. the points where the curve changes from convex to concave).

2.1.3.3 More about the normal distribution

The normal distribution has been found to be of great use. Apart from describing the distribution of measurement errors, it occurs in the study of many physical phenomena such as heat flow, diameters of machined parts, lengths of tobacco leaves and also in the study of many 'natural' measurements such as the heights and weights of human beings, IQ scores, etc. For example, Fig. 2.3 illustrates the distributions of heights

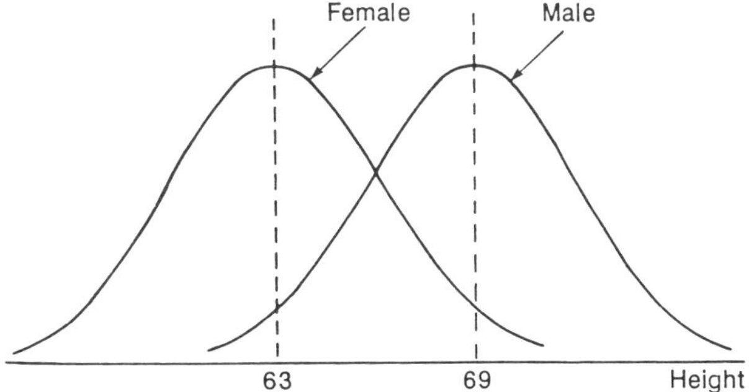

FIG. 2.3. Two normal distributions with equal standard deviations but different means.

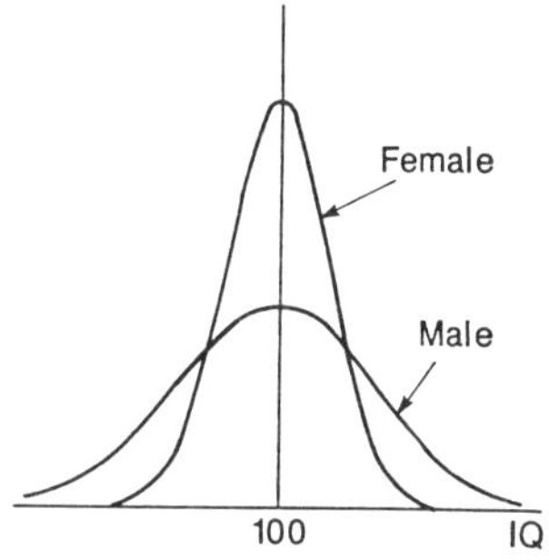

FIG. 2.4. Two normal distributions with equal means but different standard deviations.

for males and females which are described as normal with the same standard deviation (3) but with different means (63 inches for females and 69 inches for males). Figure 2.4 illustrates the distributions of IQs for males and females which are described as normal with the same mean but with different standard deviations; the distribution of IQ for males is more varied than that for females. Note that the narrower of the two distributions is also taller. This must be so because the *areas* under the two curves are equal.

The idea of 'area' is very important when dealing with probability distributions of continuous variables. The total area under the curve corresponds to a probability of 1, the highest value for a probability. Any subsection of the total area corresponds to a probability value between 0 and 1. The main usefulness of the normal distribution arises from the existence of readily available 'normal tables' (see Table T3, Appendix E) which make the calculation of these probabilities extremely easy. For example, we can make use of Table T3 to calculate the shaded area in Fig. 2.5 which is the proportion of females with heights between 66 and 68 inches, or equivalently, the probability that if a female is chosen at random, she will have a height of between 66 and 68 inches.

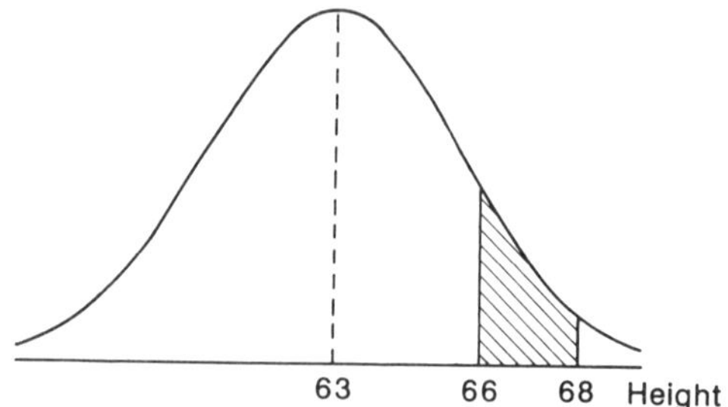

FIG. 2.5. Females with heights between 66 inches and 68 inches.

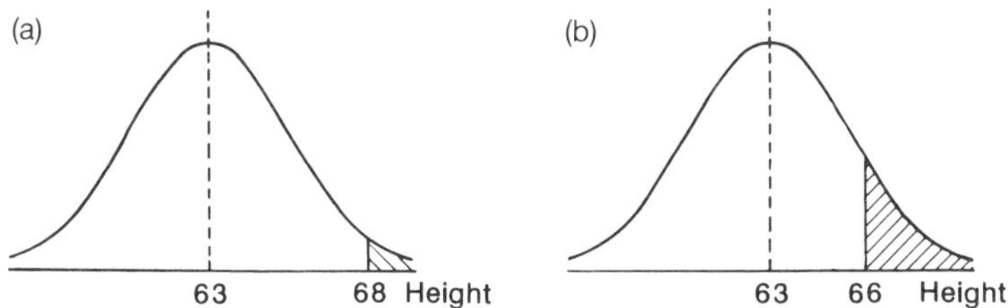

FIG. 2.6. Females with heights greater than (a) 68 inches, and (b) 66 inches.

Obviously all we have to do is find the proportion of all the females taller than 68 inches and subtract this from the proportion of all the females taller than 66 inches. To find the proportion taller than 68 inches, i.e. the shaded area of Fig. 2.6a, we first convert the height (x) into a *standardized* height (z), using

$$z = \frac{x - \mu}{\sigma},$$

where μ = the population mean, σ = the population standard deviation, and so

$$z = \frac{68 - 63}{3} = 1.67.$$

Using Table T3 we find that a z value of 1.67 corresponds to a probability of 0.0475. So we might say that 4.75% of adult females are taller than 68 inches or equivalently that, the probability of a randomly chosen female having a height exceeding 68 inches is 0.0475. Similarly, we can find that the proportion with heights exceeding 66 inches (see Fig. 2.6b) is 15.87%. Therefore there is a probability of $0.1587 - 0.0475 = 0.1112$ that a randomly chosen female will have a height between 66 and 68 inches. Note that when we calculate a standardized value (z), we are, in fact, jumping *to* a standardized normal distribution with a mean of zero and a standard deviation of 1, *from* the Normal distribution which has a mean μ and a standard deviation σ.

Table T3 is a representation of this 'standard normal distribution' and using it we can draw general conclusions for all variables which have a normal distribution. For example, we can see that a z value of 1.96 corresponds to a probability of 0.025, i.e. 2.5% of the area lies to the right of 1.96. This is displayed in Fig. 2.7a which also shows that 2.5% of the area lies to the left of -1.96 by symmetry. Thus we can conclude that 95% of z values will lie between -1.96 and 1.96. Transferring our attention to Fig. 2.7b which represents a normal distribution with mean μ and standard deviation, σ, we can say that 95% of values will lie between

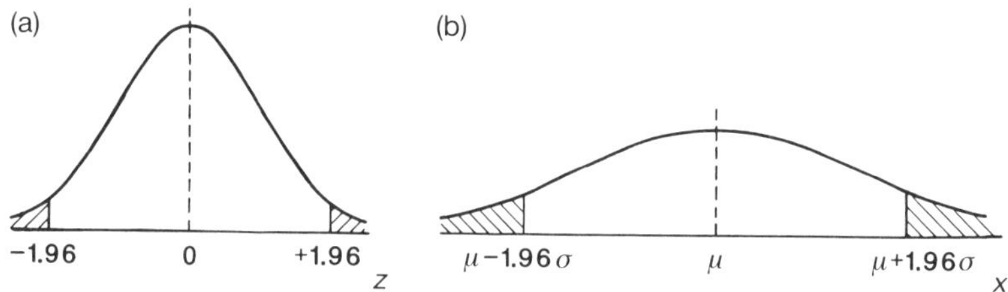

FIG. 2.7. The central 95% of a normal distribution.

$(\mu - 1.96\sigma)$ and $(\mu + 1.96\sigma)$. In general, for any variable with a normal distribution, 95% of values will be within 1.96 standard deviations of the mean. Similarly, 99% of values will be within 2.58 standard deviations of the mean.

2.1.3.4 Checking for normality

The calculation of probabilities above might appear relatively easy and straightforward. However, although the calculation can be applied to any set of data, it will only give valid conclusions when assumptions underlying the calculations are reasonably upheld. Such assumptions are:

(a) the sample is selected at random;

(b) the sample is taken from a normal distribution.

Although it is impossible to prove this second assumption we can follow the method of 'median ranks' described below, to check that it is reasonable. We shall use the data of Table 2.2.

In Table 2.3 the 50 Breakdown Voltages (BV) have been *ranked* in order of magnitude. Alongside each value for the voltage is a 'Median Rank' (MR) which is a crude estimate of the cumulative percentage of cables having a breakdown voltage up to the value under consideration. For example, based purely on the median rank of the 2.55 kV breakdown voltage, it is estimated that there are 9.3% cables with breakdown voltage up to the 2.55 value. The median rank is calculated by

$$MR = \frac{\text{rank} - 0.3}{n + 0.4} \times 100.$$

For example, the 9th highest rank in the sample of $n = 50$ has a median rank calculated as

$$MR = \frac{9 - 0.3}{50.4} \times 100 = 17.3\%.$$

TABLE 2.3. Median ranks (MR), for breakdown voltages (BV)

BV	MR	BV	MR	BV	MR	BV	MR	BV	MR
2.30	1.4	2.70	21.2	2.86	41.1	2.96	60.9	3.16	80.8
2.40	3.4	2.72	23.2	2.87	43.1	2.98	62.9	3.17	82.7
2.45	5.4	2.73	25.2	2.90	45.0	3.01	64.9	3.20	84.7
2.50	7.3	2.75	27.2	2.90	47.0	3.02	66.9	3.22	86.7
2.55	9.3	2.75	29.2	2.92	49.0	3.05	68.8	3.25	88.7
2.62	11.3	2.77	31.2	2.93	51.0	3.09	70.8	3.28	90.7
2.65	13.3	2.78	33.1	2.94	53.0	3.11	72.8	3.30	92.7
2.68	15.3	2.80	35.1	2.94	55.0	3.13	74.8	3.32	94.6
2.68	17.3	2.80	37.1	2.95	56.9	3.13	76.8	3.40	96.6
2.69	19.2	2.85	39.1	2.96	58.9	3.15	78.8	3.50	98.6

The median ranks are then plotted against breakdown voltages on normal probability paper as shown in Fig. 2.8.

The scale on normal probability paper has been chosen so that if the data formed a perfect normal distribution the points would exactly lie on a straight line. Due to variation that always exists, sample data will never lie exactly on a straight line; however, it is evident in Fig. 2.8 that a straight line provides a *reasonable fit* and so we can safely conclude that the normal distribution is a reasonable model for the variability in breakdown voltages.

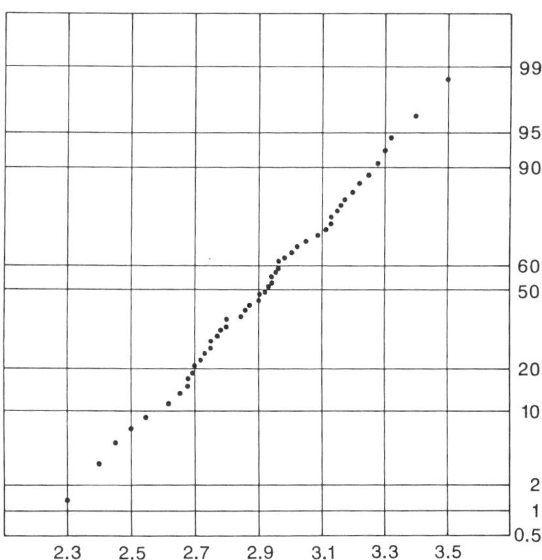

FIG. 2.8. Median rank plot for breakdown voltages.

2.1.4 Regression analysis

There are many occasions when we have an 'independent' variable or factor x that we can control, and a response or dependent variable Y that we observe at many different levels of x. There is a simple method, known as 'regression analysis', by which we can investigate the relationship between Y and x through a 'linear model' which can also be used for prediction of future values of Y. Inferences can then be made on how close this relationship is, and on the predicting uncertainty and capability of the model.

The general form of the model is that of a straight line equation

$$Y_i = \alpha + \beta x_i + \varepsilon_i, \qquad i = 1, \ldots, n,$$

where x_i is the setting of the independent variable on the ith observation; Y_i is the corresponding response; 'α' and 'β' are, respectively, the intercept and the gradient; and ε_i are random error terms, assumed to be independent, normally distributed with mean zero and variance σ^2. Note that the linear form of the model might be the result of an appropriate transformation of the available data, for example the log-transformation.

As an example, we can refer to the following situation: a range of small voltages is applied to the gate of a MOSFET (Metal Oxide Silicon Field Effect Transistor) operating in the subthreshold mode. The corresponding drain currents (I_D) are measured with a certain degree of experimental error. The subthreshold slope is to be evaluated and the standard error in it, on the basis of n observations (x_i, Y_i) $i = 1, \ldots, n$.

From transistor theory, $\log(I_D)$ against gate voltage (V_G) is expected to give a straight line. Errors in the gate voltage values are assumed negligible. We need to calculate the intercept $\hat{\alpha}$ (sample intercept) and the slope $\hat{\beta}$ (sample regression coefficient) so that the line (see Fig. 2.9)

$$\hat{Y} = \hat{\alpha} + \hat{\beta}x$$

has the best possible fit, i.e. so that the errors ε_i, in the 'least square' sense are minimal; $\hat{\alpha}$ and $\hat{\beta}$ are then the best (unbiased) estimators for α and β.

By minimizing

$$R^2 = \sum_{i=1}^{n} \varepsilon_i^2 = \sum_{i=1}^{n} (Y_i - \alpha - \beta x_i)^2 \text{ (method of least squares)},$$

we can obtain the α and β which minimize R^2, and these are given by

$$\hat{\beta} = \frac{\sum_i (x_i - \bar{x})(Y_i - \bar{Y})}{\sum_i (x_i - \bar{x})^2} = \frac{\sum_i x_i Y_i - n\bar{x}\bar{Y}}{\sum_i x_i^2 - n\bar{x}^2}$$

and

$$\hat{\alpha} = \bar{Y} - \hat{\beta}\bar{x},$$

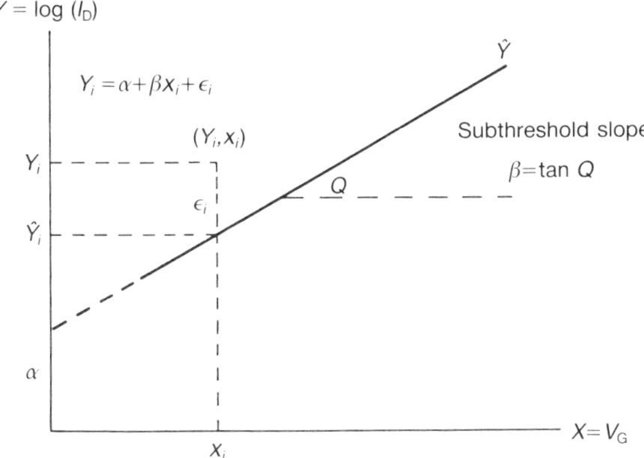

FIG. 2.9. Model for drain currents.

where

$$\bar{x} = \frac{1}{n}\sum_i x_i, \qquad \bar{Y} = \frac{1}{n}\sum_i Y_i.$$

To find the significance of β, i.e. to find whether Y is significantly linearly dependent on x, we need the standard error (standard deviation) of the estimated slope $\hat{\beta}$ which can be shown to be given by

$$SE(\hat{\beta}) = \left\{\frac{s^2}{\sum_i (x_i - \bar{x})^2}\right\}^{\frac{1}{2}},$$

where s^2 is an 'unbiased' estimator of σ^2 (the residual variance) given by

$$s^2 = \frac{RSS}{(n-2)} = \sum_i \frac{(Y_i - \hat{Y}_i)^2}{(n-2)} = \frac{1}{(n-2)}\left\{\sum_i (Y_i - \bar{Y})^2 - \hat{\beta}\sum_i (x_i - \bar{x})(Y_i - \bar{Y})\right\},$$

where RSS is the residual sum of squares.

We can now test for statistical significance of β (and so for significant effect of x on Y) by calculating the 't-ratio'

$$T = \frac{|\hat{\beta}|}{SE(\hat{\beta})},$$

and comparing it with a t-value (from the t-distribution Tables T1) depending on $n-2$ degrees of freedom at some predetermined level of significance α. That is, if T is found to be larger than $t(n-2, \alpha/2)$ where $\alpha = 0.05$ (say) we can accept that changes in the voltage significantly affect the drain current of a MOSFET linearly (through a logarithmic transformation) and by stating this there is a chance of 5% of being wrong.

Note that T represents the 'test-statistic' (TS) for testing the null hypothesis

$$H_0\text{: } \beta = 0$$

against the alternative

$$H_1\text{: } \beta \neq 0 \text{ (two-sided test).}$$

The general form of the TS is

$$TS = \frac{|\hat{\beta} - \beta|}{SE(\hat{\beta})},$$

and this becomes equal to T under the assumption that H_0 is true (see also Section 2.1.2).

Example: It has been established that the lifetime of a MOSFET-transistor can be predicted by measuring the flicker noise equivalent current. Some data were obtained under accelerated conditions (for example, the transistors were subjected to hot environment). The data are as follows:

Lifetime (Y)	Current (x)
245	17.9
220	23.6
215	30.9
211	56.1
161	61.0
135	77.0

We wish to find the best linear relationship between x and Y, by finding the best estimates of α and β in the relationship

$$Y = \alpha + \beta x.$$

Applying the simple regression formulae would result in $\hat{\beta} = -1.5931$ and $\hat{\alpha} = 268.59$ and this is represented by the line in Fig. 2.10. The residual standard error is found to be $s = 18.72$. The standard error (standard deviation) of $\hat{\beta}$ is

$$SE(\hat{\beta}) = \left\{\frac{s^2}{\sum (x_i - \bar{x})^2}\right\}^{\frac{1}{2}} = \left\{\frac{(18.72)^2}{2792.35}\right\}^{\frac{1}{2}} = 0.3543$$

and so, testing for significance of β,

$$T = \frac{|\hat{\beta}|}{SE(\hat{\beta})} = \frac{1.5931}{0.3543} = 4.5,$$

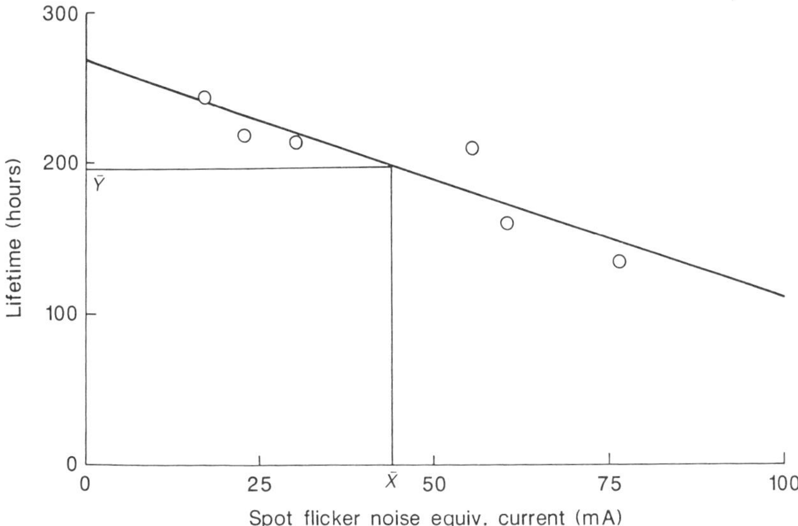

FIG. 2.10. Fitted regression line for lifetime of MOSFE transistors.

and this is highly significant, which leads to the rejection of the (null) hypothesis of a non-significant linear relationship between Y and x. Acceptance of significance for the regression coefficient β indicates that a large amount of variability in Y can be 'explained' by x. The total variability in Y can be represented through the Total Sum of Squares (TSS) of the deviations of each Y_i from the mean $\bar{Y}$, i.e.

$$TSS = \sum_{i=1}^{n} (Y_i - \bar{Y})^2.$$

In terms of areas, these squared deviations can be shown in Fig. 2.11.

By fitting the 'best' straight line $\hat{Y}$ using regression, we try to find a straight line *other* than $\bar{Y}$, which will *reduce* the area of the squared deviations from the line. Comparing Fig. 2.11 with Fig. 2.12 we can observe the tremendous reduction in area (of squared residuals) achieved by fitting the regression line. The relative reduction in area is called 'percent fit' and may be obtained as

$$PF = \frac{TSS - RSS}{TSS} \times 100 = \frac{\sum (Y - \bar{Y})^2 - \sum (Y - \hat{Y})^2}{\sum (Y - \bar{Y})^2} \times 100.$$

The difference $TSS - RSS$ is called the 'regression sum of squares' ($RegSS$) which, if $\hat{\beta}$ has been calculated, can be found from $RegSS =$

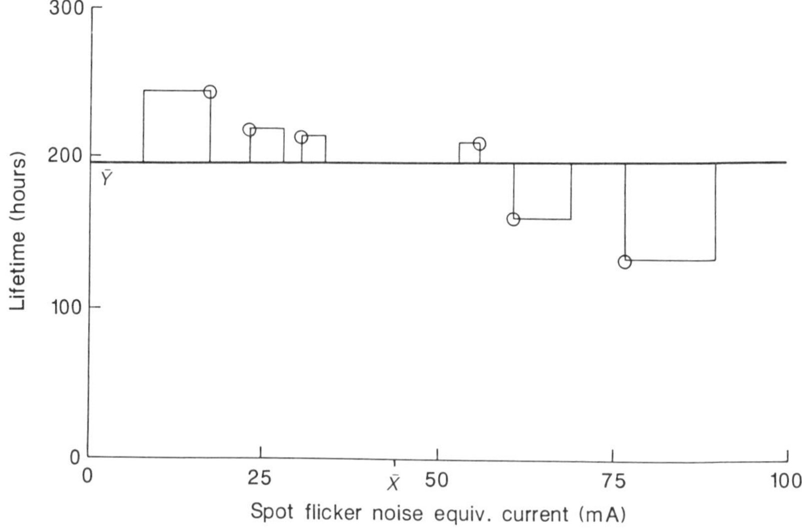

FIG. 2.11. Squared deviations from the mean.

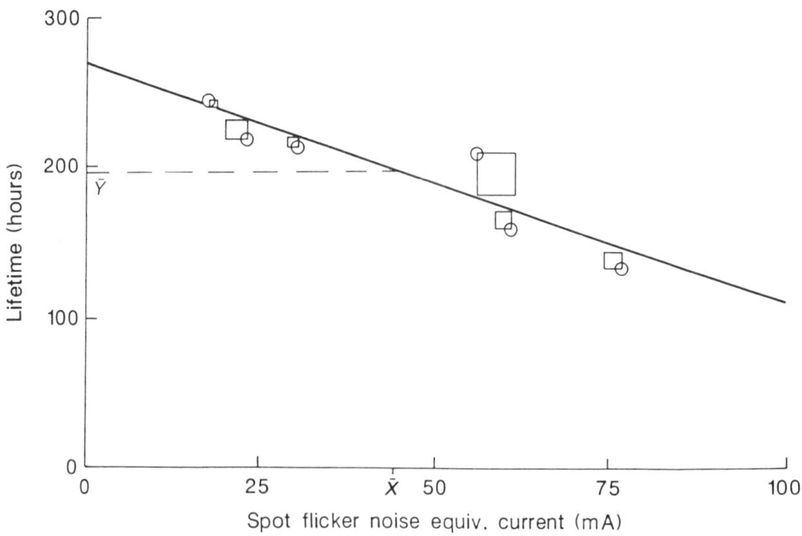

FIG. 2.12. Squared deviations from the fitted line.

$\hat{\beta}^2 \Sigma (x - \bar{x})^2$. From our example,

$$PF = \hat{\beta}^2 \frac{\Sigma (x - \bar{x})^2}{\Sigma (Y - \bar{Y})^2} \times 100 = (-1.5931)^2 \cdot \frac{2792.35}{8488.83} \times 100 = 83.5.$$

By fitting the regression line in the way we did, we have explained 83.5% of the variation in Y. The larger the PF the better the fit; it can range between 0 and 100 inclusively. The value

$$\mathbf{R}^2 = PF/100$$

is known as the 'coefficient of determination'.

The 'simple regression analysis' of fitting a simple linear model

$$Y = \alpha + \beta x$$

can be generalized to 'multiple regression analysis' by fitting a model of the form

$$Y = \alpha + \beta_1 x_1 + \cdots + \beta_k x_k$$

to investigate the relationship of k independent variables $x_1, \ldots, x_k$ to a response of interest Y. The least squares method can be used again and the same principles apply. Of course, calculations in such a case are performed today on a computer.

2.2 Simple models

In engineering and science we build simple models (theoretical equations) representing our product or process. These models have two features:

1. Certain parts of the model are unknown.
2. The observations from which we construct the model have a random part to them.

The simplest and certainly the most used framework for 1 and 2 is to say that

$$\text{observation} = \text{model} + \text{error}.$$

A third feature, which is a very common assumption, is that *different* observations are *statistically independent* or uncorrelated.

An experiment is the process of extracting observations in a particular way. There are good and bad experiments, small and large experiments, experiments conducted over long and short periods of time, or complicated and simple experiments. It is a home truth in science that one should bring some kind of order, through a model, to the experiment

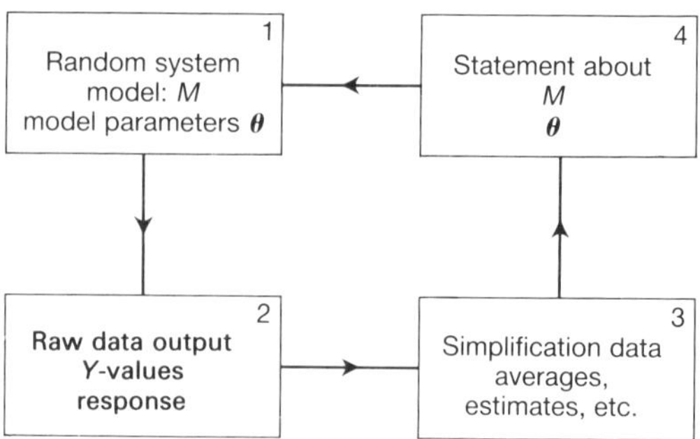

FIG. 2.13. The statistical method.

otherwise we will learn nothing. Indeed, learning itself can be thought of as a process of changing, rejecting, updating, improving (and so forth) our model. At a simple level we may be merely estimating some of the unknown quantities in the model. This process will make the model more useful, for example for forecasting or interpolating. If one of the quantities in the model is estimated as being large it may tell us that, for example, one variable is very important. Conversely, if a quantity is estimated as close to zero, it may mean that a variable is very unimportant.

A broad idea of the purpose of good data collection through experimentation is to *obtain information* of a *particular kind* about a system or phenomenon whose behaviour is partly unknown. The ignorance divides into three types: (a) that due to the random or 'stochastic' behaviour of the system (b) that due to ignorance about the model for the system (c) that due to ignorance about the parameter values in the model.

Figure 2.13 represents the view that the purpose of statistics is to find out about or make statements about the system. This is a little vague. The objective may be far more precise, for example:

1. Find the parameter θ values to give the maximum output Y.
2. Estimate the difference in two parameter values: $\theta_1 - \theta_2$.
3. Decide between alternative models M_1, M_2.
4. Find out whether a change in a variable x *internal to* the system has an effect on Y.

Where does experimentation fit into Fig. 2.13? We may decide to test, observe, experiment on the system in a more or less active or passive way. There is a very large, often complex, mathematical and philosophical literature across the entire spectrum between controlled experimentation and passive observation. Another critical feature is that when we experiment we must distinguish carefully between:

1. What can be controlled and what cannot.
2. The environment that the system operates in normally and the experimental environment.

The system is usually dependent on variables x which, depending on who you are talking to, can be called: independent variables, (input) control variables, factors, auxiliary variables and so on. In this chapter we shall use the word *factors.*

Here is an example which, though hypothetical, will fill out some of these ideas.

2.2.1 Weighing Experiment No. 1

Imagine a weighing machine on which one or more objects can be placed at the same time. Suppose that we have two objects A and B with unknown weights θ_A and θ_B. The model for a single weighing will be

$$\text{observed weight } (Y) = \text{total true weight } (\theta) + \text{error } (\varepsilon).$$

Here we can say that the total weight is the sum of the weights on the weighing machine. Suppose we carry out three weighings

	Object	Total weight	Error
Weighting 1	A	θ_A	ε_1
Weighting 2	B	θ_B	ε_2
Weighting 3	A and B	$\theta_A + \theta_B$	ε_3

We obtain the observations Y_1, Y_2, and Y_3 as

$$Y_1 = \theta_A + \varepsilon_1$$
$$Y_2 = \theta_B + \varepsilon_2,$$
$$Y_3 = \theta_A + \theta_B + \varepsilon_3.$$

Here the model is quite well defined. It is additive in that the weights add up, but there are unknown quantities to estimate, namely θ_A and θ_B, which we call the parameters and which could be part of an actual forecasting model. For example, if we wanted to know what happened if 3 perfect copies of object A and 2 perfect copies of object B were placed on the weighing machine, we would probably forecast the results as $3\hat{\theta}_A + 2\hat{\theta}_B$, where $\hat{\theta}_A$ and $\hat{\theta}_B$ are our estimates of θ_A and θ_B respectively from the experiment.

2.2.2 Choosing the best estimates

Returning to the experiment, we need to find good estimates of θ_A and θ_B. If all the errors were zero, then we could solve exactly and set $\theta_A = Y_1$, $\theta_B = Y_2$ and the third weighing would be redundant. The presence of the errors makes this inappropriate. The solution proposed in statistical theory is to minimize

$$R^2 = (Y_1 - \theta_A)^2 + (Y_2 - \theta_B)^2 + (Y_3 - [\theta_A + \theta_B])^2,$$

with respect to θ_A and θ_B. We do this by setting $\dfrac{\partial R^2}{\partial \theta_A} = \dfrac{\partial R^2}{\partial \theta_B} = 0$. Thus

$$\frac{\partial R^2}{\partial \theta_A} = -2(Y_1 - \theta_A) - 2(Y_3 - [\theta_A + \theta_B]) = 0,$$

$$\frac{\partial R^2}{\partial \theta_B} = -2(Y_2 - \theta_B) - 2(Y_3 - [\theta_A + \theta_B]) = 0.$$

Solving these simultaneous equations for θ_A and θ_B gives

$$\left.\begin{aligned} \hat{\theta}_A &= \tfrac{1}{3}(2Y_1 - Y_2 + Y_3), \\ \hat{\theta}_B &= \tfrac{1}{3}(-Y_1 + 2Y_2 + Y_3). \end{aligned}\right\} \qquad (2.8)$$

We refer to such estimates as 'least squares estimates' (*LSE*s) because the quantity R^2 is the sum of the squares of the differences between the actual observations and the model. This is the method of least squares used in Section 2.1.4. These *LSE*s have some attractive properties under suitable conditions on the ε_1, ε_2, ε_3. The assumptions we make about the errors are of a statistical kind and are discussed in more detail in Appendix B. They are as follows:

1. Each ε_i has *zero expectation*; that is, over many repetitions of the experiment we would expect the long-run averages to be zero.
2. Separate observations are 'statistically unrelated'; that is to say, uncorrelated in a well-defined sense.
3. The standard deviation of the errors (σ) is constant. This means that the dispersion of the errors is about the same for all the observations.

On the basis of the above we can now list three desired properties that the estimates $\hat{\theta}_A$ and $\hat{\theta}_B$ given by (2.8) can be shown to have:

1. They are linear in Y_1, Y_2, Y_3. This is obviously the case for the *LSE*s as shown in (2.8).
2. They are *unbiased*, i.e. $E(\hat{\theta}_A) = \theta_A$ and $E(\hat{\theta}_B) = \theta_B$. Indeed, by taking expectations for the *LSE*s (see Appendix B)

$$E(\hat{\theta}_A) = \tfrac{1}{3}(2E(Y_1) - E(Y_2) + E(Y_3)).$$

Now since each ε_i have zero expectation

$$E(Y_1) = \theta_A + E(\varepsilon_1) = \theta_A,$$
$$E(Y_2) = \theta_B + E(\varepsilon_2) = \theta_B,$$
$$E(Y_3) = \theta_A + \theta_B + E(\varepsilon_3) = \theta_A + \theta_B.$$

Substituting these into $E(\hat{\theta}_A)$ we have

$$E(\hat{\theta}_A) = \tfrac{1}{3}(2\theta_A - \theta_B + \theta_A + \theta_B) = \theta_A.$$

Similarly

$$E(\hat{\theta}_B) = \theta_B.$$

3. They have minimum variance. Indeed, using the formula for calculating variances of uncorrelated quantities (see Appendix B):

$$\begin{aligned}\mathrm{var}(\hat{\theta}_A) &= \tfrac{1}{9}\{4\,\mathrm{var}(\hat{Y}_1) + \mathrm{var}(\hat{Y}_2) + \mathrm{var}(\hat{Y}_3)\}\\ &= \tfrac{1}{9}\{4\,\mathrm{var}(\varepsilon_1) + \mathrm{var}(\varepsilon_2) + \mathrm{var}(\varepsilon_3)\}\\ &= \tfrac{2}{3}\sigma^2\end{aligned}$$

since each $\mathrm{var}(\varepsilon_i) = [\text{standard deviation } (\varepsilon_i)]^2 = \sigma^2$.

The proof of point 3 is harder, but we can verify it here by looking at some other estimators. For example, suppose that we set

$$\hat{\theta}'_A = \tfrac{1}{2}(Y_1) + \tfrac{1}{2}(Y_3 - Y_2).$$

We can easily see that this is linear and unbiased since

$$E(\hat{\theta}'_A) = \tfrac{1}{2}\theta_A + \tfrac{1}{2}(\theta_A + \theta_B - \theta_A) = \theta_A.$$

But for this $\hat{\theta}_A$ we have

$$\mathrm{var}(\hat{\theta}'_A) = \tfrac{1}{4}\sigma^2 + \tfrac{1}{4}\sigma^2 + \tfrac{1}{4}\sigma^2 = \tfrac{3}{4}\sigma^2,$$

which is larger than the value of $\tfrac{2}{3}\sigma^2$ for the *LSE* $\hat{\theta}_A$.

2.2.3 *Definitions*

If the original model is written as

$$Y_1 = x_{A,1}\theta_A + x_{B,1}\theta_B + \varepsilon_1,$$
$$Y_2 = x_{A,2}\theta_A + x_{B,2}\theta_B + \varepsilon_2,$$
$$Y_3 = x_{A,3}\theta_A + x_{B,3}\theta_B + \varepsilon_3,$$

where

$$(x_{A,1}, x_{B,1}) = (1, 0),$$
$$(x_{A,2}, x_{B,2}) = (0, 1),$$
$$(x_{A,3}, x_{B,3}) = (1, 1),$$

i.e. in general

$$Y_j = x_{A,j}\theta_A + x_{B,j}\theta_B + \varepsilon_j \qquad j = 1, 2, 3, \ldots,$$

we can consider the experiment as assigning values of two variables (X_A, X_B) with

$$X_A = \begin{cases} = 1 \text{ if A is in the experiment,} \\ = 0 \text{ if A is not in the experiment,} \end{cases}$$

and similarly for X_B and B. We are now in a position to define several terms used in experimental design and analysis.

Factor: A variable selected for experimentation; in the above example, X_A and X_B are factors. These act as 'input control' (independent) variables, their values for each weighing determining *what we actually do*. The characteristic of interest, in this case the observed weight, is also a variable called the 'response' (or 'dependent') variable. One of the purposes of fitting a model to the available data is to identify those independent variables or factors which significantly affect the response variables.

Levels: The different values of a factor which can be used in an experiment. In the above example, each factor has just two levels ('1' or '0', i.e. 'in' or 'out' of the weighing). The levels of a factor could be quantitative, i.e. set at measured values such as temperatures 50 °C, 65 °C, or 80 °C, etc; or they could be qualitative, i.e., with no numerical scale attached to them, as in the cases of different machines, operators, 'yes' or 'no', 'in' or 'out', etc. According to their level type, the factors can be distinguished into quantitative and qualitative factors. One or both types of factors can be involved in an experiment. A 2-level qualitative factor (e.g. with levels 'in' or 'out') can be represented in a binary form, i.e. as having the levels '1' or '0'.

Trial: A particular setting of levels of the different factors. In the weighing experiment there were three trials, i.e. 3-level combinations of the factors (X_A, X_B), namely $(1, 0)$, $(0, 1)$ and $(1, 1)$. Historically the name 'treatment combination' has been used to mean 'trial'.

Notice that we distinguish 'trial' from 'observation' in that we can have repeated observations of the *same* trial; these are referred to as *replications*.

If every trial has only one observation, the experiment is referred to as a *single replicate* experiment.

Sample size: The total number of observations (experimental results) in an experiment.

Factorial experiment: An experiment involving the study of more than one factor. Such an experiment is called a '*full-factorial*' if all possible factor-level combinations are being experimented upon; otherwise, it is called a '*fractional-factorial*' or a '*fractional-replicate*'.

Contrast: A linear combination of n observed values $Y_1, \ldots, Y_n$ of the form

$$L = c_1 Y_1 + c_2 Y_2 + \cdots + c_n Y_n$$

is termed a 'contrast' if the sum of the coefficients c_i, $i = 1, \ldots, n$ is zero.

If L' is another contrast expressed as

$$L' = c_1' Y_1' + c_2' Y_2' + \cdots + c_n' Y_n',$$

then L and L' are termed '*orthogonal contrasts*' if

$$c_1 c_1' + c_2 c_2' + \cdots + c_n c_n' = 0,$$

i.e. if the 'inner product' of their coefficients is zero.

The Y_i's could represent single observations, totals of observations or even averages. For example, if A_i, $i = 1, \ldots, k$ is the sum total of the experimental values for all the trials corresponding to the ith level of the k-level factor X_A, then assuming an equal number of data values per level,

$$L(\mathrm{A}) = c_1 \mathrm{A}_1 + \cdots + c_k \mathrm{A}_k$$

is a contrast provided that

$$c_1 + \cdots + c_k = 0.$$

For the case of unequal number of observations per level, then $L(\mathrm{A})$ is a contrast only when

$$r_1 c_1 + r_2 c_2 + \cdots + r_k c_k = 0,$$

where r_i, $i = 1, \ldots, k$ is the number of observations in the ith level of X_A.

2.3 Factors and factorial experiments

2.3.1 Weighing experiment No. 2

Consider a weighing experiment on a scale with an indicator as shown in Fig. 2.14. We can place objects in either pan; the dial indicates the total in the right-hand-side pan *minus* the total in the left-hand-side pan *plus* an unknown correction θ_0. With three objects A, B, and C, whose unknown weights are θ_A, θ_B, and θ_C, consider the full-factorial experiment outlined in Table 2.4.

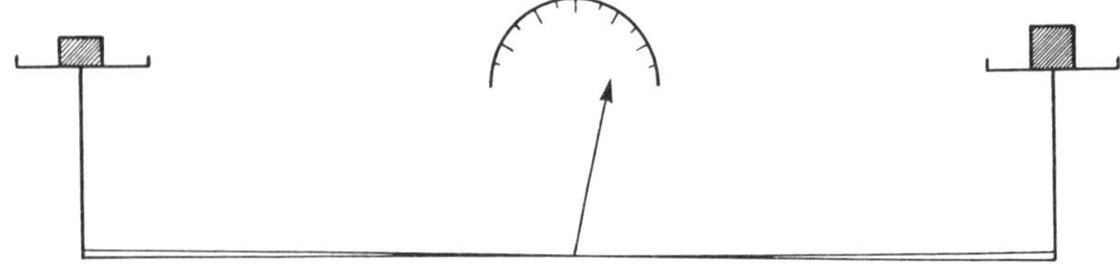

FIG. 2.14. The weighing scale.

Measurement errors ε_j are associated with every dial reading Y_j, $j = 1, \ldots, 8$ and, remembering what the dial display represents every time, we have

$$
\begin{aligned}
Y_1 &= \theta_0 + \theta_A + \theta_B + \theta_C + \varepsilon_1 \\
Y_2 &= \theta_0 - \theta_A + \theta_B + \theta_C + \varepsilon_2 \\
Y_3 &= \theta_0 + \theta_A - \theta_B + \theta_C + \varepsilon_3 \\
Y_4 &= \theta_0 + \theta_A + \theta_B - \theta_C + \varepsilon_4 \\
Y_5 &= \theta_0 - \theta_A - \theta_B + \theta_C + \varepsilon_5 \\
Y_6 &= \theta_0 - \theta_A + \theta_B - \theta_C + \varepsilon_6 \\
Y_7 &= \theta_0 + \theta_A - \theta_B - \theta_C + \varepsilon_7 \\
Y_8 &= \theta_0 - \theta_A - \theta_B - \theta_C + \varepsilon_8
\end{aligned}
$$

and in general

$$Y_j = \theta_0 + x_{A,j}\theta_A + x_{B,j}\theta_B + x_{C,j}\theta_C + \varepsilon_j \qquad j = 1, \ldots, 8, \qquad (2.9)$$

where the $x_{i,j}$'s are ± 1 values attached to the θ_i's for every trial j, with

TABLE 2.4

Trial	Weights in LH pan	Weights in RH pan	Observation on dial
1	—	A, B, C	Y_1
2	A	B, C	Y_2
3	B	A, C	Y_3
4	C	A, B	Y_4
5	A, B	C	Y_5
6	A, C	B	Y_6
7	B, C	A	Y_7
8	A, B, C	—	Y_8

$i = \text{A, B, C}; j = 1, \ldots, 8$. Note that for every i,

$$\sum_{j=1}^{8} x_{i,j} = 0. \tag{2.10}$$

We can find the 'least squares estimates' $\hat{\theta}_0$, $\hat{\theta}_A$, $\hat{\theta}_B$, $\hat{\theta}_C$ for θ_0, θ_A, θ_B and θ_C respectively, by minimizing.

$$\sum_{j=1}^{8} \varepsilon_j^2 = \sum_{j=1}^{8} \{Y_j - (\theta_0 + x_{A,j}\theta_A + x_{B,j}\theta_B + x_{C,j}\theta_C)\}^2.$$

It turns out that the *LSE*s are

$$\hat{\theta}_0 = \tfrac{1}{8}(Y_1 + \cdots + Y_8), \tag{2.11}$$

$$\hat{\theta}_A = \tfrac{1}{8}\left\{\sum_{+1}^{(A)} Y_j - \sum_{-1}^{(A)} Y_j\right\}, \tag{2.12}$$

where $\sum_{+1}^{(A)} Y_j$ represents the sum over all those observations Y_j for which object A was in the right-hand pan (i.e. when $x_{A,j} = +1$), and $\sum_{-1}^{(A)} Y_j$ represents the sum over all Y_j's for which object A was in the left-hand pan (i.e. when $x_{A,j} = -1$). Similarly for $\hat{\theta}_B$ and $\hat{\theta}_C$.

Clearly, eqns (2.11) and (2.12) are equivalent to

$$\hat{\theta}_0 = \frac{1}{n}\sum_{j=1}^{n} Y_j$$

and

$$\hat{\theta}_A = \frac{1}{n}\sum_{j=1}^{n} x_{A,j}Y_j, \tag{2.13}$$

or

$$\hat{\theta}_A = \frac{\sum_j x_{A,j}Y_j}{\sum_j x_{A,j}^2}, \tag{2.14}$$

where n is the sample size.

Note that θ_0 represents the grand average (mean) of all the experimental data values. Notice also that the numerator in eqns (2.12), (2.13), or (2.14) is a contrast of the two levels A(1) and A(2) of A, where A(1) represents all the Y_j's with $x_{A,j} = -1$, and A(2) represents all the Y_j's with $x_{A,j} = +1$. This contrast can be expressed as

$$L_A = c_1 A_1 + c_2 A_2$$

where $c_1 = -1$, $c_2 = +1$, and A_i is the sum total of the Y_j's in level A(i), $i = 1, 2$. So (2.12) can be expressed as

$$\hat{\theta}_A = \tfrac{1}{8}(A_2 - A_1)$$

or, in general,

$$\hat{\theta}_A = \frac{L_A}{m \cdot c} \tag{2.15}$$

where L_A is the contrast of level totals for A ($L_A = A_2 - A_1$), m is the number of observations in each level ($m = 4$) and c is the sum of squares of the coefficients in the contrast (in our case $c = c_1^2 + c_2^2 = (-1)^2 + (+1)^2 = 2$). If there were different number of observations for each level, say m_1 and m_2, a more general formula appropriate for a 2-level factor A is given by

$$\hat{\theta}_A = \frac{L_A}{m_1 c_1^2 + m_2 c_2^2} \tag{2.16}$$

The usefulness of the general eqns (2.14) to (2.16) will become apparent when we consider factors with more than 2 levels (Section 2.5) and when we apply data analysis techniques (Section 2.6).

We will conclude the current section with another generality: in vector and matrix notation, let us have

$$\boldsymbol{\theta} = (\theta_0, \theta_A, \theta_B, \theta_C),$$
$$\mathbf{Y} = (Y_1, \ldots, Y_8)^T,$$

and let $\mathbf{X}$ be the matrix of the $x_{i,j}$'s, $i =$ A, B, C; $j = 1, \ldots, 8$ including a column of '1''s for θ_0, i.e.

$$\mathbf{X} = \begin{bmatrix} 1 & 1 & 1 & 1 \\ 1 & -1 & 1 & 1 \\ 1 & 1 & -1 & 1 \\ 1 & 1 & 1 & -1 \\ 1 & -1 & -1 & 1 \\ 1 & -1 & 1 & -1 \\ 1 & 1 & -1 & -1 \\ 1 & -1 & -1 & -1 \end{bmatrix}$$

Then, the standard matrix formula for $\hat{\boldsymbol{\theta}}$ is

$$\hat{\boldsymbol{\theta}} = (\mathbf{X}^T\mathbf{X})^{-1}\mathbf{X}^T\mathbf{Y},$$

where $\mathbf{X}^T$ represents the transpose matrix of $\mathbf{X}$.

Note that in the current experiment, we used levels of ± 1 for the X-variables/factors, rather than the levels of $(0, 1)$ used in the first weighing experiment (Section 2.2.1).

The advantage of the ± 1 level notation becomes apparent when we have experiments involving many 2-level factors, as in the case described in Section 2.3.2. In the sequel, capital letters A, B, C, etc., will be reserved for representing factors, where $A(i)$ will mean the ith level (or the ith 'cell') of factor A; moreover A_i and $\bar{A}_i$ will respectively represent the total sum and the average of the data values corresponding to level A(i).

2.3.2 2^k-factorial designs

A simple way of understanding the structure of more complicated experiments with several factors is to look at a standard 'tableau'. Here is an example with four controllable 2-level factors and one response variable (taken from Pignatiello and Ramberg 1985).

Process: the heat treatment of leaf springs for trucks.

Response: free height of spring in unloaded position. Deviations below target value (8 inches) are considered undesirable.

Factors

A: Furnace temperature.

B: Heating time.

C: Transfer time (length of time to transfer a part from the heat furnace to the camber former).

D: Hold-down time (length of time that the camber former is closed on a hot part).

In the experiment the factors are set at high and low levels as follows:

Factors	Levels	
	1	2
A Furnace temp. (°F)	A(1) = 1840	A(2) = 1920
B Heating time (sec.)	B(1) = 25	B(2) = 23
C Transfer time (sec.)	C(1) = 12	C(2) = 10
D Hold-down time (sec.)	D(1) = 2	D(2) = 3

Underlying this experiment may be a simple 'linear model'

$$Y = \theta_0 + \theta_A \cdot (A) + \theta_B \cdot (B) + \theta_C \cdot (C) + \theta_D \cdot (D) + \varepsilon. \tag{2.17}$$

We allowed to 'reparametrize' the factors A, B, C, and D into variables which take the values ±1. For example, if *TEMP* is the value of the first factor A, then instead of A we can use in the model

$$X_A = \frac{2\{TEMP - \frac{1}{2}[A(1) + A(2)]\}}{A(2) - A(1)} = \frac{2\{TEMP - \frac{1}{2}(1840 + 1920)\}}{1920 - 1840},$$

so that $TEMP = 1840$ implies $X_A = -1$ and $TEMP = 1920$ implies $X_A = +1$. This recording makes no difference to the statistical analysis. So, the model can now be expressed as

$$Y = \theta_0 + \theta_A X_A + \theta_B X_B + \theta_C X_C + \theta_D X_D + \varepsilon, \tag{2.18}$$

TABLE 2.5

Trial	Factors: Constant	x_A	x_B	x_C	x_D	Y-values					
1	1	−1	−1	−1	−1	7.78	7.78	7.81	7.50	7.25	7.12
2	1	+1	−1	−1	+1	8.15	8.18	7.88	7.88	7.88	7.44
3	1	−1	+1	−1	+1	7.50	7.56	7.50	7.50	7.56	7.50
4	1	+1	+1	−1	−1	7.59	7.56	7.75	7.63	7.75	7.56
5	1	−1	−1	+1	+1	7.94	8.00	7.88	7.32	7.44	7.44
6	1	+1	−1	+1	−1	7.69	8.09	8.06	7.56	7.69	7.62
7	1	−1	+1	+1	−1	7.56	7.62	7.44	7.18	7.18	7.25
8	1	+1	+1	+1	+1	7.56	7.81	7.69	7.81	7.50	7.59

whereas the experiment can now be presented in a 'tableau-form' as shown in Table 2.5. The response data (Y = height values) are also given.

The 'tableau' of Table 2.5 is a 'factorial design' and is usually presented without the 'constant' column. This is a factorial experiment with eight trials, each with six replications. A very important point is that only eight out of the possible $16 = 2^4$ trials were used; a fractional factorial experiment. We say that we have a $\frac{1}{2}$-fraction of a 2^4 full-factorial. In general, for k 2-level factors, there are 2^k possible level combinations (trials) that can be experimented upon, and so a full-factorial experiment requires 2^k trials. Usually, a $1/2^p$-fraction of the full-factorial is utilized $(1 < p < k)$, thus leading to experiments requiring only $(1/2^p) \times 2^k$ or 2^{k-p} trials. We will revert to 'fractional experiments' on numerous occasions throughout this book.

Returning now to the leaf spring experiment, we may set up our model as

$$Y_{j,k} = \theta_0 + \theta_A x_{A,j} + \theta_B x_{B,j} + \theta_C x_{C,j} + \theta_D x_{D,j} + \varepsilon_{j,k}, \qquad (2.19)$$

where the suffix j runs over the 8 trials, whereas the suffix k runs over the six replications within each trial.

The $x_{i,j}$'s ($i = A, B, C, D$; $j = 1, \ldots, 8$) are the ± 1 coded values of the factors X_i, ($i = A, B, C, D$) and these can easily be determined from Table 2.5. For example,

$$Y_{1,1} = 7.78 = \theta_0 + \theta_A(-1) + \theta_B(-1) + \theta_C(-1) + \theta_D(-1) + \varepsilon_{1,1},$$

or

$$Y_{3,5} = 7.56 = \theta_0 + \theta_A(-1) + \theta_B(+1) + \theta_C(-1) + \theta_D(+1) + \varepsilon_{3,5},$$

etc. Note that for every $i = A, B, C, D$ we have $\sum_{j=1}^{8} x_{i,j} = 0$ [as in

(2.10)]. The model (2.19) is similar to the model (2.9) (see Section 2.3.1) except in that, we are now dealing with a total error which includes the error between trials (inter-experimental error) as well as the error within trials (replication error). This extra complication by no means affects the procedure for estimating $\theta_A, \ldots, \theta_D$. The least-squares approach can still be followed, and equations similar to (2.11) to (2.16) can be obtained. In fact, we have

$$\hat{\theta}_0 = \frac{1}{nr} \sum_{j=1}^{n} \sum_{k=1}^{r} Y_{j,k}$$

where n is the number of trials (8) and r is the number of replications in each trial (6).

For our case,

$$\begin{aligned}\hat{\theta}_0 &= \frac{1}{8 \times 6} (\text{sum of all observations}) \\ &= \frac{1}{48} (7.78 + 7.78 + \cdots + 7.50 + 7.59) = 7.636.\end{aligned}$$

Also, similarly as in (2.12)

$$\begin{aligned}\hat{\theta}_A &= \frac{1}{nr} \left\{ \sum_{+1}^{(A)} Y_{j,k} - \sum_{-1}^{(A)} Y_{j,k} \right\} \\ &= \frac{1}{nr} \{(\text{sum of observations at top level of A}) \\ &\quad - (\text{sum of observation at bottom level of A})\} \\ &= \frac{1}{nr} (A_2 - A_1) \quad \text{(in terms of level totals)} \qquad (2.20) \\ &= \tfrac{1}{2}(\bar{A}_2 - \bar{A}_1) \quad \text{(in terms of level averages)},\end{aligned}$$

or similarly as in (2.13)

$$\hat{\theta}_A = \frac{1}{nr} \left(\sum_{j=1}^{n} \sum_{k=1}^{r} x_{A,j} Y_{j,k} \right), \qquad (2.21)$$

or

$$\hat{\theta}_A = \frac{1}{n} \left(\sum_{j=1}^{n} x_{A,j} \bar{Y}_j \right), \qquad (2.22)$$

where $\bar{Y}_j = (1/r) \sum_{k=1}^{r} Y_{j,k}$ is the jth trial mean, i.e. thc average value of the observations at trial j. Equation (2.22) suggests that, for purposes of estimation, it is sufficient to deal only with the average of the observations at each trial as if we had the case of a single replicate experiment.

For our example we have

$$\hat{\theta}_A = \tfrac{1}{48}\{(8.15 + 8.18 + \cdots + 7.44 + 7.59 + \cdots + 7.56 + 7.69 + \cdots + 7.59) - (7.78 + \cdots + 7.25)\}$$
$$= \tfrac{1}{48}(185.92 - 180.61) = 0.1106.$$

We can similarly find $\hat{\theta}_B$, $\hat{\theta}_C$, and $\hat{\theta}_D$.

The simplicity in the estimation procedures for the unknown parameters $\theta_A, \ldots, \theta_D$ is a consequence of some attractive properties that the factor-columns (of the factorial design of Table 2.5) have: for every pair of these columns, each possible combination of levels appears, and it appears the same number of times. These are the properties of 'orthogonality' and 'balance'. For example, in the two columns corresponding to factors X_A and X_B (see Table 2.5), all possible level combinations $(-1, -1)$, $(+1, -1)$, $(-1, +1)$, and $(+1, +1)$ appear, and they appear equally often (twice). If we view the design of Table 2.5 as an array without the first, 'constant', column, it is called an 'orthogonal' design. Much use is made of designs with the orthogonality property in this book. In particular, a detailed description of orthogonal arrays can be found in Chapter 3.

2.3.3 *Interactions*

Let us consider an extended model of the leaf spring example:

$$Y = \theta_0 + \theta_A X_A + \theta_B X_B + \theta_C X_C + \theta_D X_D + \theta_{AB} X_A X_B + \varepsilon. \quad (2.23)$$

The term $\theta_{AB} X_A X_B$ is an 'interaction term'. We can accommodate this term in the experimental design shown in Table 2.5, by extending the 'tableau' to include an extra column, which can be obtained by literally multiplying the X_A and X_B columns term by term, i.e.

X_A	X_B	$X_A X_B$
−1	−1	+1
+1	−1	−1
−1	+1	−1
+1	+1	+1
−1	−1	+1
+1	−1	−1
−1	+1	−1
+1	+1	+1

This procedure of obtaining interactions columns can be applied to any case involving 2-level factors. The $X_A X_B$ column happens to be orthogonal to all the factor columns. We can treat $X_A X_B$ as if it is a new factor in its own right and we can estimate the associated parameter θ_{AB} as before, i.e. $\hat{\theta}_{AB} = 1/nr\{$(sum of observations for which $X_A X_B$ is at top level) – (sum of observations for which $X_A X_B$ is at bottom level)$\}$. Moreover, the model (2.23) can also be expressed as

$$Y_{j,k} = \theta_0 + \theta_A x_{A,j} + \theta_B x_{B,j} + \theta_C x_{C,j} + \theta_D x_{D,j} + \theta_{AB} x_{A,j} x_{B,j} + \varepsilon_{jk}, \tag{2.24}$$

where $j = 1, \ldots, n$; $k = 1, \ldots, r$, and

$$\hat{\theta}_{AB} = \frac{1}{nr}\left\{\sum_{j=1}^{n}\sum_{k=1}^{r}(x_{A,j}x_{B,j})Y_{j,k}\right\} = \frac{1}{n}\left\{\sum_{j=1}^{n}(x_{A,j}x_{B,j})\bar{Y}_j\right\}.$$

Generally, an interaction between factor A and factor B exists, when the effect of A (on the response) differs according to the levels of B or, in other words, when the effect of one factor (on the response) depends upon the value of the other. For example, suppose $\overline{A_1B_i}$ and $\overline{A_2B_i}$ are the averages of all the observations corresponding to levels A(1) and A(2) respectively (for factor A) when factor B is on its ith level, $i = 1, 2$. If the value of $(\overline{A_2B_1} - \overline{A_1B_1})$ differs significantly from the value of $(\overline{A_2B_2} - \overline{A_1B_2})$, then we can say that the interaction effect between factors A and B, is significant. With respect to our example, $\overline{A_1B_1}$ is the average of the observations corresponding to $(X_A, X_B) = (-1, -1)$, i.e. of trials 1 and 5, whereas $\overline{A_2B_1}$ is the average of the observations corresponding to $(X_A, X_B) = (+1, -1)$, i.e. of trials 2 and 6 (see Table 2.5). Therefore

$$\overline{A_1B_1} = \frac{1}{2\times 6}(7.78 + \cdots + 7.12 + 7.94 + \cdots + 7.44) = 7.605,$$

and

$$\overline{A_2B_1} = \frac{1}{2\times 6}(8.15 + \cdots + 7.44 + 7.69 + \cdots + 7.62) = 7.843,$$

and so

$$(\overline{A_2B_1} - \overline{A_1B_1}) = 0.238.$$

Similarly,

$$(\overline{A_2B_2} - \overline{A_1B_2}) = 7.65 - 7.446 = 0.204.$$

Whether or not the two values 0.238 and 0.204 are 'significantly' different, is a matter of 'significance testing' through the 'Analysis of Variance' technique to be discussed in Section 2.6.

We can now distinguish between two types of effects.

Main effects: these are represented by terms such as $\theta_A X_A$, $\theta_B X_B, \ldots,$ in the model. Historically, the 'effect of A' has been used both for θ_A and $\hat{\theta}_A$ and also for $2\theta_A$ and $2\hat{\theta}_A$ indicating the 'total effect' of X_A on the model as X_A changes from '-1' to '$+1$'. Theoretically, the main effect of a certain factor, is the *mean of the effect* by that factor on the experimental values, *taken over* the various levels of the other factors.

Interaction effects: these are represented by terms such as $\theta_{AB}X_AX_B$. We may also have $\theta_{ABC}X_AX_BX_C$ and so forth. Again, there have been slightly different historical usages. Theoretically an interaction effect is a *measure of the extent* to which the value of the response associated with changes in the level of one factor, depends on the level(s) of one or more other factors.

2.4 Aliasing

If we consider again the model (2.23), we can include in it the term

$$\theta_{BC}X_BX_C$$

representing the interaction between factors B and C. The column in the factorial design corresponding to X_BX_C can be a column distinct from all the others and orthogonal to all the factor columns and to the X_AX_B interaction column.

However, if we also try to introduce in the model the term

$$\theta_{AD}X_AX_D$$

we will discover that the associated X_AX_D-column is identical to the X_BX_C-column. This phenomenon is called 'aliasing' (or 'co-linearity' more generally in statistics), and it should always be avoided. We say that the effects $B \times C$ and $A \times D$ form an 'alias group'; only one member of an alias group should be allowed in a model. 'Aliasing' causes confounding of effects and is the consequence of working with fractional designs rather than with full-factorials; in fact, the fewer the trials the greater the aliasing. For example, in the spring leaf experiment, we cannot study both interactions $B \times C$ and $A \times D$, because their effects cannot be distinguished. Of course, if from past experience and/or engineering judgement, it is believed that, say, the $A \times D$ interaction effect is not likely to be significant, then we can proceed with the study of $B \times C$ and attribute the effect only to $B \times C$ and not to $A \times D$.

Therefore a rough prescription for choosing a design is:

(i) write down which terms in a model you think are likely to be near zero;

(ii) make sure that your (fractional) experimental design is such that no terms in the model are aliased.

TABLE 2.6

Trial	Sources						
	A X_A	B X_B	C X_C	A × B X_AX_B	A × C X_AX_C	B × C X_BX_C	A × B × C $X_AX_BX_C$
1	−1	−1	−1	+1	+1	+1	−1
2	−1	−1	+1	+1	−1	−1	+1
3	−1	+1	−1	−1	+1	−1	+1
4	−1	+1	+1	−1	−1	+1	−1
5	+1	−1	−1	−1	−1	+1	+1
6	+1	−1	+1	−1	+1	−1	−1
7	+1	+1	−1	+1	−1	−1	−1
8	+1	+1	+1	+1	+1	+1	+1

2.4.1 Constructing fractional designs

In the simple case of an experiment with only three 2-level factors, the full-factorial design is shown in Table 2.6. The word 'sources' is intended to include factor as well as interaction effects.

The factors A, B and C were assigned on the first three columns. Note that:

(a) All possible combinations of levels ($2^3 = 8$ of them) do appear in the first three columns.

(b) All possible interactions (three 2-way interactions and one 3-way interaction) were assigned to a column which is simply the product of the columns corresponding to the factors in the interaction.

(c) The design of Table 2.6 is an orthogonal and balanced design.

Suppose now that there is an experimental capacity of only four trials (rather than eight). We seek a $\frac{1}{2}$-fraction of the full-factorial with which to estimate the effects of A, B, C interpreted as the parameters θ_A, θ_B, θ_C in a model $Y = \theta_0 + \theta_A X_A + \theta_B X_B + \theta_C X_C$. The design should be such that, at least the effects A, B, and C are not aliased (not confounded). A simple way of achieving this, is to use the third-order interaction A × B × C as a 'knife' to do surgery on the full-factorial.

We can split the full-factorial experiment into two parts:

Part 1: All those trials with A × B × C at bottom level, i.e. with $X_AX_BX_C = -1$. So only the trials 1, 4, 6, and 7 will be included.

Part 2: All those trials with A × B × C at top level, i.e. with $X_AX_BX_C = +1$. This will include trials, 2, 3, 5, and 8.

TABLE 2.7a. Part 1

Trial	A	B	C	A × B	A × C	B × C	A × B × C
1	−1	−1	−1	+1	+1	+1	−1
4	−1	+1	+1	−1	−1	+1	−1
6	+1	−1	+1	−1	+1	−1	−1
7	+1	+1	−1	+1	−1	−1	−1

TABLE 2.7b. Part 2

Trial	A	B	C	A × B	A × C	B × C	A × B × C
2	−1	−1	+1	+1	−1	−1	+1
3	−1	+1	−1	−1	+1	−1	+1
5	+1	−1	−1	−1	−1	+1	+1
8	+1	+1	+1	+1	+1	+1	+1

The two parts are represented in Tables 2.7a and 2.7b. A glance in the designs of Table 2.7 reveals the following:

(a) There is no way of studying the effect of the 3-way interaction A × B × C within any of the two parts.

(b) For each part, every 2-way interaction column is identical to a factor column (except perhaps for a sign change in Part 1). For example in Part 2, the (A × B)-column is identical to the C-column; or, in Part 1, the (A × C)-column is identical to the B-column (up to a sign change), etc.

The above indicates that, if one uses the design of Part 1 or of Part 2, every two-way interaction effect is aliased with a main effect. In fact,

A is aliased with B × C
B is aliased with A × C
C is aliased with A × B

We *must not* include both aliased terms (such as A *and* B × C) in a model. So, in order to use any of the 4-trial designs, for the study of the three factors of interest A, B, and C, it must be assumed that all interaction effects are insignificant. Note, of course, that no main effect is

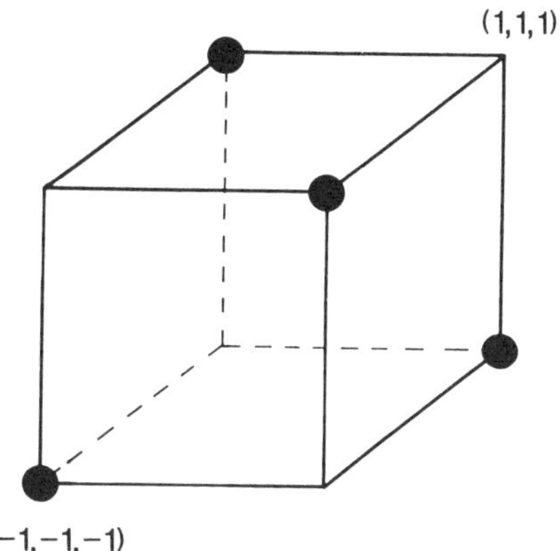

FIG. 2.15. $\frac{1}{2}$-fraction of 2^3 factorial.

aliased with another main effect (such a design is called a 'resolution III design', see Section 3.3.3).

Each 'part-experiment' of Table 2.7 is called a 'block'. Of course, we could still perform the whole experiment of 8 trials; say, by doing block 1 on day 1 and block 2 on day 2. However, if there is a hidden day-effect, then this will be aliased with the A × B × C interaction. This leads to a piece of terminology: 'the interaction A × B × C is confounded with blocks'. It is worth looking at the geometry of the blocking. The 8 corners of the cube of Fig. 2.15 represent the 8 trials of the 2^3-experiment. The dotted corners represent the trials of block 1.

In the leaf spring experiment, it can be easily seen that the design (see Tableau of Table 2.5 without the constant column) was obtained by using the interaction A × B × C × D as the 'knife' column. The resulting design includes only the trials corresponding to the 'top level' of A × B × C × D (it can be easily established that for all trials $X_A X_B X_C X_D = +1$). The aliased pairs for this case are as follows:

A, B × C × D
B, A × C × D
C, A × B × D
D, A × B × C
A × B, C × D
A × C, B × D
A × D, B × C

Note that no main effect is aliased with another main effect or a 2–way interaction: however, every main effect is aliased with a third-order interaction and some two-way interactions are aliased with each other (such a design is otherwise called a 'Resolution IV design' see Section 3.3.3).

So, in general we can split a full-factorial design into half-fractions by using a high-order interaction as a 'knife'. Furthermore, we can split a design into quarter fractions by using two high-order interactions, into eighth-fractions by using three, and so on. In general, using p interactions we can obtain a

$$\frac{1}{2^p} \times 2^k = 2^{k-p} \text{ fractional design}$$

or a 2^{k-p} experiment or namely a $(1/2^p)$-fraction of a 2^k experiment (i.e. of an experiment needing normally 2^k trials for the study of k 2-level factors). A high-order interaction used in such a way for the construction of fractions is otherwise known as a 'defining contrast'.

2.4.2 Recognizing aliasing

There is an algebraic method to establish what aliasing results when a high-order interaction is used for the construction of certain fractions and blocks. We can demonstrate this by using the 2^3 example of Section 2.4.1, with A, B, and C being the factors of interest. It similarly applies to any 2^k factorial experiment. If we define 'I' to represent a 'constant term' or an 'identity factor' in the algebraic sense, i.e. as having the property

$$\mathrm{A} \times I = \mathrm{A}, \qquad \mathrm{B} \times I = \mathrm{B} \quad \text{and} \quad \mathrm{C} \times I = C, \tag{2.25}$$

and if we assume the additional property that

$$I = \mathrm{A} \times \mathrm{A} = \mathrm{B} \times \mathrm{B} = \mathrm{C} \times \mathrm{C},$$

or

$$I = \mathrm{A}^2 = \mathrm{B}^2 = \mathrm{C}^2, \tag{2.26}$$

then we can set up the following 'aliasing rule': Assume that I is identical to the 'defining contrast', i.e. in our case assume

$$I = \mathrm{A} \times \mathrm{B} \times \mathrm{C}. \tag{2.27}$$

Now generate alias groups by 'multiplying' both sides of (2.27) by each effect. The result indicates what is aliased with each effect.

For example, if we multiply both sides of (2.27) by the effect A we have

$$\mathrm{A} \times I = \mathrm{A} \times (\mathrm{A} \times \mathrm{B} \times \mathrm{C}) \Leftrightarrow \mathrm{A} = (\mathrm{A} \times \mathrm{A}) \times \mathrm{B} \times \mathrm{C} \Leftrightarrow \mathrm{A} = \mathrm{B} \times \mathrm{C}.$$

This indicates that the interaction $\mathrm{B} \times \mathrm{C}$ is aliased with the main effect A; in other words, $(\mathrm{A}, \mathrm{B} \times \mathrm{C})$ is an alias group. We can similarly find that

(B, A × C) and (C, A × B) are also alias groups. The aliasing rule based on (2.27), is otherwise known as a rule based on 'confounding A × B × C'.

The above simple method of recognizing aliasing can be generalized to m^k-experiments, i.e. to experiments involving k factors with m-levels each. As far as the 'identity factor' I is concerned, properties (2.25) and (2.27) will still hold, with property (2.26) being generalized to

$$I = \mathrm{A}^{\mathrm{m}} = \mathrm{B}^{\mathrm{m}} \ldots .$$

In experiments involving many multi-level factors, constructing highly fractionated designs and recognizing all alias groups is more complicated. However, many 'ready-made' designs exist which are adequate for the majority of the experimental requirements, and these are described in Chapter 3 (see also Appendices C and D).

2.5 Investigating curvature

For investigating the linear effects of quantitative factors, 2^k full or fractional factorial designs can be used.

In order to assess curvature (non-linear effects) for a particular factor, it is necessary to study more than two levels for that factor. For example, linear as well as quadratic effects for a quantitative factor can be extracted by studying this factor at three equispaced levels; if cubic effects need also to be extracted, four levels should be studied, and so forth.

2.5.1 *Extraction of non-linear effects*

To see first how to extract linear and quadratic effect, let us consider a quantitative factor A with 3 equispaced levels A(1), A(2) and A(3). If A_i represents the response total at level A(i), $i = 1, 2, 3$, we might find the results of Fig. 2.16. If the factor A produces a linear response, the total linear effect from A(1) to A(2) is

$$(\mathrm{A}_2 - \mathrm{A}_1),$$

and the total linear effect from A(2) to A(3) is

$$(\mathrm{A}_3 - \mathrm{A}_2).$$

Therefore the total linear effect for factor A is given by

$$L(\mathrm{A}) = (\mathrm{A}_2 - \mathrm{A}_1) + (\mathrm{A}_3 - \mathrm{A}_2) = \mathrm{A}_3 - \mathrm{A}_1.$$

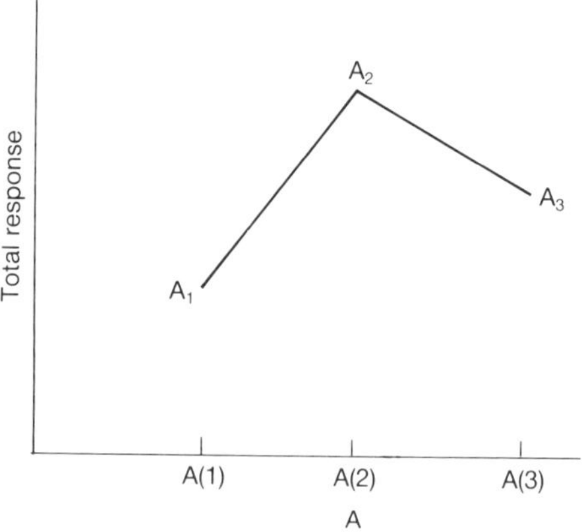

FIG. 2.16. Total linear effect of factor A.

Note that $L(\text{A})$ is a contrast since it can be represented as

$$L(\text{A}) = c_1\text{A}_1 + c_2\text{A}_2 + c_3\text{A}_3,$$

with

$$c_1 = -1, \qquad c_2 = 0, \qquad c_3 = 1,$$

so that

$$\sum_{i=1}^{3} c_i = 0.$$

Now, if the factor A produces a quadratic effect on the response, the slope from A(1) to A(2) will be different from the slope between A(2) and A(3); the difference in slopes will provide an estimate of the quadratic effect $Q(\text{A})$ of factor A, i.e.

$$Q(\text{A}) = (\text{A}_3 - \text{A}_2) - (\text{A}_2 - \text{A}_1) = \text{A}_3 - 2\text{A}_2 + \text{A}_1.$$

Note that $Q(\text{A})$ is also a contrast with

$$c_1' = 1, \qquad c_2' = -2, \qquad c_3' = 1$$

which is orthogonal to $L(\text{A})$, since

$$\sum_{i=1}^{3} c_i c_i' = (-1) \times (+1) + (0) \times (-2) + (1) \times (1) = 0.$$

Similarly, by proper choice of coefficients for the response totals, linear, quadratic as well as cubic effects could be extracted for a quantitative factor which would then need to be studied at four equidistant levels. It can be shown that the appropriate coefficients (in the orthogonal contrasts) would be

Linear:	$C_1 = -3$	$C_2 = -1$	$C_3 = +1$	$C_4 = +3$
Quadratic:	$C_1' = +1$	$C_2' = -1$	$C_3' = -1$	$C_4' = +1$
Cubic:	$C_1'' = -1$	$C_2'' = +3$	$C_3'' = -3$	$C_4'' = +1$

To represent different effects for multi-level quantitative factors, tables of 'orthogonal polynomials' are available which give the proper coefficients depending on the number of the factor levels. Such a table can be found in Appendix E (see Table T4). The theoretical background behind orthogonal polynomials is outlined in Appendix B.

2.5.2 3^n-factorial designs

We have already seen that when a quantitative factor A has three levels, there is a linear as well as a quadratic effect to be considered in any model. For example, in a single factor experiment, a model including terms to represent both effects for A is given by

$$Y = \theta_0 + \theta_{A_L}(A_L) + \theta_{A_Q}(A_Q). \tag{2.28}$$

As in Section 2.3.2, we can re-parameterize A_L and A_Q into variables whose values sum up to zero when varied over all trials [as in (2.10)]. These values are simply the coefficients of the orthogonal contrasts $L(A)$ and $Q(A)$ of Section 2.5.1. So we have

TABLE 2.8

	A	
Levels	X_{A_L}	X_{A_Q}
A(1)	−1	+1
A(2)	0	−2
A(3)	+1	+1

The above two columns are orthogonal in a 'vector sense' since

$$(-1) \times (+1) + (0) \times (-2) + (+1) \times (+1) = 0.$$

So, with actual observations the model (2.28) can be expressed as

$$Y_j = \theta_0 + \theta_{A_L} x_{A_L,j} + \theta_{A_Q} x_{A_Q,j} + \varepsilon_j, \qquad j = 1, 2, \ldots, n,$$

where $x_{A_L,j}$ and $x_{A_Q,j}$ are the values of the variables X_{A_L} and X_{A_Q} respectively (see Table 2.8) for the estimation of θ_{A_L} and θ_{A_Q}.

For example, if we consider the simplest possible case of three observations ($n = 3$), one for each level of A, we have

$$Y_1 = \theta_0 + \theta_{A_L}(-1) + \theta_{A_Q}(+1) + \varepsilon_1$$
$$Y_2 = \theta_0 + \theta_{A_L}(0) + \theta_{A_Q}(-2) + \varepsilon_2$$
$$Y_3 = \theta_0 + \theta_{A_L}(+1) + \theta_{A_Q}(+1) + \varepsilon_3.$$

Through the least squares method we can find the *LSE*s for the θ's as

$$\left.\begin{aligned} \hat{\theta}_0 &= \tfrac{1}{3}(Y_1 + Y_2 + Y_3) \\ \hat{\theta}_{A_L} &= \frac{(Y_3 - Y_1)}{2} \\ \hat{\theta}_{A_Q} &= \frac{(Y_1 - 2Y_2 + Y_3)}{6}. \end{aligned}\right\} \tag{2.29}$$

The above can be easily generalized to n observations as

$$\hat{\theta}_0 = \bar{Y} \text{ (grand average of data values)}$$
$$\hat{\theta}_i = \frac{\sum_{j=1}^{n} x_{i,j} Y_j}{\sum_{j=1}^{n} x_{i,j}^2}, \qquad i = A_L, A_Q \tag{2.30}$$

(note the similarity of (2.30) and (2.14)), or, in terms of level totals,

$$\left.\begin{aligned} \hat{\theta}_{A_L} &= \frac{(A_3 - A_1)}{m \times 2} \\ \hat{\theta}_{A_Q} &= \frac{(A_1 - 2A_2 + A_3)}{m \times 6}, \end{aligned}\right\} \tag{2.31}$$

where A_i is the total sum for level A(i) $i = 1, 2, 3$, and m the number of observations in each level (assuming m is the same at each level).

We can generalize the above formulae even further as

$$\left.\begin{aligned} \hat{\theta}_0 &= \bar{Y} \\ \hat{\theta}_{A_L} &= \frac{L(A)}{m \times C_L} \\ \hat{\theta}_{A_Q} &= \frac{Q(A)}{m \times C_Q}, \end{aligned}\right\} \tag{2.32}$$

where

$$C_L = C_1^2 + C_2^2 + C_3^2 = (-1)^2 + (0)^2 + (+1)^2 = 2$$
$$C_Q = (C_1')^2 + (C_2')^2 + (C_3')^2 = (+1)^2 + (-2)^2 + (+1)^2 = 6.$$

Note the similarity of (2.32) with (2.15). A similar formula to (2.16) can be obtained in place of (2.32) for the case of unequal numbers of observations in each level.

TABLE 2.9

Trial	Factors A	Factors B	Data
1	A(1)	B(1)	Y_1
2	A(1)	B(2)	Y_2
3	A(1)	B(3)	Y_3
4	A(2)	B(1)	Y_4
5	A(2)	B(2)	Y_5
6	A(2)	B(3)	Y_6
7	A(3)	B(1)	Y_7
8	A(3)	B(2)	Y_8
9	A(3)	B(3)	Y_9

Therefore, more generally, if the effect $\mathrm{A_e}$ of a quantitative factor A (linear, quadratic or even cubic, etc., for factors with 4 levels or more) is represented by a contrast of k-level totals of the form

$$L(\mathrm{A_e}) = C_1A_1 + C_2A_2 + \cdots + C_kA_k,$$

then the estimate of the parameter in the model associated with that effect can be calculated by

$$\hat{\theta}_{\mathrm{A_e}} = \frac{L(\mathrm{A_e})}{m \times c}, \tag{2.33}$$

where $c = C_1^2 + \cdots + C_k^2$, and m is the number of data values in each level (with the obvious extension to the case when there is a different m_i for each A(i), $i = 1, \ldots, k$). All the above can be easily extended to an experiment involving two quantitative factors A and B, both at 3 levels: a 3^2-experiment requires 9 trials, which can take place according to the full factorial design of Table 2.9. The two columns of the design of Table 2.9 are balanced and orthogonal to each other because every combination of levels occurs and exists the same number of times (once). Each column represents a main factor effect which can split up into a linear and a quadratic effect thus generating two columns according to Table 2.8. We can also generate interaction columns by simply multiplying the columns corresponding to the effects in the interaction. Altogether, eight columns can be generated according to the tableau of Table 2.10. The values in each column of the tableau in Table 2.10 represent the x_{ij}'s associated with each effect; applying (2.30) we can easily estimate $\theta_{\mathrm{A_L}}$ and similarly $\theta_{\mathrm{A_Q}}$, $\theta_{\mathrm{B_L}}$, $\theta_{\mathrm{B_Q}}$, $\theta_{\mathrm{A_L \times B_L}}$, etc.

TABLE 2.10

Trial	Source								Data
	A_L	A_Q	B_L	B_Q	$A_L \times B_L$	$A_L \times B_Q$	$A_Q \times B_L$	$A_Q \times B_Q$	
1	−1	+1	−1	+1	+1	−1	−1	+1	Y_1
2	−1	+1	0	−2	0	2	0	−2	Y_2
3	−1	+1	+1	+1	−1	−1	+1	+1	Y_3
4	0	−2	−1	+1	0	0	+2	−2	Y_4
5	0	−2	0	−2	0	0	0	+4	Y_5
6	0	−2	+1	+1	0	0	−2	−2	Y_6
7	+1	+1	−1	+1	−1	+1	−1	+1	Y_7
8	+1	+1	0	−2	0	−2	0	−2	Y_8
9	+1	+1	+1	+1	+1	+1	+1	+1	Y_9

Note that every column in the tableau is orthogonal to any other 'in the vector sense'. However, the tableau of Table 2.10 is not an 'orthogonal' design, because, if one considers the x_{ij}'s as 'factor levels' the properties of balance and orthogonality 'in the matrix sense' are frequently violated. For example, the balance (equal number of occurrences for each level) is already broken for A_Q; a new kind of 'level' is created on $A_Q \times B_Q$ ('level' 4) which appears only once; the column of A_L is not orthogonal to A_Q nor to $A_L \times B_Q$ and so forth. However, there are some exceptions: some orthogonal pairs are the following

$$(A_L, B_L), (A_Q, B_L), (A_L, B_Q).$$

Nevertheless, the tableau of Table 2.10 should not be considered as an experimental design. It is only for purposes of model-parameter estimation. The actual experimental design to follow is that of Table 2.9.

Full- and fractional-factorial designs involving many 3-level factors are discussed again in Chapter 3 (see also Appendices C and D). For complex experiments with factors with multiple levels and factors with complicated constraints on their levels, there are two main approaches developed over the last few decades:

(i) 'purely combinational constructions' to be described in full in Chapter 3, and

(ii) response surface methods to be discussed in Chapter 5.

2.6 Analysis of variance

A group of techniques for the analysis of experiments comprises various names, such as analysis of variance, regression analysis, linear hypothesis testing, the F-test, t-test, multiple comparisons, and so on.

A simple way of understanding the subject is the numerical decomposition of the data into various components.

2.6.1 *A simple case*

Returning to the weighing experiment No. 1 (Section 2.2.1) consider the sum of the squares of the observations

$$Y_1^2 + Y_1^2 + Y_3^2.$$

We can write the 'fitted' observations as

$$\begin{aligned} \hat{Y}_1 &= \hat{\theta}_A \\ \hat{Y}_2 &= \hat{\theta}_B \\ \hat{Y}_3 &= \hat{\theta}_A + \hat{\theta}_B, \end{aligned}$$

by merely 'hatting' the original model. Using (2.8) we obtain

$$\begin{aligned} \hat{Y}_1 &= \tfrac{1}{3}(2Y_1 - Y_2 + Y_3) \\ \hat{Y}_2 &= \tfrac{1}{3}(-Y_1 + 2Y_2 + Y_3) \\ \hat{Y}_3 &= \tfrac{1}{3}(Y_1 + Y_2 + 2Y_3). \end{aligned}$$

We may calculate the *residuals* as the difference of the $\hat{Y}$'s from the original data, i.e.

$$\begin{aligned} R_1 &= Y_1 - \hat{Y}_1 = \tfrac{1}{3}(Y_1 + Y_2 - Y_3) \\ R_2 &= Y_2 - \hat{Y}_2 = \tfrac{1}{3}(Y_1 + Y_2 - Y_3) \\ R_3 &= Y_3 - \hat{Y}_3 = -\tfrac{1}{3}(Y_1 + Y_2 - Y_3). \end{aligned}$$

It can readily be verified that

$$(Y_1^2 + Y_2^2 + Y_3^2) = (\hat{Y}_1^2 + \hat{Y}_2^2 + \hat{Y}_3^2) + (R_1^2 + R_2^2 + R_3^2). \tag{2.34}$$

The left-hand side of (2.34) is called 'the (uncorrected) total sum of squares'. The quantity $R_1^2 + R_2^2 + R_3^2$ is called the 'residual sum of squares' (*RSS*).

We may estimate the 'error variance' or 'residual variance' σ_R^2 by

$$S_R^2 = \frac{RSS}{n - p}.$$

where n is the number of observations (sample size) and p the number of parameters in the model. In the example, $n - p = 3 - 2 = 1$. The term $RegSS = \hat{Y}_1^2 + \hat{Y}_2^2 + \hat{Y}_3^2$ is called the regression sum of squares and is the amount of 'variability' in the model (see also Section 2.1.4). It is possible to decompose *RegSS* still further: suppose that we introduce the assumption that $\theta_A = \theta_B = \theta$. Then our model would become

$$Y_1 = \theta + \varepsilon_1$$
$$Y_2 = \theta + \varepsilon_2$$
$$Y_3 = 2\theta + \varepsilon_3.$$

The *LSE* of θ is

$$\hat{\theta} = \tfrac{1}{6}(Y_1 + Y_2 + 2Y_3),$$

which may be verified by solving

$$\frac{\partial R^2}{\partial \theta} = \frac{\partial}{\partial \theta}\{(Y_1 - \theta)^2 + (Y_2 - \theta)^2 + (Y_3 - 2\theta)^2\} = 0.$$

The *RegSS* for this model, which we call $RegSS_H$ (the H referring to the 'hypothesis' that $\theta_A = \theta_B$), is

$$RegSS_H = \hat{\theta}^2 + \hat{\theta}^2 + 4\hat{\theta}^2 = 6\hat{\theta}^2$$
$$= \tfrac{1}{6}(Y_1 + Y_2 + 2Y_3)^2.$$

Our full decomposition is then

$$Y_1^2 + Y_2^2 + Y_3^2 = RegSS_H + (RegSS - RegSS_H) + RSS.$$

The term $(RegSS - RegSS_H)$ is called the 'test sum of squares', i.e.

$$TestSS_H = RegSS - RegSS_H,$$

and represents the *increase* in regression sum of squares as we move from the 'smaller' model (under the assumption H) to the 'larger' model (without the assumption H). If this quantity is large relative to the estimate of σ_R^2, (namely S_R^2), we shall reject the hypothesis H.

The general test, called the F-test, is to calculate the following

$$F\text{-ratio} = \frac{TestSS_H/q}{S_R^2},$$

where q = number of parameters in large model minus number of parameters in small model. In our case, $q = 2 - 1 = 1$.

Under normal distribution assumptions and under the condition H (that the smaller model holds) the F-ratio has a distribution called the F-distribution with q and $n - p$ degrees of freedom. Thus, if the calculated F-ratio is in the tail of this distribution then we reject H. The critical values for the F-distribution are given in Tables T2 in Appendix E. These values depend on the q and $n - p$ degrees of freedom and also on a 'level of significance', α, which is the risk we are prepared to take if we decide to reject the hypothesis H (see also Section 2.1).

2.6.2 *The general case*

The kind of analysis given for the simple weighing experiment No. 1 in Section 2.6.1, is actually numerically harder than that in the case using orthogonal arrays. Indeed, for an orthogonal experimental design we can immediately write down a decomposition of the 'total (uncorrected) sum of squares' into separate terms, one for each parameter and one for the residual.

For example, considering the weighing experiment No. 2 (see Section 2.3.1) we can easily obtain

$$RegSS = \hat{Y}_1^2 + \hat{Y}_2^2 + \cdots + \hat{Y}_8^2 = 8\hat{\theta}_0^2 + 8\hat{\theta}_A^2 + 8\hat{\theta}_B^2 + 8\hat{\theta}_C^2,$$

and so

$$Y_1^2 + \cdots + Y_8^2 = 8\hat{\theta}_0^2 + 8\hat{\theta}_A^2 + 8\hat{\theta}_B^2 + 8\hat{\theta}_C^2 + RSS, \qquad (2.35)$$

where

$$RSS = R_1^2 + \cdots + R_8^2 = \sum_{i=1}^{8} (Y_i - \hat{Y}_i)^2,$$

with

$$\begin{aligned} R_1 &= Y_1 - (\hat{\theta}_0 + \hat{\theta}_A + \hat{\theta}_B + \hat{\theta}_C) \\ &\vdots \\ R_8 &= Y_8 - (\hat{\theta}_0 - \hat{\theta}_A - \hat{\theta}_B - \hat{\theta}_C). \end{aligned}$$

Each parameter θ_0, θ_A, θ_B, and θ_C in the original model (2.9) has its own sum of squares given by one of the first four terms in the right-hand side of (2.35), i.e.

$$\begin{aligned} SS(\theta_0) &= 8(\hat{\theta}_0)^2 \\ SS(\theta_i) &= 8(\hat{\theta}_i)^2 \qquad i = \text{A, B, C}. \end{aligned} \qquad (2.36)$$

Equation (2.36) can be used to find the sum of squares of any two-level factor. Note that

$$SS(\theta_0) = 8(\bar{Y})^2 = (T)^2/8, \qquad (2.37)$$

where T = total sum of all observations. The value of $SS(\theta_0)$ as given by (2.37) is called the 'correction factor' (CF) and can be used to obtain the so-called 'total corrected sum of squares' TSS as

$$TSS = [\text{total (uncorrected) sum of squares}] - [\text{correction factor}],$$

i.e.

$$TSS = \sum_i Y_i^2 - CF. \qquad (2.38)$$

As in Section 2.6.1 we can build up a '$TestSS_H$' by testing a hypothesis

$$\text{H: } \theta_A = \theta_B = \theta_C = 0$$

as

$$TestSS_H = RegSS - RegSS_H,$$

or

$$TestSS_H = SS(\theta_0) + SS(\theta_A) + SS(\theta_B) + SS(\theta_C) - RegSS_H,$$

where we can easily find

$$RegSS_H = 8(\bar{Y})^2,$$

and then calculate

$$F_H\text{-ratio} = \frac{TestSS_H/q}{S_R^2},$$

where

$$q = 4 - 1 = 3$$

and

$$\hat{\sigma}_R^2 = S_R^2 = RSS/(n-p),$$

with $n = 8$ and $p = 4$.

Another special case of the above would be to test for

$$\text{H:}\ \ \theta_0 = \theta_A = \theta_B = \theta_C = 0,$$

or even in particular

$$\text{H:}\ \ \theta_i = 0,$$

where $i = $ A, B, or C. In the last case it turns out that

$$Test\ \text{H} = SS(\theta_i), \qquad i = \text{A, B, C}.$$

In general, in order to test for significance of a particular effect A_e for factor A, it suffices to calculate the

$$F\text{-ratio for } A_e = \frac{SS(\theta_{A_e})/q(A_e)}{S_R^2}$$

and determine whether it is greater than the critical value from the one-sided F-tables (Appendix E) at some level of significance α. If this F-ratio is greater than the critical value, we say that we cannot accept the hypothesis that the effect A_e is not significant (i.e. that $\theta_{A_e} = 0$) and there is a risk of $\alpha\%$ that we are wrong in saying so.

The value $q(A_e)$ is called the degrees of freedom for A_e and depends on the effect of A_e. For example, if A_e is the linear effect of a 3-level factor A, then $q(A_e) = 1$. But if it represents the total main effect of a 3-level factor A then $q(A_e) = 2$. This is because the main effect for A includes both the linear as well as the quadratic effect of A (see Section 2.5). The degrees of freedom corresponding to a main effect of a factor is always one less than the number of levels for that factor. The degrees of freedom corresponding to an interaction is the product of the individual degrees of freedom for the effects in the interaction.

TABLE 2.11. ANOVA table

Source	df	S	$MS = S/\text{df}$	F-ratio
θ_1	df_1	$SS(\theta_1)$	$SS(\theta_1)/\text{df}_1$	MS_1/S_R^2
θ_2	df_2	$SS(\theta_2)$	$SS(\theta_2)/\text{df}_2$	MS_2/S_R^2
$\vdots$	$\vdots$	$\vdots$	$\vdots$	$\vdots$
Residual	df_R	$\sum_j (Y_j - \hat{Y})^2 = RSS$	$RSS/\text{df}_R = S_R^2$	
TSS	$n-1$	$\sum Y_j^2 - CF$		

We usually present the results of the sum of squares decomposition and F-testing in an 'Analysis of variance' (ANOVA) table in the format of Table 2.11.

In the ANOVA table, df stands for 'degrees of freedom', S for Sum of squares and MS stands for the mean sum of squares (or variance) of the effect θ_i, obtained by dividing the sum of squares for θ_i to the degrees of freedom df_i corresponding to θ_i. 'Source' covers anything from main-effects, interaction effects, to linear quadratic effects, etc.

An easy way of calculating the RSS (otherwise simply called the 'error' sum of squares) is by subtracting the sum of the squares of all the effects from the total TSS; the residual degrees of freedom can be obtained in a similar way. The mean sum of squares for the residual provides an estimate of the 'error variance' σ_e^2 (or σ_R^2). Usually, an asterisk is placed next to the value of the F-ratio if this value is significant at 5% level; and two asterisks if significant to 1% level.

2.6.3 ANOVA formulae

2.6.3.1 Formulae using contrasts

We have already seen that a simple formula for calculating the sum of squares for the main effect of a two-level factor A is given by (see (2.36)).

$$S_A = SS(\theta_A) = N(\hat{\theta}_A)^2, \tag{2.39}$$

where N is the total number of data values. Note that if A is a quantitative factor, (2.39) gives the sum of squares of the linear effect of A, which, since A has only two levels, coincides with the main effect of A. Different formulae for $SS(\theta_A)$ can be obtained by considering different formulae for $\hat{\theta}$. For example, in terms of level totals [see (2.20)]

$$S_{\mathrm{A}} = SS(\theta_{\mathrm{A}}) = \frac{(A_2 - A_1)^2}{N}, \tag{2.40}$$

or in terms of contrasts [see, for example, (2.15)], for an effect $\mathrm{A_e}$

$$S_{\mathrm{A_e}} = SS(\theta_{\mathrm{A_e}}) = \frac{[L(\mathrm{A_e})]^2}{mc} = mc[\hat{\theta}_{\mathrm{A_e}}]^2, \tag{2.41}$$

where in this case, $L(\mathrm{A_e}) = -A_1 + A_2$, m is the number of data values in each A-level and c is the sum of squares of the coefficients in the contrast, in this case $c = (-1)^2 + (+1)^2 = 2$. Equation (2.41) can be applied for the sum of squares of any factor effect $\mathrm{A_e}$ which can be represented by a contrast of the form

$$L(\mathrm{A_e}) = c_1 A_1 + \cdots + c_k A_k,$$

with

$$c = \sum_{i=1}^{k} c_i^2,$$

[see, for example, (2.32)], assuming an equal number of observations per level. For unequal m_i's for each level A(i), $i = 1, \ldots, k$, we have

$$S_{\mathrm{A_e}} = SS(\theta_{\mathrm{A_e}}) = \frac{(c_1 A_1 + \cdots + c_k A_k)^2}{m_1 c_1^2 \cdots + m_k c_k^2}. \tag{2.42}$$

So, assuming that $\mathrm{A_e}$ can be represented by a contrast of level totals, then (2.41) (or (2.42)) constitutes a *variation of 1 degree of freedom* and it is one component of the total variation due to factor A whose sum of squares is represented by S_{A}.

If $\mathrm{A_e'}$ is another effect whose contrast $L(\mathrm{A_e'})$ is orthogonal to $L(\mathrm{A_e})$, in the sense that the inner product of their coefficients is zero (see Section 2.2.3), then $SS(\theta_{\mathrm{A_e'}})$ is another component of variation for A with 1 degree of freedom; moreover $SS(\theta_{\mathrm{A_e}})$ and $SS(\theta_{\mathrm{A_e'}})$ are 'mutually independent'. In fact, if A is a k-level factor, S_{A} can be resolved to $k-1$ mutually independent components. (Recall that there are $k-1$ degrees of freedom corresponding to A.) For example, if A is a 3-level quantitative factor, we have [see (2.32)].

$$S_{\mathrm{A}} = [SS(\theta_{\mathrm{A_L}}) + SS(\theta_{\mathrm{A_Q}})] = \frac{[L(\mathrm{A})]^2}{mC_{\mathrm{L}}} + \frac{[Q(\mathrm{A})]^2}{mC_{\mathrm{Q}}},$$

where $L(\mathrm{A})$ and $Q(\mathrm{A})$ represent respectively the contrasts for the linear and quadratic effects of A (see Section 2.5.2).

Resolution of variation by the use of contrasts is not restricted only to quantitative factors. For example, if we wish to investigate the differences among five products P(1), . . . , P(5), two of which P(1) and P(2) originate from company A, and the rest from company B, one could resolve the differences among the P(i)'s as

$$S_p = SS_1 + SS_2 + SS_3,$$

i.e.

$$\begin{aligned} S_p = {} & S(\text{difference between company A and company B}) \\ & + S(\text{difference between P(1) and P(2) within A}) \\ & + S(\text{difference among P(3), P(4), and P(5) within B}). \end{aligned}$$

To compare between companies one could use the contrast

$$L_1 = 3(P_1 + P_0) - 2(P_3 + P_4 + P_5)$$

or

$$L_1 = \frac{(P_1 + P_2)}{2m} - \frac{(P_3 + P_4 + P_5)}{3m},$$

where P_i is the sum of the characteristic values for level $P(i)$, $i = 1, \ldots, 5$ and m is the number of values in each level. In either case, applying (2.41) we can obtain

$$SS_1 = \frac{[3(P_1 + P_2) - 2(P_3 + P_4 + P_5)]^2}{30m},$$

which is one component of variation for factor P with 1 degree of freedom.

A second component with one degree of freedom can be easily found to be

$$SS_2 = \frac{(P_1 - P_2)^2}{2m},$$

(using the contrast $L_2 = (+1)P_1 + (-1)P_2 + (0)(P_3 + P_4 + P_5)$ which is orthogonal to L_1), whereas the third will be a component with *two* degrees of freedom (comparison of 3 qualitative levels) and can be found by subtraction, i.e.

$$SS_3 = S_p - SS_1 - SS_2.$$

2.6.3.2 Formulae for main effects and interactions

If the total effect of a particular factor is not significant, there is little point in resolving it into components of variation using orthogonal contrasts. Usually, we first calculate the main effects of all the factors, and resolve only the significant ones. There is an easy procedure to follow for the calculation of main effects and interactions.

First calculate the correction factor:

$$CF = \frac{(\text{sum total of all observations})^2}{\text{total number of all observations}} = \frac{(\Sigma_j Y_j)^2}{N}. \tag{2.43}$$

Then, the total sum of squares (corrected):

$$TSS = [(\text{sum of all observations squared}) - (\text{correction factor})],$$

i.e.

$$TSS = \sum_{j=1}^{N} Y_j^2 - CF. \tag{2.44}$$

The total degrees of freedom is equal to $N-1$.

For the sum of squares of the main effect of a k-level factor A, with m observations for each level, we use the following formula

$$S_A = \frac{(A_1)^2 + \cdots + (A_k)^2}{m} - CF, \tag{2.45}$$

where A_i is the sum total in level $A(i)$, $i = 1, \ldots, k$. If the number of observations per level is different , m_i, say, $i = 1, \ldots, k$, then

$$S_A = \frac{(A_1)^2}{m_1} + \cdots + \frac{(A_k)^2}{m_k} - CF. \tag{2.46}$$

The degrees of freedom corresponding to a k-level factor is equal to $k-1$.

For interaction effects, say between factors A and B, similar formulae can be applied. We have already seen that when A and B are 2-level factors, an 'interaction column' can be easily constructed (see Section 2.3.3) which can be treated as a column corresponding to a 'new' 2-level factor, in which case (2.45) with $k=2$, can still be applied for the calculation of the interaction sum of squares. Of course, one should be careful to avoid aliasing (see Section 2.4), i.e. to avoid the case when a column corresponding to an interaction of interest is already assigned to another factor. For certain fractional designs, interaction tables exist which facilitate the checking for aliasing between main effects and 2-way interaction effects. These will be described in Chapter 3 (see also Appendix D). For 3-level factors, these interaction tables will show that there are *two* 3-level columns corresponding to a 2-way interaction. In such a case these two columns can be considered to correspond to two new factors, in which case (2.45) with $k=3$ can be applied for each new factor with the two results being added together to obtain the interaction sum of squares. However, there is a general formula which can be applied for the sum of squares of interaction effects, which avoids the separate calculation of the sum of squares of all the 'columns' corresponding to the interaction.

For example, if A is an a-level factor and B is a b-level factor then, assuming A × B is not aliased with another effect,

$$S_{\mathrm{A}\times\mathrm{B}} = \frac{(A_1B_1)^2 + (A_1B_2)^2 + (A_2B_1)^2 + \cdots + (A_aB_b)^2}{m} - S_{\mathrm{A}} - S_{\mathrm{B}} - CF, \tag{2.47}$$

where A_iB_j is the sum total of the observations corresponding to the combination level A(i)B(j) $i = 1, \ldots, a$; $j = 1, \ldots, b$ (i.e. of the data values obtained when both levels A(i) and B(j) were used) and m is now the number of observations in each combination level, assumed the same for all levels. If m is not the same, and there are, say, m_{ij} values in the combination level A(i)B(j), then

$$S_{\mathrm{A}\times\mathrm{B}} = \frac{(A_1B_1)^2}{m_{11}} + \cdots + \frac{(A_aB_b)^2}{m_{ab}} - S_{\mathrm{A}} - S_{\mathrm{B}} - CF, \tag{2.48}$$

with df $= (a-1)(b-1)$. Equation (2.48) [and so (2.47)] can be extended for three-way interactions (involving a third factor C with c levels) as follows:

$$S_{\mathrm{A}\times\mathrm{B}\times\mathrm{C}} = \frac{(A_1B_1C_1)^2}{m_{111}} + \cdots + \frac{(A_aB_bC_c)^2}{m_{abc}}$$
$$- S_{\mathrm{A}} - S_{\mathrm{B}} - S_{\mathrm{C}} - S_{\mathrm{A}\times\mathrm{B}} - S_{\mathrm{A}\times\mathrm{C}} - S_{\mathrm{B}\times\mathrm{C}} - CF.$$

Obvious generalizations can be obtained for higher-order interactions. Components of interaction sum of squares for quantitative factors with more than 2 levels (e.g. $S_{\mathrm{A_L}\times\mathrm{B_L}}, S_{\mathrm{A_L}\times\mathrm{B_Q}}$ etc.) can be found using orthogonal polynomials (see Section 2.5 and Appendix B). A detailed description of steps to follow can be found in Section 3.3.2.5. Otherwise, columns such as the ones depicted in the tableau of Table 2.10 (see Section 2.5.2) can easily be constructed, and a special case of eqn (2.41) can be applied.

Indeed, a special case of (2.41) occurs when dealing with a contrast of observations Y_i, i.e. when

$$L(\mathrm{A_e}) = c_1Y_1 + \cdots + c_nY_n,$$

in which case

$$S_{\mathrm{A_e}} = SS(\theta_{\mathrm{A_e}}) = \frac{(c_1Y_1 + \cdots + c_nY_n)^2}{c_1^2 + \cdots + c_n^2}. \tag{2.49}$$

For example, if we are interested on the linear effect of A in the (hypothetical) 3^2-experiment of Table 2.9 (Section 2.5.2) we can use the first column of the tableau of Table 2.10 and eqn (2.49) to obtain

$$S_{A_L} = SS(\theta_{A_L})$$
$$= \frac{[(-1)Y_1 + (-1)Y_2 + (-1)Y_3 + (0)(Y_4 + Y_5 + Y_6) + (+1)(Y_7 + Y_8 + Y_9)]^2}{(-1)^2 + (-1)^2 + (-1)^2 + (0)^2 + (0)^2 + (0)^2 + (+1)^2 + (+1)^2 + (+1)^2}.$$

We would have obtained the same result had we applied eqn (2.41) as

$$S_{A_L} = SS(\theta_{A_L}) = \frac{[(-1)A_1 + (0)A_2 + (+1)A_3]^2}{3 \times [(-1)^2 + (0)^2 + (+1)^2]}.$$

But, eqn (2.49) is particularly useful for other component effects in the tableau of Table 2.10. For example, using the last column of the tableau, the interaction between the quadratic effects of A and B can be found as

$$S_{A_Q \times B_Q} = SS(\theta_{A_Q \times B_Q}) = \frac{(Y_1 - 2Y_2 + Y_3 - 2Y_4 + 4Y_5 - 2Y_6 + Y_7 - 2Y_8 + Y_9)^2}{4 \times (+1)^2 + 4 \times (-2)^2 + (4)^2}.$$

2.6.3.3 Inter-experimental and replication error

We have seen that an estimate of the error variance σ_e^2 is given by the 'mean residual sum of squares'

$$S_R^2 = MS_e = \frac{RSS}{df_R}$$

against which every source (effect) can be tested for significance. The error variance represents the error between experimental trials ('inter-experiment error' or 'first-order variance') and (in the case when there are replications in each trial) the error within experimental trials (otherwise known as 'error among replications' or 'second-order variance' or simply 'replication error').

In the case of a replicated factorial experiment, the *RSS* can be resolved into the 'inter-experiment' error sum of squares or first-order error sum of squares S_{e_1} and the 'replication' or second-order error sum of squares S_{e_2}, so that

$$RSS = S_{e_1} + S_{e_2}.$$

In such a case, the inter-experiment error e_1 can be tested (for significance) against the second-order variance using

$$F\text{-ratio for } e_1 = \frac{MSe_1}{MSe_2},$$

where $MSe_1 = S_{e_1}/\mathrm{df}_{e_1}$ (first-order variance), and $MS_{e_2} = S_{e_2}/\mathrm{df}_{e_2}$ (second-order variance). If e_1 is found to be non-significant, then it can be 'pooled' together with e_2, and RSS will then be used for the testing of the effects under consideration in the usual way. However, if e_1 is found to be significant, then every effect under consideration (main effect, interaction effect, etc.) should be tested against the first-order error variance MSe_1.

Assuming that n is the number of trials with r replications each, S_{e_1} can be calculated by first considering every trial to be a level of an 'experimental-trial-factor', Tr, whose sum of squares can be calculated in the usual way of applying (2.45), i.e.

$$S_{\mathrm{Tr}} = \frac{(\text{sum total in trial } 1)^2 + \cdots + (\text{sum total in trial } n)^2}{r} - CF.$$

Note that if $r = 1$, then $S_{\mathrm{Tr}} = TSS$. The trial-factor Tr has n levels and corresponds to $n - 1$ degrees of freedom; it accounts for the total variation among all trial runs of the experiment. Some parts of this variation are accounted for by the sources under consideration; the remaining part represents the first-order error variation e_1. So, S_{e_1} can be found by subtracting the sum of squares of all the sources from S_{Tr}. So

$$S_{e_1} = S_{\mathrm{Tr}} - \left\{\sum_{\mathrm{Ef}} S_{\mathrm{Ef}}\right\}, \tag{2.50}$$

where the summation symbol is over all the effects (Ef) considered. By the same principle

$$\mathrm{df}_{e_1} = (n-1) - \sum_{\mathrm{Ef}} \mathrm{df}_{\mathrm{Ef}}.$$

Clearly, if

$$\sum_{\mathrm{Ef}} \mathrm{df}_{\mathrm{Ef}} = n - 1,$$

then e_1 is not retrievable ($S_{e_1} = 0$) and only e_2 can be considered. S_{e_2} can be calculated either by subtraction from TSS, i.e. from

$$S_{e_2} = TSS - \sum_{\mathrm{Ef}} S_{\mathrm{Ef}} - S_{e_1} = TSS - S_{\mathrm{Tr}}, \tag{2.51}$$

or by the following formula

$$S_{e_2} = S_{\mathrm{Tr}(1)} + S_{\mathrm{Tr}(2)} + \cdots + S_{\mathrm{Tr}(n)},$$

where $S_{\mathrm{Tr}(i)}$ is the total (corrected) sum of squares for the data values *within* the ith trial, $i = 1, \ldots, n$. For example, if $Y_{i1}, \ldots, Y_{ir}$ are the r replications in the ith trial then

$$S_{\mathrm{Tr}(i)} = (Y_{i1}^2 + \cdots + Y_{ir}^2) - \frac{(Y_{i1} + \cdots + Y_{ir})^2}{r} \qquad (\mathrm{df} = r - 1).$$

TABLE 2.12a

Trial	Design A	B	C	D	A×B	Data(*Y*)					
1	−1	−1	−1	−1	+1	7.78	7.78	7.81	7.50	7.25	7.12
2	+1	−1	−1	+1	−1	8.15	8.18	7.88	7.88	7.88	7.44
3	−1	+1	−1	+1	−1	7.50	7.56	7.50	7.50	7.56	7.50
4	+1	+1	−1	−1	+1	7.59	7.56	7.75	7.63	7.75	7.56
5	−1	−1	+1	+1	+1	7.94	8.00	7.88	7.32	7.44	7.44
6	+1	−1	+1	−1	−1	7.69	8.09	8.06	7.56	7.69	7.62
7	−1	+1	+1	−1	−1	7.56	7.62	7.44	7.18	7.18	7.25
8	+1	+1	+1	+1	+1	7.56	7.81	7.69	7.81	7.50	7.59

Note that an estimate for the variance σ_i^2 at trial i, is given by $S_i^2 = [1/(r-1)]S_{\mathrm{Tr}(i)}$. In general,

$$S_{e_2} = \sum_{i=1}^{n}\left\{\sum_{j=1}^{r} Y_{ij}^2 - \frac{(\sum_{j=1}^{r} Y_{ij})^2}{r}\right\}, \tag{2.52}$$

with

$$\mathrm{df}_{e_2} = n(r-1).$$

If the number of replications is different from trial to trial, say r_i for the ith trial, the ANOVA calculations differ slightly. The procedure to follow is outlined in Chapter 4 (Section 4.5.1).

2.6.3.4 Numerical example

Using the ANOVA principles we will analyse the data of the case-study mentioned in Section 2.3.2, concerning the heat treatment of leaf springs for trucks. We need to investigate the effects of four factors, furnace temperature (A), heating time (B), transfer time (C) and hold-down time (D) on the spring height which, in unloaded position, is required to be over 8 inches. The interaction effect between A and B was also to be investigated. The experimental design and data is presented in Table 2.12a.

Analysis

We have $n = 8$, $r = 6$ and $N = n \times r = 48$. First we calculate the correction factor (eqn (2.43))

$$CF = \frac{(\sum Y)^2}{N} = \frac{(7.78 + \cdots + 7.59)^2}{48} = \frac{(366.53)^2}{48} = 2798.838.$$

Then, the total sum of squares (applying (2.44))

$$TSS = \sum Y^2 - CF = 7.78^2 + \cdots + 7.59^2 - CF = 2.966 \qquad (\text{df} = 47).$$

For the main effect of A (applying (2.45))

$$S_A = \frac{A_1^2 + A_2^2}{4 \times 6} - CF$$

$$= \frac{(7.78 + \cdots + 7.12 + 7.50 + \cdots + 7.25)^2 + (8.15 + \cdots + 7{\cdot}59)^2}{24} - CF$$

$$= \frac{(180.61)^2 + (185.92)^2}{24} - 2798.838 = 0.587 \qquad (\text{df} = 2 - 1 = 1).$$

Similarly

$$S_B = \frac{B_1^2 + B_2^2}{24} - CF = \frac{(185.38)^2 + (181.15)^2}{24} - CF = 0.373 \qquad (\text{df} = 1)$$

$$S_C = 0.01, \qquad S_D = 0.129 \qquad (\text{df} = 1).$$

Since we are dealing with two-level factors, the interaction sum of squares can be calculated by considering A × B as a factor in its own right and by applying (2.45). Instead we will apply the more general formula (2.47).

$$S_{A \times B} = \frac{(A_1B_1)^2 + (A_2B_1)^2 + (A_1B_2)^2 + (A_2B_2)^2}{12} - S_A - S_B - CF$$

$$= \tfrac{1}{12}\{(7.78 + \cdots + 7.12 + 7.94 + \cdots + 7.44)^2 + (8.15 + \cdots + 7.62)^2$$
$$+ (7.5 + \cdots + 7.25)^2 + (7.59 + \cdots + 7.59)^2\}$$
$$- 0.587 - 0.373 - 2798.838 = 0.004 \qquad (\text{df} = 1).$$

Since there are six replications per trial we need to assess the between-trials error e_1 and the within-trials (replication) error e_2. Applying (2.50)

$$S_{e_1} = S_{Tr} - \{S_A + S_B + S_C + S_D + S_{A \times B}\},$$

where

$$S_{Tr} = \tfrac{1}{6}\{(\text{sum of data in Trial 1})^2 + \cdots + (\text{sum in Trial 8})^2\} - CF$$

$$= \frac{(7.78 + \cdots + 7.12)^2 + \cdots + (7.56 + \cdots + 7.59)^2}{6} - CF = 1.122,$$

and so

$$S_{e_1} = 1.122 - [0.587 + 0.373 + 0.01 + 0.129 + 0.004] = 0.019,$$

with

$$\text{df}_{e_1} = (n - 1) - [\text{df}_A + \text{df}_B + \text{df}_C + \text{df}_D + \text{df}_{A \times B}] = 2.$$

TABLE 2.12b. ANOVA for the leaf spring experiment

Source	df	S	MS	F-ratio
A	1	0.587	0.587	13.3**
B	1	0.373	0.373	8.5
C	1	0.01	0.01	0.2
D	1	0.129	0.129	2.93
A × B	1	0.004	0.004	0.1
e_1	2	0.019	0.0095	
e_2	40	1.844	0.0461	
$(e_{1,2})$	(42)	(1.863)	(0.044)	
Total	47	2.966		

Finally, by subtraction (applying (2.51))

$$S_{e_2} = TSS - \{S_A + S_B + S_C + S_D + S_{A\times B}\} - S_{e_1} = TSS - S_{Tr} = 1.844,$$

$$\text{with } df_{e_2} = df_{tot} - \{df_A + \cdots + df_{A\times B}\} - df_{e_1} = 47 - 5 - 2 = 40.$$

The ANOVA Table is as shown in Table 2.12b. Note that e_1, when tested against e_2 is not significant, and so it is pooled together with e_2 to create $e_{1,2}$ against which all the effects are tested.

The F-ratios of all the effects are compared with a critical value from the F-tables (Table T2.2—Appendix E) with 1 and 42 degrees of freedom; we have

$$cv = F(1, 42; 5\%) = 4.08 \quad \text{and} \quad F(1, 42; 1\%) = 7.31.$$

This means that factors A and B are significant at $\alpha = 1\%$ level. Since $F(1, 42; 10\%) = 2.84$, factor D is significant at the 10% level.

2.6.3.5 Multiple comparisons

If the experimental study is concerned with only one factor, then we are dealing with a 'one-way ANOVA', which can be considered as a generalization of the 2-sample t-test (see Section 2.1.2.1). In fact, if the single factor of interest has only two levels then the 1-way ANOVA F-test is equivalent to the 2-sample t-test.

When the factor has more than two levels, the ANOVA technique is a convenient way of testing the general (null) hypothesis that there is no difference among that level-means. A significant F-ratio indicates that *there is* a difference among the level means (i.e. that there is a 'factor-effect') but it does not indicate *which* of the level means are

TABLE 2.13

Batch	Breakdown voltages (kV)				
B(1)	2.70	2.80	2.45	2.62	2.90
B(2)	3.01	2.90	3.20	3.13	2.77
B(3)	2.72	3.05	2.68	3.11	2.78
B(4)	2.50	2.68	3.13	2.40	2.92
B(5)	2.30	2.80	2.93	2.69	3.16
B(6)	2.94	2.75	3.15	3.22	3.30
B(7)	3.32	3.25	2.95	2.85	3.02
B(8)	2.86	2.65	3.09	2.94	2.96
B(9)	2.55	2.73	2.96	2.75	2.87
B(10)	2.98	3.28	3.17	3.40	3.50

different and we would need to carry out two-sample t-tests on all possible pairs of levels. However, in the case of multi-level factors, this would require too many significance tests, and an alternative approach is to use a multiple comparison test such as the 'Duncan's test', due to D. B. Duncan (1956). This can be demonstrated using the data of Table 2.2 of Section 2.1.3, which is reproduced in Table 2.13.

An analysis of variance of these data (see Table 2.13a) has indicated a strong 'batch effect' i.e. evidence of differences between the 10 batches (i.e. the 10 levels of the batch factor). In order to assess which of the batches are different using only one test we first calculate and then arrange the batch means (level means) in order of magnitude as shown in Table 2.13b.

Using Table T5 of Appendix E, we can now obtain coefficients, using a 5% significance level, for comparing between 2 and 10 batch means, on the basis of 40 degrees of freedom (the residual df). These coefficients are shown in Table 2.13c.

The expression 'spanned batch means' refers to all level means of the batch factor which are spanned by any two means inclusive of the two

TABLE 2.13a

Source	df	S	MS	F-ratio
Between batches	9	1.5438	0.1715	3.63**
Within batches (residual)	40	1.8881	0.0472	
Total	49	3.4318		

TABLE 2.13b

Batch level	B(1)	B(4)	B(9)	B(5)	B(3)	B(8)	B(2)	B(6)	B(7)	B(10)
Mean	2.694	2.726	2.772	2.776	2.868	2.900	3.002	3.072	3.078	3.266

TABLE 2.13c

No. of spanned batch means	2	3	4	5	6	7	8	9	10
Coefficients from Table T5	2.86	3.01	3.10	3.17	3.22	3.27	3.30	3.33	3.35

TABLE 2.13d

No. of spanned means	2	3	4	5	6	7	8	9	10
5% LSD	0.278	0.292	0.301	0.308	0.313	0.318	0.321	0.324	0.326

TABLE 2.13e

Batch	B(1)	B(4)	B(9)	B(5)	B(3)	B(8)	B(2)	B(6)	B(7)	B(10)
Mean	2.694	2.726	2.772	2.776	2.868	2.900	3.002	3.072	3.078	3.266

means; thus, from Table 2.13b, B(1) and B(9) have a span of 3 batch means, or, B(4) and B(2) have a span of 6 batch means, etc.

We can now obtain a measure called 'least significant difference' (*LSD*) for the 5% level using the formula

$$5\%\ LSD = (\text{coefficient from Table T5}) \times \left\{\frac{\text{residual mean sum of squares}}{\text{number of observations in each level}}\right\}^{\frac{1}{2}}.$$

Therefore with a residual *MS* of 0.0472 and 5 observations per batch, the *LSD*s are as shown in Table 2.13d.

We can first test the difference between the means of levels B(1) and B(4) (see Table 2.13b) whose difference is

$$2.726 - 2.694 = 0.032.$$

Since B(4) and B(1) have a span of 2, their 5% *LSD* is (see Table 2.13d) 0.278 which is larger than 0.032, so that B(4) and B(1) are considered non-significantly different. We can similarly see that B(1) and B(9) are not different. However, the difference between the means of B(6) and B(1) is

$$3.072 - 2.694 = 0.378,$$

which is larger than the 5% *LSD* = 0.321 (see Table 2.13d) corresponding to a span of 8 [8 is the span between B(6) and B(1)—see Table 2.13b]; thus B(6) and B(1) are said to be significantly different at the 5% level. When no significant difference is found between two means, these means and all intervening means are classed as non-significantly different and underscored with a line. The final results are shown in Table 2.13e.

2.6.4 *Pooling*

The purpose of pooling is to obtain a more accurate estimate of the error variance. For cases when no estimate of the error variance is available, pooling together the sum of squares of the smallest magnitude seems to be the only way to test for significance of the remaining effects whose sum of squares are comparatively large. Taguchi recommends that pooling should take place even in a case when a retrievable error variance estimate exists but corresponds to only a few degrees of freedom. In fact, he suggests that enough 'small' sources should be pooled together until an estimate of the error variance, corresponding to *almost half* of the available degrees of freedom, is obtained. Even if this results in an overestimated error variance, he advocates that to overlook small sources 'is not a great loss'. Also sources which are significant even with an overly great error variance are indeed significant. For example, a main effect that is still significant relative to an error variance which includes interactions can certainly be regarded as a reliable effect.

Taguchi believes that in industrial experiments, when the experimentation capability is limited, to carry out additional trial runs just for information on error variance, is 'highly questionable from the standpoint of efficiency of the experiment'. To pool small sources and substitute them for the error variance is a convenient method of estimating the inter-experimental error (first-order variance) without conducting an excessive number of experiments.

2.6.4.1 Methods of pooling

A source could be considered 'small', if the associated sum of squares constitute less than 4% of the total (corrected) sum of squares. The sum of squares of this source can then be 'pooled with the error', by adding it to the existing error sum of squares; the source's degrees of freedom are also added to the existing residual degrees of freedom. As already mentioned, Taguchi's approach is to continue pooling until the 'residual degrees of freedom' constitute almost half the total number of degrees of freedom. He recommends this method as practical and yet safe. He actually proposes that a source which has been added to error can be tested by the error which includes itself!

2.6.4.2 Testing without pooling

The 'pooling' of small effects for the construction of an estimate for the error variance, has attracted much criticism as a method which can induce extreme bias in a statistical analysis and which can result to spurious conclusions (see Box and Ramirez 1986).

A technique which avoids invalid pooling and allows for selection of significant effects even in the cases when an estimate of the error variance is not available, is that recommended by C. Daniel (1959). The technique involves plotting the ordered absolute values of the single-degree-of-freedom contrasts against the quantities of the half-normal distribution using half-normal probability paper (i.e. using half the scale of ordinary normal probability paper). A 'half-normal probability paper' can be easily prepared as follows (see also Daniel 1959).

We 'fold' a normal probability paper in half, i.e. we delete the printed probability scale P for the range of P less than 50%. For the range of P over 50% we replace P by

$$P^1 = 2P - 100.$$

Having calculated n, say, contrasts (of level totals) for n single-degree-of-freedom effects (i.e. in the case of multi-level quantitative factors the linear, quadratic etc. components) we rank them in terms of absolute magnitude, and for each one we calculate a corresponding empirical (probability) rank (ER) given by

$$ER = \frac{(\text{Rank} - \frac{1}{2})}{n} \times 100 \qquad \text{Rank} = 1, 2, \ldots, n.$$

The empirical ranks are then plotted against the absolute contrasts on the half-normal probability paper (a similar procedure was followed when 'testing for normality'—see Section 2.1.3.4).

Effects which are composed only of random variation would be approximately on a straight line through the origin. Effects which are significant will not conform to this linearity, and will tend to fall towards the top right-hand corner above the linear configuration determined by the rest of the contrasts. If a linear pattern does not go through the origin, this is an indication of the possible presence of outliers. Plots showing two distinct lines rather than just one, indicate that perhaps proper randomization procedures have not been adhered to.

The 'half-normal plots' approach is a significant step towards greater objectivity in deciding what is random and what is systematic, and it is strongly recommended especially in situations when an estimate of the error variance is not retrievable.

2.7 Bibliography

There are countless texts on elementary probability and statistics but few with an engineering slant. One such example is Chatfield (1983). A popular introduction to regression analysis is Draper and Smith (1981). Recent 'diagnostic' methods in regression analysis are contained in Cook and Weisberg (1982) and Atkinson (1986). A range of standard methods are implemented in computer packages and MINITAB is very widely used in education. SPSS, GENSTAT, and SAS are larger-scale packages with SAS containing excellent data-handling facilities. GLIM is a more specialist package of particular use for categorical and non-normal data. Learning is best carried out with a teacher, and straightforward textbooks supplemented by package and manual.

3
DESIGNING EXPERIMENTS

In this chapter the history to date of experimental designs is outlined, from the 'change-one-factor-at-a-time' method, to the 'change-many-factors-at-a-time' method through (fractional) multi-factorial designs.

Fisher's original simple designs ('randomized block designs', 'Latin squares', etc.) are briefly mentioned (Section 3.2) as an introduction to multi-factorial designs (Section 3.3) capable of dealing with numerous multi-level variables and their interactions, through a systematic and cost-effective manner. Such designs include Box's 2^{k-p} 'resolution' designs (Section 3.3.3), Plackett and Burman's screening designs (Section 3.3.4), and Taguchi's orthogonal and non-orthogonal arrays (Sections 3.3.5 and 3.3.6).

Taguchi's designs are by no means new designs. In fact, all the designs covered in Section 3.3.5 are identical or equivalent to the 'Plackett and Burman' designs outlined in Section 3.3.4. The novelty of Taguchi's designs lies in the straight forward manner in which they are presented which allows not only the easy assignment of sources (factors and interactions—see also Appendix D), but also the handling of complicated situations, such as those when there are restrictions on randomization (see, for example, Section 3.3.5.4), when the type of factor changes according to the levels of another factor (see Section 3.3.5.3), when multi-level factors need to be studied using only small arrays with 2- or 3-level columns (see Section 3.3.6.2), etc. Another of Taguchi's novelties is the straight forward method of constructing complicated designs starting from a relatively small number of simple arrays (see, for example, Sections 3.3.5.1, 3.3.5.2, and 3.3.6.3).

Procedures for analysing the resulting experimental data for all the designs covered in this chapter are also outlined.

3.1 Fundamentals of experimental design

3.1.1 Original developments

A brief history of experimental design was given in Chapter 1. In this chapter we describe in greater detail the design and analysis of experiments, starting with standard methods and moving on to various suggestions of Taguchi.

Development of design of experiments in its early period began in the 1920s with R. A. Fisher at the Rothamsted Agricultural Experimental Station, England. Fisher successfully compared the quality of plant varieties by first dividing the land into several blocks of homogeneous fertility conditions, and then randomly planting the plant varieties within each one of these blocks. This allowed comparisons to be made among the plant varieties and the fertility conditions (Fisher 1935).

Fisher's clear awareness of the need for block construction and randomization (hence the 'randomized block designs', 'Latin squares', etc.), constituted the basis of all subsequent developments of experimental design and led to the birth and advancement of the concept of factorial analysis. It is possible to consider a very large part of recent work in experimental design as a continuation of Fisher's work on agricultural experimentation.

Initially, all the effort was devoted into improving and further developing data-analysis techniques; development of the design methods was rather slow, mainly because the duration of agricultural and biological experiments was usually long. It was not until the early 1950s when the design of experiments began to be used widely in technological studies that its progress and development was very rapid. In particular, the use of fractional factorial designs throughout Japanese industry repeatedly demonstrated how reliably the effects of numerous controllable factors can be determined with the least expense and effort.

3.1.2 'One-factor-at-a-time' method

However, in the majority of Western industries the most popular method of multi-factorial experimentation was (and still is, on many occasions) the 'change-one-factor-at-a-time' method: this is when the experimenter decides to vary only one factor, while keeping all the other factors fixed at a specific set of conditions. For example, the only difference between the first and the second experimental trial according to the design of Table 3.1, is the change of factor A from level 1 to level 2, all the other factors B, C, . . . , G being kept fixed at their level 1. The experiment continues until each factor has changed once, while everything else is kept constant in its most recent state.

The popularity of this method (which is also called the 'combination design') is due to its simplicity; assignment of the controllable factors according to the combination design of Table 3.1 allows the experimenter to get by with changing the level of only one factor in each trial, needing only $k + 1$ trial runs, where k is the number of factors.

TABLE 3.1. A simple version of the 'one-factor-at-a-time' method

Trial run	A	B	C	D	E	F	G
1	1	1	1	1	1	1	1
2	2	1	1	1	1	1	1
3	2	2	1	1	1	1	1
4	2	2	2	1	1	1	1
5	2	2	2	2	1	1	1
6	2	2	2	2	2	1	1
7	2	2	2	2	2	2	1
8	2	2	2	2	2	2	2

The simplicity of the method is, however, misleading and can lead to unreliable results and wrong conclusions. Indeed, the difference e.g. between A(1) and A(2) (i.e. between level 1 and level 2 of factor A) and so the 'effect of A', is estimated from data obtained by holding the conditions of all other factors B, . . . , G constant, in this case at B(1)C(1)D(1)E(1)F(1)G(1). So, no matter how good is the precision with which this effect has been estimated, this effect will be correct only for the case when all other factors are at their fixed levels B(1), . . . , G(1); there is no guarantee whatsoever that A will have the same effect when other factor conditions change. But conditions, for example in the user's environment, are most of the time different from laboratory conditions where the characterization experiment took place.

3.1.3 'Reliable' factor effects

A factorial effect, therefore, is a 'reliable' effect only if it is 'an effect of high reproducibility', i.e. only when its influence on the experimental values holds consistently *even if* the conditions of the other factors change. A reliable experiment, therefore, is an experiment which allows the determination of factorial effects when everything else varies, and not when everything else is kept constant. For a 'one-factor-at-a-time design' to be a reliable experimental design, all possible combinations of level settings should be experimented on. However, the effort and experimental cost required for such a design, otherwise termed 'a full-factorial' design, could be prohibitively large and unrealistic. For example in the case of the seven 2-level factors *A, B, . . . , G* of Table 3.1, a full-factorial design would require $2^7 = 128$ trial runs.

The search therefore should be on for designs which allow the determination of reliable (consistent) effects, with a fraction of the cost

TABLE 3.2. Orthogonal array $OA_8(2^7)$. (We can use this design to study seven two-level factors with only eight experimental trials. The full-factorial design would require $2^7 = 128$ experimental trials. This is a much more efficient design than the one depicted in Table 3.1.)

	Factors						
Trial	A	B	C	D	E	F	G
1	1	1	1	1	1	1	1
2	1	1	1	2	2	2	2
3	1	2	2	1	1	2	2
4	1	2	2	2	2	1	1
5	2	1	2	1	2	1	2
6	2	1	2	2	1	2	1
7	2	2	1	1	2	2	1
8	2	2	1	2	1	1	2

and effort required for a full-factorial design. Fractional 'orthogonal arrays' satisfy these requirements.

3.1.4 'Orthogonal' designs—an example

'Orthogonal arrays' are factorial (usually, highly fractional) designs, for which stress is laid on factorial effects of high reproducibility. A factor is considered 'significant' (or as having effect), only if it has a consistent effect when the conditions of other factors differ. In every pair of columns of such an array every combination of levels appears the same number of times; this guarantees that the averaged effect of each factor can be determined while the levels of all other factors are varied.

Table 3.2 represents the orthogonal array $OA_8(2^7)$, with which one can study seven 2-level factors with only 8 trial runs. Note that in every pair of columns, all possible combinations (11), (12), (21), and (22) appear with the same frequency. In finding the effects of the two levels A(1) and A(2) of factor A, we compare the mean of the data from four runs of experiments, No. 1, No. 2, No. 3 and No. 4, performed under the conditions of A(1), and the mean of four runs Nos. 5, 6, 7, and No. 8 performed under the conditions of A(2). To compare B(1) and B(2), we only need to compare the mean of runs Nos. 1, 2, 5, and 6 which were performed with B(1), and the mean of the runs No. 3, 4, 7, and 8 performed with B(2). Similarly for the other factors. If a comparison, for example, between A(1) and A(2) yields a significantly large difference, this can be translated as evidence of the existence of an 'effect' due to factor A, consistent even when the conditions of other factors change, and which has a good chance of being reproduced in the user environment even if the single condition of scale changes.

Since only consistent factorial effects will be classified as significant with the use of orthogonal arrays, the reliability and efficiency of these fractional experimental designs (for example, of the design of Table 3.2. rather than that of Table 3.1) can be appreciated. We will revert to the concept of orthogonal arrays later in this chapter.

3.2 Simple designs

3.2.1 Completely randomized design

This is a design in which all levels of a factor (treatments) are assigned to the experimental units in a completely random manner. For example, if we want to test for differences among four types of phospor A, B, C, and D, with respect to the light-output in a TV tube, we could follow a randomized procedure in order to experiment with the phosphor-types on 4 TV tubes, perhaps according to the following design No. 1.

Design No. 1. (The numbers in brackets show the measured light output in microamperes) (μA).

	TV tubes			
	1	2	3	4
Phosphor-type distribution	A [25]	C [4]	D [9]	A [28]
	B [22]	A [12]	C [2]	B [11]
	C [18]	B [3]	A [10]	D [20]
	C [22]	D [10]	B [5]	D [15]

The objective of complete randomization here is to average out the differences among TV tubes which might affect the light-output results. The above data can be analysed considering the phosphor-type as the main factor of interest.

Analysis of data for Design No 1

First we find the correction factor

$$CF = \frac{(\Sigma\, Y)^2}{16} = \frac{(25 + 4 + \ldots + 5 + 22)^2}{16} = 2916.$$

Then the total sum of squares

$$TSS = \sum Y^2 - CF = 25^2 + \ldots + 22^2 - 1916 = 1010.$$

For the sum of squares due to the phosphor type S_p, we find the sum of the observations within the A-cell ($\Sigma\, Y_A$) and also $\Sigma\, Y_B$, $\Sigma\, Y_C$, $\Sigma\, Y_D$; then

$$S_p = \frac{(\Sigma\, Y_A)^2 + (\Sigma\, Y_B)^2 + (\Sigma\, Y_C)^2 + (\Sigma\, Y_D)^2}{4} - CF$$

$$= \frac{(25 + 12 + 10 + 28)^2 + \cdots + (10 + \cdots + 15)^2}{4} - 2916$$

$$= [(75)^2 + (41)^2 + (46)^2 + (54)^2]/4 - 2916 = 168.5.$$

The error sum of squares is found by subtraction

$$S_E = TSS - S_p = 1010 - 168.5 = 841.5.$$

So we have

ANOVA for Design No. 1

Source	df	S	MS	F-ratio
Phosphor type	3	168.5	56.17	0.8
Error (residual)	12	841.5	70.13	
Total	15	1010		

The F-ratio value of 0.8, compared with the critical value at 5% level (from the F-tables) $F(3, 12; 5) = 3.49$ shows insignificance. The insignificant value of the F-ratio indicates that there is no reason to reject the hypothesis of equal light-output among the four phosphor-types.

In the ANOVA table, note that a relatively large value for the error sum of squares produces a large value for the error variance. But this residual variation includes the experimental error as well as the variance among the TV tubes whose assessment and elimination (from the error variance) is not possible if one applies Design No. 1. Any variation, for example, within phosphor-type C, may reflect variation between TV tubes 1, 2, and 3. Another disadvantage of this completely randomized design is that phosphor-type D is never used in tube 1, and C is never used in tube 4.

3.2.2 Randomized block design

This is an experimental design in which all levels of a factor are randomized within a 'block' (homogeneous group) and many blocks are run. Such a design would require that each phosphor-type be used once on each TV tube (block) and randomization is now restricted within tubes. This will make the assessment and elimination of the variation among tubes possible. A small readjustment of Design No 1 would make it a randomized block design (assuming that the phosphor-type distribution was random) with column-wise 'blocking'.

Design No. 2

	Tube			
	1	2	3	4
Phosphor-type distribution	A [25] B [22] C [18] D [15]	C [4] A [12] B [3] D [10]	D [9] C [2] A [10] B [5]	A [28] B [11] D [20] C [22]

Note that each phosphor-type is used once on each tube. Better comparisons can now be made between phosphor-types since they are *all* applied (randomly) over homogeneous terrains, and the error variation is not associated with any variation deliberately introduced in the experimental conditions.

Analysis of data for Design No. 2

Since the data is the same, as before we have

$$CF = 2916$$

$$TSS = 1010$$

and

$$S_{\mathrm{p}} = 168.5.$$

Now, the tube effect may also be isolated and its variation eliminated from the error variance. If $\sum Y_i$ is the sum of the data values in tube i, $i = 1, \ldots, 4$ we have for the tube sum of squares (block sum of squares – S_{B})

$$S_{\mathrm{B}} = \frac{(\sum Y_1)^2 + \ldots + (\sum Y_4)^2}{4} - CF$$

$$= \frac{(25 + 22 + 18 + 15)^2 + \ldots + (28 + \ldots + 22)^2}{4} - 2916$$

$$= \frac{(80)^2 + (29)^2 + (26)^2 + (81)^2}{4} - 2916 = 703.5.$$

The error sum of squares can now be found by subtraction as

$$S_E = TSS - S_p - S_B = 1010 - 168.5 - 703.5 = 138 \qquad (\text{df} = 9).$$

Consequently

ANOVA for Design No. 2

Source	df	S	MS	F-ratio
Phosphor-type	3	168.5	56.17	3.7 sign at 10% level
Tube (block)	3	703.5	234.5	15.3** sign at 1% level
Error	9	138	15.33	
Total	15	1010		

To test the hypothesis

$$H_0 : \mu_A = \mu_B = \mu_C = \mu_D,$$

the F-ratio is 3.7, which is significant at the 10% level since $F(3, 9; 10) = 2.81$. So the hypothesis of equal phosphor–means (i.e. of no phosphor-type effect) is rejected. This is a different conclusion from the one produced by the analysis of the completely randomized Design No. 1. The randomized block design, which allows for removal of the block (tube) effect, reduced the error variance estimate from 70.13 to 15.33 and clearly showed the actual significance of the phosphor-type effect.

It is also possible to test for tube-effect. The F-ratio for this effect (15.3) is highly significant, which means that the light output significantly depends on the TV tube used.

Remarks: Another way of representing Design No. 2 is the following:

Design No. 2.1

Tubes	A	B	C	D	Tube-cell (row) totals
1	25	22	18	15	80
2	12	3	4	10	29
3	10	5	2	9	26
4	28	11	22	20	81
Phosphor-cell (column) Totals	75	41	46	54	216 = Grand total

This outline, greatly helps the calculations for the determination of the sum of squares of the phosphor-type factor or the tube (block) factor. This is a typical 2-way outline of any factorial experiment involving two factors. Instead of phosphor-types vs. TV tubes, one could have tyre-brands vs. car-types, drug treatments vs. patients, different teaching methods vs. different pupils, fertilizers vs. plots of land etc. In the case of the 2-factor experiment, the above design satisfies the orthogonality property: every level combination (between the two factors) occurs the same number of times. In fact, this is a full-factorial design involving 2 factors with 4 levels each.

Another example: People as 'blocks'. In a particular manufacturing plant there are five methods for assembling. If we were to assign 8 people to each method at random, we would be dealing with a completely randomized design. In such a case, for example, more people might have been assigned to method 1 rather than method 2.

Alternatively we could have made sure that all 8 people used each of the methods. The order of usage could be random. This is dealing with a randomized block design with the 8 people as 8 blocks. The assessment of the variation due to the method as well as to the 8 people would then be possible.

Remarks: (i) In general, 'blocking' makes significant components of variation easier to detect. This is achieved with the removal of a portion from the residual variation. This portion of variation is attributed to the 'block effect' and is assigned $(b - 1)$ degrees of freedom, where b is the number of blocks. Hopefully, the amount of residual removed will be quite large, at least large enough to compensate for the reduction in the degrees of freedom for the error variance.

(ii) As a consequence of the balance that exists in such an experiment we can say that the treatments (phosphor-types, assembling methods, etc.) are orthogonal to blocks (TV tubes, people, etc.).

3.2.2.1 The case of missing data

There is always a possibility in a randomized block design that one or more observations are accidentally lost. For example, an animal may die, or a tyre may deflate during an experiment monitoring tread loss during driving, etc. In the phosphor-type experiment, a tube might accidentally be destroyed or might fail to produce a light output (if the failure were due to the phosphor-type, this would *not* be considered a 'missing value').

For a single-factor completely randomized design, a missing value would not be problem, since the ANOVA technique in this case can deal with unequal factor cells. But for a 2-way analysis this means a loss of orthogonality.

In order to deal with one or more missing values, we usually replace them with values which minimize the error sum of squares.

Example: Suppose that in the Design No. 2.1, the data value corresponding to the phosphor-type C on the second plot is missing. The resulting data, with Y in place of the missing value, are now as follows:

Design No. 2.2

T	A	B	C	D	Row total
1	25	22	18	15	80
2	12	3	Y	10	$25+Y$
3	10	5	2	9	26
4	28	11	22	20	81
Column Total	75	41	$42+Y$	54	$212+Y$

We now have $CF=(212+Y)^2/16$
Also

$$S_E = TSS - S_p - S_B$$
$$= \sum \text{(squares of the observations)} - CF - \left\{\sum \frac{(\text{column total})^2}{4} - CF\right\} - \left\{\sum \frac{(\text{row total})^2}{4} - CF\right\},$$

i.e.

$$S_E = \{25^2+22^2+\cdots+Y^2+\cdots+20^2\} - \frac{75^2+41^2+(42+Y)^2+54^2}{4} - \frac{80^2+(25+Y)^2+26^2+81^2}{4} + \frac{(212+Y)^2}{16}.$$

In order to find the value of Y which minimizes S_E, we differentiate S_E with respect to Y and set it equal to zero. Since the derivatives of all constant terms are zero, we simply have

$$\frac{d(S_E)}{dY} = 2Y - \frac{2(42+Y)}{4} - \frac{2(25+Y)}{4} + \frac{2(212+Y)}{16} = 0$$

from which we obtain

$$Y = 6.2.$$

Replacing Y with 6.2 in the Design No. 2.2, we can proceed with the ANOVA analysis which differs from the previous one only in the degrees of freedom for the error term which are reduced by one, since there are only 15 observations and Y is determined by these.

If there are more than one missing values, a similar procedure is followed: S_E is partially differentiated with respect to each missing value and is set equal to zero; this procedure provides as many equations as unknown values, from which estimates for the missing values can be obtained. In general, for the case of only one missing value Y_{ij} which corresponds to the ith treatment and the jth block, it can be shown that

$$Y_{ij} = \frac{tT_i + bB_j - G}{(t-1)(b-1)}$$

where

t = number of treatments
b = number of blocks
T_i = the sum of the observed values for the ith treatment
B_j = the sum of the observed values for the jth block
G = the grand sum of all observed values.

In our example, since the number of phosphor-types (treatments) is 4, ($t = 4$), the number of tubes (blocks) is 4 (i.e. $b = 4$) and since $T_3 = 42$, $B_2 = 25$ and $G = 212$, we have

$$Y_{ij} = Y_{32} = \frac{4 \times 42 + 4 \times 25 - 212}{3 \times 3} = 6.2.$$

The above method of estimating missing values is otherwise known as the 'Fisher–Yates' method. Clearly the method requires that there is a retrievable error sum of squares with sufficient error-degrees-of-freedom. We will revert to more general cases of missing values in Section 4.5.

3.2.2.2 Randomized incomplete block design

This is a randomized block design in which it is not possible to apply all levels of the factor of interest (i.e. all treatments) in every block. Such a design is called 'balanced' if each pair of treatments occurs together the same number of times in blocks. For example, in the 'phosphor' experiment, suppose that only 3 out of 4 phosphor-types could be used in each tube (block). A balanced design for such a case could be as follows

Design No. 2.3

	Phosphor-types			
Tubes	A	B	C	D
1	25	—	18	15
2	—	3	4	10
3	10	5	2	—
4	28	11	—	20

Note that in each tube only 3 types of phosphor are applied: however, each pair of types, (A, C) say, occurs only on tube 1 and 3; B and C occur together on plots 2 and 3, etc.

Analysis of data for Design No. 2.3

Using the 12 available data-values, we find the correction factor *CF*, the total sum of squares *TSS*, and the block (tube) sum of squares S_B in the usual way:

$$CF = \frac{(151)^2}{12} = 1900.1$$

$$TSS = \sum Y^2 - CF = (25^2 + \cdots + 20^2) - 1900.1 = 832.9$$

$$S_B = \frac{58^2 + 17^2 + 17^2 + 59^2}{3} - 1900.1 = 574.2$$

We now calculate the phosphor-factor sum of squares S_p 'adjusted for blocks' as

$$S_p = \frac{P_A^2 + P_B^2 + P_C^2 + P_D^2}{k \cdot t \cdot n}$$

where

k = number of factor levels per block ($k = 3$)
t = number of all factor levels ($t = 4$)
n = number of times each pair of factor-levels appears together throughout the experiment ($n = 2$)

and

$$P_A = k \times (\text{total of A-cell}) - (\text{sum of all block totals which contain level } A)$$
$$= 3 \times (63) - (58 + 17 + 59) = 55$$

Similarly

$$P_B = 3 \times (19) - (17 + 17 + 59) = -36$$

$$P_C = 3 \times (24) - (58 + 17 + 17) = -20$$

$$P_D = 3 \times (45) - (58 + 17 + 59) = 1$$

(Note that $P_A + P_B + P_C + P_D = 0$, which is always the case.) Then

$$S_p = \frac{55^2 + (-36)^2 + (-20)^2 + (1)^2}{3 \times 4 \times 2} = 196.75.$$

Finally, by subtraction

$$S_{Error} = TSS - S_p - S_B = 832.9 - 196.75 - 574.2 = 61.95.$$

Then

ANOVA for Design No. 2.3

Source	df	S	MS	F-ratio
Phosphor-type (adjusted)	3	196.75	65.6	5.3
Blocks (tubes)	3	574.2	—	
Error	5	61.95	12.4	
Total	11			

The F-ratio shows evidence of phosphor-type's significance at least at 10% level.

Note that we do not usually test for a block effect. To do so, the block sum of squares needs to be 'adjusted for phosphor-types' in the same way as we 'adjusted' the S_p. But in such case, the block-factor would be considered as *the* main factor of interest (with the fertizilizer as a secondary factor for which no F-ratio would be obtained). Thus there are two different ANOVA analyses for this same experiment.

3.2.3 Latin square design

In general, this is an experimental design in which each level of each factor is combined only once with each level of two other factors. In particular this is a square design on which there is blocking not only column-wise but also row-wise. For example, if we needed to assess the effect of the four phosphor-types on the light output using not only one but four types of face-plate glass, we could have used a block design with tubes as the 4 column-wise blocks (as in Design No. 2) and the glass types as the 4 row-wise blocks, for example:

Design No. 3

Face-plate glass-type	Tube 1	2	3	4
1	A (25)	B (3)	C (2)	D (20)
2	B (22)	C (4)	D (9)	A (28)
3	C (18)	D (10)	A (10)	B (11)
4	D (15)	A (12)	B (5)	C (22)

Note that A appears once in every column and in every row; similarly for B, C, and D. Interest might still be centred on the phosphor-types, but two restrictions are placed on the randomization. Such a design is only possible when the number of levels of both restrictions (block factors) equals the number of the levels of the factor of interest. All randomization is not lost in this design, as the particular Latin square to be used may be chosen at random from many possible Latin squares of the required size. For example, from the above 4×4 standard square, another $4!3! - 1$ Latin squares, all different, may be generated by permuting all the rows, except the first, and all the columns, one of which may be chosen in random.

By a simple extension of the analysis of Design No. 2 we can analyse the data as outlined in Design No. 3. The extra calculation involves the determination of the sum of squares for the glass factor S_G (row-block factor). A 'glass-cell' consists of the data of the corresponding row. So

$$S_G = \frac{(\text{sum of data in first row})^2 + \cdots + (\text{sum of data in fourth row})^2}{\text{number of data in each row}} - CF$$

$$= \frac{(25+3+2+20)^2 + \cdots + (15 + \cdots + 22)^2}{4} - 2916 = 30.5,$$

and the new error sum of squares (with 6 degrees of freedom):

$$S_E = TSS - S_p - S_B - S_G = 1010 - 168.5 - 703.5 - 30.5 = 107.5,$$

leading to:

ANOVA for Design No. 3

Source	df	S	MS	F-ratio
Phosphor	3	168.5	56.17	3.1 sign at 10% level
Tube (column-block)	3	703.5	234.5	13.1** sign at 1% level
Glass (row-block)	3	30.5	10.2	0.6
Error	6	107.5	17.9	
Total	15	1010		

Overall, although significant differences are detected among phosphor-types and among tubes, there is no evidence to suggest a significant difference among the 4 face-plate glass varieties.

Another example: We wish to compare 5 methods a, b, c, d, and e for improving the strength of a material; the strength could depend on the environmental conditions on a day-to-day basis, and also on the operator handling the process. The experiment is to run for five days and there are 5 operators available. The simplest design is a 5×5 Latin-square blocking for the days and operators as follows

Design No. 3.1

	Days				
Operators	1	2	3	4	5
1	a	d	e	c	b
2	c	a	b	e	d
3	d	b	c	a	e
4	b	e	a	d	c
5	e	c	d	b	a

3.2.4 Graeco-Latin square

This is an experimental design in which four factors are arranged so that each level of each factor is combined only once with each level of the other three factors. For example, following the square outline of the previous sections, we could have

Design No. 4

	Columns			
Rows	1	2	3	4
1	A_α	B_β	C_γ	D_δ
2	B_γ	A_δ	D_α	C_β
3	C_δ	D_γ	A_β	B_α
4	D_β	C_α	B_δ	A_γ

This is an extended Latin square where a third restriction on randomization is imposed. This restriction is at levels α, β, γ, δ (of the fourth factor) which each appear once and only once in each row, in each column and within each level of treatment A, B, C, or D. A Graeco-Latin square can be considered as the result of superimposing two Latin squares (one being written with Greek letters and the other with Latin letters, hence the name 'Graeco-Latin'). These Latin squares are orthogonal to each other, because when they are superimposed, every letter of one square occurs once and only once with every letter of the other. For the analysis of a Graeco-Latin square, the usual 4-way ANOVA can be applied.

3.2.5 Youden square

This is a special incomplete Latin square. For example, in the phosphor–glass-types experiment, suppose that only three glass varieties are available, and we have

Design No. 5

	Tube			
Glass-type	1	2	3	4
1	A	B	C	D
2	B	C	D	A
3	C	D	A	B

Clearly the addition of the (D, A, B, C) row would make this a Latin square (assuming a fourth glass-type was available). So a method for constructing a Youden square is to omit one row from a Latin-square design. To analyse a Youden square one can follow the procedure for the incomplete block analysis (see Section 3.2.2.2) noticing that the glass-types are orthogonal to both tube (blocks) and phosphor-types, and therefore their sum of squares can be calculated in the usual way.

3.3 Multi-factorial designs

3.3.1 Design Matrix

In the previous sections we have considered up to 4-factor experimental designs; for example, if we call the TV-tubes the 'block factor', the randomized block design No. 2 is a 2-factor design and can be represented as shown at top of p. 106.

Similarly, the Latin square can be considered as a 3-factor design, and Design No. 3 can be represented as shown at bottom of p. 106.

The above array-type representation is called a 'design matrix' and is a very convenient way of representing complicated multi-factor multi-level experimental designs. Every column represents a factor, every row represents an experimental trial, and entries in the matrix represent the factor levels. For example, the first row of the above matrix represents the experimental trial which took place with tube 1, on glass variety 1, using phosphor-type A and yielding a light-output of 25 μa. A 4-factor design matrix can be obtained from the Graeco-Latin square as shown on p. 107.

Design No. 2′

Trial	Factors: Tube (block)	Phosphor-type	Data
1	1	A	25
2	1	B	22
3	1	C	18
4	1	D	15
5	2	A	12
6	2	B	3
7	2	C	4
8	2	D	10
9	3	A	10
10	3	B	5
11	3	C	2
12	3	D	9
13	4	A	28
14	4	B	11
15	4	C	22
16	4	D	20

Design No. 3′

Trial	Factors: Tube (block)	Glass-type	Phosphor	Data
1	1	1	A	25
2	1	2	B	22
3	1	3	C	18
4	1	4	D	15
5	2	1	B	3
6	2	2	C	4
7	2	3	D	10
8	2	4	A	12
9	3	1	C	2
10	3	2	D	9
11	3	3	A	10
12	3	4	B	5
13	4	1	D	20
14	4	2	A	28
15	4	3	B	11
16	4	4	C	22

Design No. 4′

Trial	Factors			
	Column-factor	Row-factor	Greek-factor	Latin-factor
1	1	1	α	A
2	1	2	γ	B
3	1	3	δ	C
4	1	4	β	D
5	2	1	β	B
6	2	2	δ	A
7	2	3	γ	D
8	2	4	α	C
9	3	1	γ	C
10	3	2	α	D
11	3	3	β	A
12	3	4	δ	B
13	4	1	δ	D
14	4	2	β	C
15	4	3	α	B
16	4	4	γ	A

3.3.2 Fractional and full-factorial designs

Note that all the design matrices mentioned so far are orthogonal arrays, because for every pair of columns, every level combination occurs the same number of times (for these cases only once). If the occurrence of every level-combination was not just for every *pair* of columns, but for *all* the columns, i.e. if *every* level-combination for all the factors occurred, then the design would be called a 'full-factorial design'. If at least one level-combination (of all the factors involved) does not appear, then we have a 'fractional-factorial design'. For example, Design No. 2′ is a full-factorial design, whereas Designs No. 3′ and No. 4′ are fractional-factorial designs.

In general, if one deals with k factors with p levels and with l factors with q levels, the number of experimental trials needed for a full-factorial design is $p^k q^l$. For example, in the Design No. 3′ there are three 4-level factors and a full-factorial would require

$$4^3 = 64 \text{ trials.}$$

Consequently, Design No. 3′ is a $\frac{1}{4}$th-fraction of a full-factorial, Design No. 4′ is a $\frac{1}{16}$th-fraction of a full-factorial, and so on.

Fractional designs can dramatically reduce the effort, expense and time required for multi-factorial experiments. Fractional orthogonal designs provide a statistically valid method of studying the effects of many factors in the same experiment, and of recognizing those effects which are of high reproducibility.

3.3.2.1 The $OA_8(2^7)$-design

The orthogonal design of Table 3.2, $OA_8(2^7)$, which can handle seven 2-level factors assuming no interaction, is a $\frac{1}{16}$th-fraction of a full-factorial normally requiring 128 trials. As with other fractional designs the only drawback of this design is the need to assume that certain interactions can be ignored. In many practical situations this is a valid assumption. Indeed, interactions among three or more factors, apart from the fact that they are difficult to explain, are usually negligible in practice. Sometimes, on the basis of past experience, even some, or all, 2-way interactions can be assumed negligible. For example, if seven 2-level factors need to be studied with only 8 experiments, $OA_8(2^7)$ can be used, but it has to be assumed that *all* interactions are insignificant. The statistical analysis for such a design produces uncorrelated main effects, but these will be confounded with 2-way, 3-way, or higher-order interactions.

Example: Find the 2-way interactions confounded with each main effect of the $OA_8(2^7)$ design.

First, recode the levels 1 and 2 into '+1' and '−1' respectively. Then, the $OA_8(2^7)$ can be represented as shown in Table 3.1a.

TABLE 3.1a

Trial	A	B	C	D	E	F	G
1	+1	+1	+1	+1	+1	+1	+1
2	+1	+1	+1	−1	−1	−1	−1
3	+1	−1	−1	+1	+1	−1	−1
4	+1	−1	−1	−1	−1	+1	+1
5	−1	+1	−1	+1	−1	+1	−1
6	−1	+1	−1	−1	+1	−1	+1
7	−1	−1	+1	+1	−1	−1	+1
8	−1	−1	+1	−1	+1	+1	−1
	(B × C)	(A × C)	(A × B)	(A × E)	(A × D)	(A × G)	(A × F)
	(D × E)	(D × F)	(D × G)	(B × F)	(B × G)	(B × D)	(B × E)
	(F × G)	(E × G)	(E × F)	(C × G)	(C × F)	(C × E)	(C × D)

At the bottom of the design matrix of Table 3.1a we indicate the 2-way interactions confounded with the main factor effect associated with a particular column. For instance, the main effect of factor A (column 1) is confounded with the 2-way interactions $B \times C$, $D \times E$, $F \times G$. This can be easily confirmed by simply multiplying the columns in pairs. For example, a multiplication of the entries of the columns 2(B) and 3(C) will yield the entries of column 1(A). Column 1 is also the product of columns 4 and 5 (D and E) and of columns 6 and 7 (F and G). This explains the confounding.

Nevertheless, the same design can be used for the study of four 2-level factors and three 2-way interactions, if from past experience it is known that all other interactions are negligible and can be ignored. For example, the following 'effects' could be assigned respectively to columns 1 to 7

$$a,\ b,\ a \times b,\ c,\ a \times c,\ b \times c,\ d.$$

This means that columns 3, 5, and 6 will remain unassigned and the experiment will be as directed by columns 1, 2, 4, and 7 which will be assigned the factors a, b, c, and d respectively. This design can then be expressed as $OA_8(2^4)$.

3.3.2.2 A case-study with 2-level factors

An experiment was performed in order to find the best conditions for the electric welding of two iron sheets, with regard to the mechanical strength of the welded part. Four main factors of interest were identified and two levels were selected for each, as follows

Factor	Level 1	Level 2
A: Brand	A(1) = J100	A(2) = B17
B: Electric current value	B(1) = 100 A	B(2) = 50 A
C: Manipulating method	C(1) = Weaving	C(2) = Single
D: preheated condition	D(1) = Not preheated	D(2) = preheated at 100 °C.

It was decided to study the 2-way interaction between the factors A, B, and C. The OA_8 design was used and the 8 trials yielded the results of Table 3.1b

Analysis

As usually we first calculate the correction factor

$$CF = \frac{(\text{grand total})^2}{8} = \frac{(120)^2}{8} = 1800$$

TABLE 3.1b

Trial	Design A	B	A × B	C	A × C	B × C	D	Strength (kg/mm^2)
1	1	1	1	1	1	1	1	15
2	1	1	1	2	2	2	2	20
3	1	2	2	1	1	2	2	4
4	1	2	2	2	2	1	1	9
5	2	1	2	1	2	1	2	25
6	2	1	2	2	1	2	1	29
7	2	2	1	1	2	2	1	10
8	2	2	1	2	1	1	2	8
								120

and the total sum of squares

$$TSS = \sum Y^2 - CF = (15^2 + \cdots + 8^2) - 1800 = 552.$$

For the sum of squares due to A

$$S_A = \frac{[\text{sum total in } A(1)]^2 + [\text{sum total in } A(2)]^2}{4} - CF = \frac{(A_1)^2 + (A_2)^2}{4} - CF$$

$$= \frac{(15 + 20 + 4 + 9)^2 + (25 + 29 + 10 + 8)^2}{4} - 1800 = 72.$$

Similarly

$$S_B = \frac{[\text{sum total in B}(1)]^2 + [\text{sum total in B}(2)]^2}{4} - CF$$

$$= \frac{(89)^2 + (31)^2}{4} - 1800 = 420.5$$

$$S_D = 18 \text{ and } S_D = 4.5.$$

For the interaction effect between A and B we have

$$S_{A\times B} = \frac{[\text{sum total in } A(1)B(1)]^2 + [\text{sum in A(1)B(2)}]^2 + [\text{sum in A(2)B(1)}]^2 + [\text{sum in A(2)B(2)}]^2}{2}$$

$$- CF - S_A - S_B = \frac{(15 + 20)^2 + (4 + 9)^2 + (25 + 29)^2 + (10 + 8)^2}{2}$$

$$- 1800 - 72 - 420.5 = 24.5.$$

Similarly

$$S_{A\times C} = 8 \text{ and } S_{B\times C} = 4.5.$$

Therefore Table 3.1c can be constructed

TABLE 3.1c. ANOVA

Source	df	Sum of squares	F-ratio
A	1	72	8.2* significant at 5% level
B	1	420.5	48.1** significant at 1% level
C	1	18p	
D	1	4.5p	
A × B	1	24.5	2.8
A × C	1	8p	
B × C	1	4.5p	
Error	—	—	
(Pooled error)	(4)	(35)	
Total	7	552	

Note that there is no retrievable error sum of squares, but we can create one by pooling together the sum of squares which individually constitute less than 4% of the total sum of squares (denoted by 'p'). Only the factors A and B seem to have a significant effect. By considering the average values on each factor level (sum total in level/4) we can easily deduce that the best levels for these factors, so that the mechanical strength of the welded part achieves its maximum, is Level 2 for A (i.e. Brand B17) and first level of factor B (i.e. electric current of 100 A). The optimal settings of the insignificant factors C and D should be decided by cost considerations or ease of handling.

3.3.2.3 Recognizing interaction columns

A simple way of recognizing which columns correspond to what (2-way) interactions in designs that allow their study, is to make use of Taguchi's upper triangular representations such as the one in Table 3.1d which corresponds to the $OA_8(2^7)$ design:

For example, entry 3 in the first row of the triangular matrix of Table 3.1d means that the interaction between columns 1 and 2 in $OA_8(2^7)$ can be assigned to column 3. Also, entry 6 in the second row means that the interaction between the factors of columns 2 and 4 in $OA_8(2^7)$ can be assigned to column 6, and so forth. (For interaction tables of other

TABLE 3.1d

	Col. No.						
Col. No.	1	2	3	4	5	6	7
1	—	3	2	5	4	7	6
2		—	1	6	7	4	5
3			—	7	6	5	4
4				—	1	2	3
5					—	3	2
6						—	1
7							—

2-level orthogonal arrays, see Appendix D.) Such triangular representations can also be used for showing the interactions between 3-level columns. But since the interactions between 3-level factors requires 4 degrees of freedom, two 3-level columns have to be allocated for such interactions; in other words, two 3-level columns have to be left unassigned (free) so that an interaction effect can be calculated.

3.3.2.4 3-level designs and interaction matrices

Table 3.2a shows the orthogonal array $OA_{27}(3^{13})$ which is a design capable of dealing with 13 3-level factors (assuming no interactions) at the expense of only 27 experimental trials rather than 1 594 323 ($=3^{13}$) required for a full factorial.

When there is a need to study 2-way interactions, the triangular matrix of Table 3.2b can indicate which columns to leave unassigned. For example, for studying the interaction between 2 factors assigned to columns 1 and 2, columns 3 and 4 will have to be left free; or, for the interaction between the factors of columns 4 and 7, columns 9 and 11 will have to be left unassigned; and so on.

As far as the analysis is concerned, the calculation of the interaction sum of squares proceeds in the usual way by considering the interaction-cell totals; where by interaction cells, we mean the cells corresponding to combinations (1, 1), (1, 2), (1, 3), (2, 1), (2, 2), (2, 3), (3, 1), (3, 2), and (3, 3). So, for 3-level factors (see also Chapter 2)

$$S_{A\times B} = \sum_{i,j=1}^{3} \frac{[\text{sum total in cell A}(i)\text{B}(j)]^2}{\text{number of values in cell}} - CF - S_A - S_B.$$

In practice, when A and B are multi-level quantitative factors, it is sufficient to concentrate only on the interaction components $A_L \times B_L$,

TABLE 3.2a. $OA_{27}(3^{13})$

Trial	Columns 1	2	3	4	5	6	7	8	9	10	11	12	13
1	1	1	1	1	1	1	1	1	1	1	1	1	1
2	1	1	1	1	2	2	2	2	2	2	2	2	2
3	1	1	1	1	3	3	3	3	3	3	3	3	3
4	1	2	2	2	1	1	1	2	2	2	3	3	3
5	1	2	2	2	2	2	2	3	3	3	1	1	1
6	1	2	2	2	3	3	3	1	1	1	2	2	2
7	1	3	3	3	1	1	1	3	3	3	2	2	2
8	1	3	3	3	2	2	2	1	1	1	3	3	3
9	1	3	3	3	3	3	3	2	2	2	1	1	1
10	2	1	2	3	1	2	3	1	2	3	1	2	3
11	2	1	2	3	2	3	1	2	3	1	2	3	1
12	2	1	2	3	3	1	2	3	1	2	3	1	2
13	2	2	3	1	1	2	3	2	3	1	3	1	2
14	2	2	3	1	2	3	1	3	1	2	1	2	3
15	2	2	3	1	3	1	2	1	2	3	2	3	1
16	2	3	1	2	1	2	3	3	1	2	2	3	1
17	2	3	1	2	2	3	1	1	2	3	3	1	2
18	2	3	1	2	3	1	2	2	3	1	1	2	3
19	3	1	3	2	1	3	2	1	3	2	1	3	2
20	3	1	3	2	2	1	3	2	1	3	2	1	3
21	3	1	3	2	3	2	1	3	2	1	3	2	1
22	3	2	1	3	1	3	2	2	1	3	3	2	1
23	3	2	1	3	2	1	3	3	2	1	1	3	2
24	3	2	1	3	3	2	1	1	3	2	2	1	3
25	3	3	2	1	1	3	2	3	2	1	2	1	3
26	3	3	2	1	2	1	3	1	3	2	3	2	1
27	3	3	2	1	3	2	1	2	1	3	1	3	2

$A_L \times B_Q$ and $A_Q \times B_L$, where the subscripts L and Q indicate the linear and quadratic effect respectively. Indeed, these parts of the interaction are enough to give a good estimate of the interaction between multi-level factors, and, in the majority of cases, the other interaction components (say $A_Q \times B_Q$, $A_C \times B_Q$ etc.) are negligible and usually difficult to interpret.

3.3.2.5 Interactions between multi-level factors—a case-study

A pigment is dispersed in a liquid by ball milling in a drum on a set of rollers. The effectiveness of the milling is judged by passing the slurry

TABLE 3.2b. Interactions between two columns for $OA_{27}(3^{13})$

1	2	3	4	5	6	7	8	9	10	11	12	13
(1)	3 4	2 4	2 3	6 7	5 7	5 6	9 10	8 10	8 9	12 13	11 13	11 12
	(2)	1 4	1 3	8 11	9 12	10 13	5 11	6 12	7 13	5 8	6 9	7 10
		(3)	1 2	9 13	10 11	8 12	7 12	5 13	6 11	6 10	7 8	5 9
			(4)	10 12	8 13	9 11	6 13	7 11	5 12	7 9	5 10	6 8
				(5)	1 7	1 6	2 11	3 13	4 12	2 8	4 10	3 9
					(6)	1 5	4 13	2 12	3 11	3 10	2 9	4 8
						(7)	3 12	4 11	2 13	4 9	3 8	2 10
							(8)	1 10	1 9	2 5	3 7	4 6
								(9)	1 8	4 7	2 6	3 5
									(10)	3 6	4 5	2 7
										(11)	1 13	1 12
											(12)	1 11

through a filter for a fixed period of time and then measuring the pressure drop across the filter. The study of the effect of the two main variables was required. These variables along with the levels to be studied are

R: Speed measured in number of revolutions per minute:

V: Volume of liquid in the drum (in litres):

Levels
$R(1) = 10$ $R(2) = 20$ $R(3) = 30$ $R(4) = 40$
$V(1) = 100$ $V(2) = 120$ $V(3) = 140.$

The pressure drop for every combination of the above factors was measured with the results of Table 3.3a.

Analysis

As usually we calculate the correction factor

$$CF = \frac{(135.9)^2}{12} = 1539.0675$$

and the total sum of squares

$$TSS = \sum Y^2 - CF = 209.0825 \qquad (\text{df} = 11).$$

The total effect for V can be found through the sum of squares for V as

$$\begin{aligned} S_V &= \frac{(V_1)^2 + (V_2)^2 + (V_3)^2}{4} - CF \\ &= \frac{(41.2)^2 + (56.8)^2 + (37.9)^2}{4} - 1539.0675 = 50.955 \quad (\text{df} = 2). \end{aligned}$$

These sum of squares consist of the sum of squares for the linear effect S_{V_L} and the sum of squares for the quadratic effect S_{V_Q}, each with 1 degree of freedom. Since we are dealing with equidistant levels, we can use the orthogonal polynomials of Table 3.4 (see also Chapter 2) to calculate the components S_{V_L} and S_{V_Q}. Thus

$$S_{V_L} = \frac{[(-1) \times V_1 + (0) \times V_2 + (1) \times V_3]^2}{r \times ((-1)^2 + 0^2 + 1^2)} = \frac{(-41.2 + 37.9)^2}{4 \times 2} = 1.36125$$

$$\begin{aligned} S_{V_Q} &= \frac{[(1) \times V_1 + (-2) \times V_2 + (1) \times V_3]^2}{r \times (1 + (-2)^2 + 1^2)} \\ &= \frac{[41.2 - (2 \times 56.8) + 37.9)^2}{4 \times 6} = 49.59375. \end{aligned}$$

TABLE 3.3a

V	R 10	20	30	40	Total
100	15.2	12.1	8.1	5.8	$41.2 = V_1$
120	20.3	16.2	11.5	8.8	$56.8 = V_2$
140	13.1	10.8	7.7	6.3	$37.9 = V_3$
Total	$48.6 = R_1$	$39.1 = R_2$	$27.3 = R_3$	$20.9 = R_4$	135.9

Note that

$$S_{V_L} + S_{V_Q} = 1.36125 + 49.59375 = 50.955 = S_V.$$

We already note that the quadratic effect of V is the dominant component of the effect due to V. Similarly we have

$$S_R = \frac{(48.6)^2 + (39.1)^2 + (27.3)^2 + (20.9)^2}{3} - 1539.0675 = 151.889 \qquad (\text{df} = 3).$$

The linear, quadratic, and cubic components of S_R can be found using again the orthogonal polynomials of Table 3.4. Thus

$$S_{R_L} = \frac{[(-3) \times 48.6 + (-1) \times 39.1 + (1) \times 27.3 + (3) \times 20.9]^2}{3 \times ((-3)^2 + (-1)^2 + (1)^2 + (3)^2)} = 150.1 \qquad (\text{df} = 1)$$

$$S_{R_Q} = \frac{[(1) \times 48.6 + (-1) \times 39.1 + (-1) \times 27.3 + (1) \times 20.9]^2}{3 \times (1^2 + (-1)^2 + (-1)^2 + 1^2)} = 0.801 \qquad (\text{df} = 1)$$

$$S_{R_C} = \frac{[(-1) \times 48.6 + (3) \times 39.1 + (-3) \times 27.3 + (1) \times 20.9]^2}{3 \times [1^2 + 3^2 + (-3)^2 + 1^2]} = 0.988 \qquad (\text{df} = 1).$$

Note that

$$S_{R_L} + S_{R_Q} + S_{R_C} = S_R$$

The dominant component is the linear one. The interaction between the linear effects of V and R can be found by calculating the sum of squares due to $V_L \times R_L$ as follows:

First, we calculate the following contrasts for each level of V

$$L_{R_L}(V_i) = C_1 \times (V_iR_1) + C_2 \times (V_iR_2) + C_3 \times (V_iR_3) + C_4 \times (V_iR_4) \qquad i = 1, 2, 3,$$

where $C_1, \ldots, C_4$ are the orthogonal coefficients corresponding to the linear effect for the 4-level factor R, i.e. -3, -1, 1, and 3 (see Table 3.4), and V_iR_j ($i = 1, 2, 3$; $j = 1, 2, 3, 4$) represents the data value (or their sum if there are more than one) for each level combination (i, j) of V and R. Thus

$$L_{R_L}(V_1) = (-3) \times (15.2) + (-1) \times (12.1) + (1) \times (8.1) + (3) \times 5.8 = -32.2$$

$$L_{R_L}(V_2) = (-3) \times (20.3) + (-1) \times (16.2) + (1) \times (11.5) + (3) \times 8.8 = -39.2$$

$$L_{R_L}(V_3) = (-3) \times (13.1) + (-1) \times (10.8) + (1) \times (7.7) + (3) \times 6.3 = -23.5$$

and

$$S_{V_L \times R_L} = \frac{[C'_1 \times L_{R_L}(V_1) + C'_2 \times L_{R_L}(V_2) + C'_3 \times L_{R_L}(V_3)]^2}{n[C_1^2 + C_2^2 + C_3^2 + C_4^2] \times [(C'_1)^2 + (C'_2)^2 + (C'_3)^2]},$$

where n: the number of observations in each cell $V(i)R(j)$ $(n = 1)$, and C'_1, C'_2, C'_3 are the orthogonal coefficients corresponding to the linear effect of the 3-level factor V, i.e. -1, 0, and 1.

So

$$S_{V_L \times R_L} = \frac{[(-1) \times (-32.2) + 0 \times (-39.2) + (1) \times (-23.5)]^2}{1 \times 20 \times 2} = 1.89225.$$

Similarly

$$S_{V_Q \times R_L} = \frac{[(1) \times (-32.2) + (-2) \times (-39.2) + (1) \times (-23.5)]^2}{1 \times 20 \times (1^2 + (-2)^2 + 1^2)} = 4.2941.$$

(1, -2, and 1 are the orthogonal coefficients corresponding to the quadratic effect of the 3-level factor V).

Note that the sum

$$S_{V_L \times R_L} + S_{V_Q \times R_L} = 6.1863$$

is the sum of squares due to $V \times R_L$, i.e. due to the interaction of factor V with the linear effect of R, and is associated with 2 degrees of freedom.

This can otherwise be calculated by

$$S_{V \times R_L} = \frac{[L_{R_L}(V_1)]^2 + [L_{R_L}(V_2)]^2 + [L_{R_L}(V_3)]^2}{[C_1^2 + C_2^2 + C_3^2 + C_4^2]} - \frac{[L_{R_L}(V_1) + L_{R_L}(V_2) + L_{R_L}(V_3)]^2}{r \times [C_1^2 + C_2^2 + C_3^2 + C_4^2]}$$

where r is the number of observations in each R-cell $(r = 3)$. Thus

$$S_{V \times R_L} = \frac{[(-32.2)^2 + (-39.2)^2 + (-23.5)^2]}{[(-3)^2 + (-1)^2 + (1)^2 + (3)^2]} - \frac{[(-32.2) + (-39.2) + (-23.5)]^2}{3 \times (20)} = 6.1863.$$

(This is what needs to be calculated if V were a qualitative factor.) A similar procedure can be followed by calculating the sum of squares for $V_L \times R_Q$, $V_Q \times R_Q$ etc.

For example, for the interaction with R_Q we need the contrasts

$$L_{R_Q}(V_i) = b_1 \times (V_i R_1) + b_2 \times (V_i R_2) + b_3 \times (V_i R_3) + b_4 \times (V_i R_4) \quad i = 1, 2, 3$$

where $b_1, \ldots, b_4$ are the orthogonal coefficients corresponding to the quadratic effect for four-level factors, i.e. 1, −1, −1, and 1 (see Table 3.4). We then calculate $S_{V_L \times R_Q}$ and $S_{V_Q \times R_Q}$ as before. In fact, we have

$$L_{R_Q}(V_1) = 0.8 \qquad L_{R_Q}(V_2) = 1.4 \qquad \text{and} \qquad L_{R_Q}(V_3) = 0.9,$$

and so

$$S_{V_L \times R_Q} = \frac{[(-1) \times (0.8) + (0) \times (1.4) + (1) \times 0.9]^2}{1 \times 4 \times 2} = 0.00125,$$

and

$$S_{V_Q \times R_Q} = \frac{[(1) \times (0.8) + (-2) + (1.4) + (1) \times 0.9]^2}{1 \times 4 \times 6} = 0.05042.$$

We similarly find $S_{V_L \times R_C} = 0.00025$ and $S_{V_Q \times R_C} = 0.000083$.

TABLE 3.3b. Outlines of all the variation components

Source			df	Sum of squares		
V	V_L		1	1.36125	50.955	
	V_Q		1	49.59375		
R	R_L		1	150.1	151.889	
	R_Q		1	0.801		
	R_C		1	0.988		
$V \times R_L$	$V_L \times R_L$	$V \times R$	1	1.89225	6.18635	6.23835
	$V_Q \times R_L$		1	4.2941		
$V \times R_Q$	$V_L \times R_Q$		1	0.00125	0.05167	
	$V_Q \times R_Q$		1	0.05042		
$V \times R_C$	$V_L \times R_C$		1	0.00025	0.00033	
	$V_Q \times R_C$		1	0.00083		
Total			11	209.082		

One can construct a 'residual sum of squares' by pooling together the components sum of squares each accounting for less than 4% of the total sum of squares. This will then show that the linear effect of R (speed) and the quadratic effect of V (volume), are the only significant effects.

3.3.3 2^{k-p} 'resolution' designs

These are 2-level factorial designs capable of dealing with k 2-level factors; they are $\frac{1}{2^p}$-fractions of the full-factorial (and therefore needing

TABLE 3.4. Orthogonal polynomials

	No. of levels: K					
	2	3		4		
Coefficients	linear	linear	quadratic	linear	quadratic	cubic
W_1	−1	−1	+1	−3	+1	−1
W_2	+1	0	−2	−1	−1	+3
W_3		+1	+1	+1	−1	−3
W_4				+3	+1	+1
$\sum_i W_i^2 = S$	2	2	6	20	4	20

2^{k-p} trials), and are classified (Box and Hunter 1961) according to their degree of fractionation as follows:

(i) *Resolution III designs,* in which no main effect is confounded with another main effect, but main effects are confounded with two factor interactions, and 2-factor interactions are confounded with each other. For example, the $OA_4(2^3)$ design of Table 3.5a is a Resolution III design and can be called a 2_{III}^{3-1} design

TABLE 3.5a. A Resolution III design

Trial	A	B	C
1	1	1	1
2	1	2	2
3	2	1	2
4	2	2	1
	(B × C)	(A × C)	(A × B)

(ii) *Resolution IV designs,* in which no main effect is confounded with any other main effect or two factor interaction, but 2-factor interactions are confounded with each other. For example, the $OA_8(2^4)$ design shown in Table 3.5b is a Resolution IV design and can be called a 2_{IV}^{4-1} design.

(iii) *Resolution V designs,* in which no main effect or 2-factor interaction is confounded with any other main effect or 2-factor interaction, but 2-factor interactions are confounded with 3-factor interactions. For example, the $OA_{16}(2^5)$ design of Table 3.5c is a Resolution V design and can be called a 2_V^{5-1} design.

TABLE 3.5b. A Resolution IV design

Trial	A	B	A×B	C	A×C	B×C	D
1	1	1	1	1	1	1	1
2	1	1	1	2	2	2	2
3	1	2	2	1	1	2	2
4	1	2	2	2	2	1	1
5	2	1	2	1	2	1	2
6	2	1	2	2	1	2	1
7	2	2	1	1	2	2	1
8	2	2	1	2	1	1	2
			(C×D)		(B×D)	(A×D)	

Note that a 2_{V}^{5-1} design is equivalent to a 2_{III}^{15-11} design both requiring 16 experimental trials for the study of 15 effects. Also, a 2_{IV}^{4-1} design is equivalent to a 2_{III}^{7-4} design both requiring 8 trials for the study of 7 effects. Whenever interactions can tentatively be assumed non-existent, the Resolution III designs are most useful in 'screening' situations, i.e. when it is desirable to determine which one of a subset of a large number of variables is important.

3.3.4 *Plackett and Burman $(P–B)_{L,N}$ designs*

These are designs suitable for studying $k = (N-1)/(L-1)$ factors, each with L-levels, with N experimental trials. They were first introduced by Plackett and Burman (1946) and are most useful for 'screening' experiments. We will denote them as $(\mathrm{P–B})_{L,N}$ designs. Their construction is described below.

3.3.4.1 The case of $L = 2$

These are Resolution III designs (i.e. designs which provide uncorrelated estimates of main effects) suitable for exploring $K = N - 1$ 2-level factors ($L = 2$), in N trials, where N is a multiple of 4. Matrices for these designs exist for $N = 4$ up to 100 (except for the isolated case of $N = 92$).

When N is a power of two, the $(\mathrm{P} - \mathrm{B})_{2,N}$ designs are identical to fractional orthogonal designs shown in Appendix D. As their associated triangular interaction arrays show, these designs can also be used for the study of interactions between factors. Of course, in such cases, the number of factors for study need to be less than $N - 1$ [see, for example, Section 3.3.2.2 where the equivalent to a $(\mathrm{P–B})_{2,8}$ design was studied].

TABLE 3.5c. A Resolution V design

Trial	A	B	A×B	C	A×C	B×C	D×E	D	A×D	B×D	C×E	C×D	B×E	A×E	E
1	1	1	1	1	1	1	1	1	1	1	1	1	1	1	1
2	1	1	1	1	1	1	1	2	2	2	2	2	2	2	2
3	1	1	1	2	2	2	2	1	1	1	1	2	2	2	2
4	1	1	1	2	2	2	2	2	2	2	2	1	1	1	1
5	1	2	2	1	1	2	2	1	1	2	2	1	1	2	2
6	1	2	2	1	1	2	2	2	2	1	1	2	2	1	1
7	1	2	2	2	2	1	1	1	1	2	2	2	2	1	1
8	1	2	2	2	2	1	1	2	2	1	1	1	1	2	2
9	2	1	2	1	2	1	2	1	2	1	2	1	2	1	2
10	2	1	2	1	2	1	2	2	1	2	1	2	1	2	1
11	2	1	2	2	1	2	1	1	2	1	2	2	1	2	1
12	2	1	2	2	1	2	1	2	1	2	1	1	2	1	2
13	2	2	1	1	2	2	1	1	2	2	1	1	2	2	1
14	2	2	1	1	2	2	1	2	1	1	2	2	1	1	2
15	2	2	1	2	1	1	2	1	2	2	1	2	1	1	2
16	2	2	1	2	1	1	2	2	1	1	2	1	2	2	1
							(A×B×C)				(A×B×D)		(A×C×D)	(B×C×D)	

When N is not a power of 2, the $(\text{P–B})_{2,N}$ designs do not generally allow the study of interactions, but they allow useful gaps to be filled in the area of 'screening' designs for 2-level factors.

Each (P–B) design can be easily constructed using a 'generating vector' which, for example, in the case of $N = 8$ has the form

$$(2\ 2\ 2\ 1\ 2\ 1\ 1).$$

The design matrix is formed by arranging the vector as the first column and off-setting by one vector element for each new column; in other words, a new column is obtained from the previous one by moving the elements of the previous column down once, and placing the last element in first position. This procedure occurs $N - 2$ times yielding a $(N - 1) \times (N - 1)$ array; the matrix is finally completed by a row of ones. For example, from the above generating vector we obtain the design of Table 3.6a.

TABLE 3.6a. $(\text{P–B})_{2,8}$ design

2	1	1	2	1	2	2
2	2	1	1	2	1	2
2	2	2	1	1	2	1
1	2	2	2	1	1	2
2	1	2	2	2	1	1
1	2	1	2	2	2	1
1	1	2	1	2	2	2
1	1	1	1	1	1	1

This is equivalent to the $OA_8(2^7)$ design (this can be easily seen after a readjustment of the rows and columns of the $(\text{P–B})_{2,8}$ design).

As another example we can refer to the $(\text{P–B})_{2,12}$ design whose generating vector is

$$(2\ 2\ 1\ 2\ 2\ 2\ 1\ 1\ 1\ 2\ 1).$$

Then, by taking this as the first column, shifting it cyclically one place $k = 11$ times and adding a final row of ones, we have the design of Table 3.6b. This is equivalent to the $OA_{12}(2^{11})$ design which can be found in Appendix D. If we replace '1' and '2' by '+1' and '−1' respectively, it can be seen that no column in the above matrix can be considered as the 'product' of two other columns in the same matrix. So the $(\text{P–B})_{2,12}$ design should not be used to analyse interactions. Generating vectors for other $(\text{P–B})_{2,N}$ designs can be found in Appendix C.

TABLE 3.6b. $(P\text{–}B)_{2,12}$ design

2	1	2	1	1	1	2	2	2	1	2
2	2	1	2	1	1	1	2	2	2	1
1	2	2	1	2	1	1	1	2	2	2
2	1	2	2	1	2	1	1	1	2	2
2	2	1	2	2	1	2	1	1	1	2
2	2	2	1	2	2	1	2	1	1	1
1	2	2	2	1	2	2	1	2	1	1
1	1	2	2	2	1	2	2	1	2	1
1	1	1	2	2	2	1	2	2	1	2
2	1	1	1	2	2	2	1	2	2	1
1	2	1	1	1	2	2	2	1	2	2
1	1	1	1	1	1	1	1	1	1	1

3.3.4.2 The case $L \geq 3$

For factors with more than two levels, Plackett and Burman suggest other 'generating vectors' which can be found in Appendix C. The construction procedure is similar to the one used when $L = 2$. For example, if $L = 3$ and $N = 9$, the design for studying $k = (9 - 1)/(3 - 1) =$ four 3-level factors can be generated by the vector

$$(1\ 2\ 3\ 3\ 1\ 3\ 2\ 2).$$

By taking this as the first column, shifting it cyclically one place $K = 4$ times and adding a final row of ones, the design of Table 3.6c is obtained.

TABLE 3.6c. $(P\text{–}B)_{3,9}$ design

1	2	2	3
2	1	2	2
3	2	1	2
3	3	2	1
1	3	3	2
3	1	3	3
2	3	1	3
2	2	3	1
1	1	1	1

By a simple readjustment of the rows of this design we obtain the design of Table 3.6d which is equivalent to 'Taguchi's $OA_9(3^4)$ design' (see Appendix D).

TABLE 3.6d

1	1	1	1
1	2	2	3
1	3	3	2
2	1	2	2
2	2	3	1
2	3	1	3
3	1	3	3
3	2	1	2
3	3	2	1

3.3.5 Taguchi's orthogonal arrays

The 2-level and 3-level orthogonal arrays that Taguchi recommends, are often identical or equivalent to 2-level or 3-level Plackett–Burman designs (Section 3.3.4) and can be found in Appendix D, where triangular interaction matrices for those designs which allow the study of interactions, e.g., for OA_8, OA_{16}, OA_{27}, etc., are also presented.

Taguchi recommends that experiments be designed with emphasis on main effects. Past engineering experience should be used for the selection of characteristics with minimal interactions as much as possible. Taguchi believes that interaction is not a matter of assignment but should be dealt with by changing to characteristic values possessing additivity or monotonicity, and by consideration of interrelationships between the levels chosen for the different factors, prior to experimentation. The designs OA_{12}, OA_{18}, OA_{36}, and OA_{54} are among a group of specially designed arrays in which interactions are distributed more or less uniformly to all columns; this enables the practitioner to focus on main effects and helps increase the efficiency and reproducibility of small-scale experimentation. In Appendix D, orthogonal designs can also be found for dealing with both 2-level and 3-level factors [such as the $OA_{36}(2^{11} \times 3^{12})$].

It is also possible to perform complicated assignments of experiments in which different numbers of levels exist together by using techniques such as the 'multi-level formation' technique described below.

3.3.5.1 'Multi-level formation' in a two-level orthogonal array

If we need to study a 4-level factor, a 4-level column can be created in a 2-level array by sacrificing three 2-level columns, consisting of any two columns and the column corresponding to their interaction. Consequently, this formation will have to take place in a design which allows the study of interactions such as the OA_8, OA_{16}, etc.

Example: the $OA_8(4 \times 2^4)$ formed from $OA_8(2^7)$.

It is known for the $OA_8(2^7)$, that, column 3 is the interactive column for the factors associated with columns 1 and 2. Therefore, the columns 1, 2, and 3 are removed from the orthogonal array and in their place we create a new column at four levels (see Table 3.7a).

TABLE 3.7 (a) $OA_8(2^7)$. (b) $OA_8(4 \times 2^4)$

	Columns								Columns				
Trial	1	2	3	4	5	6	7		(123)	4	5	6	7
1	1	1	1	1	1	1	1		1	1	1	1	1
2	1	1	1	2	2	2	2		1	2	2	2	2
3	1	2	2	1	1	2	2		2	1	1	2	2
4	1	2	2	2	2	1	1	→	2	2	2	1	1
5	2	1	2	1	2	1	2		3	1	2	1	2
6	2	1	2	2	1	2	1		3	2	1	2	1
7	2	2	1	1	2	2	1		4	1	2	2	1
8	2	2	1	2	1	1	2		4	2	1	1	2

Note that the 4-level column of $OA_8(4 \times 2^4)$ was constructed by substituting 1, 2, 3, and 4 for the four combinations (11), (12), (21), and (22) respectively, formed from the levels of the columns 1 and 2 of $OA_8(2^7)$. The interaction column 3 of $OA_8(2^7)$ was erased. Any two columns and their interaction from $OA_8(2^7)$ can be used. So, three columns with 1 degree of freedom are 'used' to create a new column with 3 degrees of freedom.

Example: the $OA_{16} \times (8 \times 2^8)$ formed from $OA_{16}(2^{15})$.

In order to study an 8-level factor we need 7 degrees of freedom, and therefore 7 2-level columns will have to be used for the construction of an 8-level column. This is a simple extension of the technique used for the creation of four levels corresponding to all possible level-combinations (4) of any two columns. Now, the eight levels of the new column will correspond to all possible level-combinations of any three columns of $OA_{16}(2^{15})$, namely $(1, 1, 1)$, $(1, 1, 2)$, $(1, 2, 1)$, $(1, 2, 2)$, $(2, 1, 1)$, $(2, 1, 2)$ $(2, 2, 1)$, and $(2, 2, 2)$; these three columns should be chosen so that each one is *not* an interaction column for the other two. This is because the columns corresponding to any (2-way or 3-way) interactions among those three columns, will have to be erased from the original array. This will leave us with one 8-level column, and eight 2-level columns.

Let us use columns 1, 2, and 4 from $OA_{16}(2^{15})$. Note that each one of those columns is not an interaction column for the other two (see triangular matrix for $OA_{16}(2^{15})$ in Appendix D). Every possible combination of their levels will create a distinct level of an 8-level column. Thus the $OA_{16}(8^1 \times 2^8)$ of Table 3.7b is obtained. Note the columns from $OA_{16}(2^{15})$ that have been erased, (apart from 1, 2, 4) are the interaction columns between 1 and 2 (column 3), between 1 and 4 (column 5), between 2 and 4 (column 6), and between 1, 2, and 4 (column 7). Note that using the $OA_{16}(2^{15})$ and following similar procedures, we can create up to five 4-level columns resulting to the $OA_{16}(4^5)$ array which can be seen in Appendix D.

TABLE 3.7b. $OA_{16}(8 \times 2^8)$ formed from OA_{16} (2^{15})

	Columns								
Trial	1–7	8	9	10	11	12	13	14	15
1	1	1	1	1	1	1	1	1	1
2	1	2	2	2	2	2	2	2	2
3	2	1	1	1	1	2	2	2	2
4	2	2	2	2	2	1	1	1	1
5	3	1	1	2	2	1	1	2	2
6	3	2	2	1	1	2	2	1	1
7	4	1	1	2	2	2	2	1	1
8	4	2	2	1	1	1	1	2	2
9	5	1	2	1	2	1	2	1	2
10	5	2	1	2	1	2	1	2	1
11	6	1	2	1	2	2	1	2	1
12	6	2	1	2	1	1	2	1	2
13	7	1	2	2	1	1	2	2	1
14	7	2	1	1	2	2	1	1	2
15	8	1	2	2	1	2	1	1	2
16	8	2	1	1	2	1	2	2	1

3.3.5.2 'Multi-level formation' in a 3-level orthogonal array

We can create multi-level columns in a 3-level orthogonal array, by following similar procedures to those in Section 3.3.5.1. The resulting array will still be orthogonal provided that the number of levels of the new columns are multiples of 3.

Of course, we have to bear in mind that for the creation of an n-level column we need $n-1$ degrees of freedom, and so in a 3-level orthogonal array we need to sacrifice $(n-1)/2$ columns (each with 2 degrees of freedom).

Example: the $OA_{27}(9 \times 3^9)$ formed from $OA_{27}(3^{13})$.

To create a 9-level column in a 3-level orthogonal array, we need any two columns and their (two) interaction columns. So, if we consider the columns 1 and 2 of $OA_{27}(3^{13})$, their interactions columns 3 and 4 [see triangular matrix for $OA_{27}(3^{13})$ in Appendix D] will also have to be erased from the original array. The nine levels of the new column will correspond to every combination between the levels of the two columns 1 and 2, namely

$$(1,1),\ (1,2),\ (1,3),\ (2,1),\ (2,2),\ (2,3),\ (3,1),\ (3,2),\ \text{and}\ (3,3).$$

In this way one new column with 8 degrees of freedom will be formed at the expense of four columns, each with 2 degrees of freedom. The new orthogonal array is as shown in Table 3.7c.

TABLE 3.7c. $OA_{27}(9 \times 3^9)$ formed from $OA_{27}(3^{13})$

	Columns									
Trial	1, 2, 3, 4	5	6	7	8	9	10	11	12	13
1	1	1	1	1	1	1	1	1	1	1
2	1	2	2	2	2	2	2	2	2	2
3	1	3	3	3	3	3	3	3	3	3
4	2	1	1	1	2	2	2	3	3	3
5	2	2	2	2	3	3	3	1	1	1
6	2	3	3	3	1	1	1	2	2	2
7	3	1	1	1	3	3	3	2	2	2
8	3	2	2	2	1	1	1	3	3	3
9	3	3	3	3	2	2	2	1	1	1
10	4	1	2	3	1	2	3	1	2	3
11	4	2	3	1	2	3	1	2	3	1
12	4	3	1	2	3	1	2	3	1	2
13	5	1	2	3	2	3	1	3	1	2
14	5	2	3	1	3	1	2	1	2	3
15	5	3	1	2	1	2	3	2	3	1
16	6	1	2	3	3	1	2	2	3	1
17	6	2	3	1	1	2	3	3	1	2
18	6	3	1	2	2	3	1	1	2	3
19	7	1	3	2	1	3	2	1	3	2
20	7	2	1	3	2	1	3	2	1	3
21	7	3	2	1	3	2	1	3	2	1
22	8	1	3	2	2	1	3	3	2	1
23	8	2	1	3	3	2	1	1	3	2
24	8	3	2	1	1	3	2	2	1	3
25	9	1	3	2	3	2	1	2	1	3
26	9	2	1	3	1	3	2	3	2	1
27	9	3	2	1	2	1	3	1	3	2

3.3.5.3 Trans-factor technique

This technique is applied to the construction of designs appropriate for the study of distinct classes of factors for different levels of a certain factor. These designs are also known as nested-factorial designs and cover the case when the type of factor to be studied differs depending on the levels of a certain factor.

For example, suppose that B is a factor whose type becomes B′ when A is at level 1, and B″ when A is at level 2. Suppose that B′ and B″ are to be studied at two levels each. Under the assumption that the interaction A × B is insignificant we can assign A and B in an orthogonal array, the $OA_8(2^7)$ say, as shown in Table 3.8.

TABLE 3.8

Factors:	A	B					
Columns:	1	2 3	4	5	6	7	
Trial		(B′)					
1	1	1	1	1	1	1	
2	1	1	2	2	2	2	
3	1	2	1	1	2	2	
4	1	2	2	2	1	1	
		(B″)					
5	2	1	1	2	1	2	
6	2	1	2	1	2	1	
7	2	2	1	2	2	1	
8	2	2	2	1	1	2	

Note that the interaction column 1 × 2 (= 3) has been eliminated from the design, and this made possible the following resolution of the degrees of freedom:

Source	df
A	1
B (for A(1))	1
B (for A(2))	1
Total	3

The factor B can be considered to be a ‘pseudo-factor’ defined to be B′ when A is at level 1, and B″ when A is at level 2. Interest is concentrated not on the main effect of B, but separately on the main effects of B′ [within A(1)] and B″ [within A(2)]; these are found by separately analysing the data corresponding to A(1) (for B′), and the data corresponding to A(2) (for B″). Other factors, or even other pseudo-factors can of course be assigned to the other available columns of the design of Table 3.8.

A case-study

An experiment took place to investigate the factors affecting the orientation of crystal resonators in piezoelectronic devices. The initial phase of the experiment concentrated on the cutting stage of the production process.

Of interest was whether or not the cutting blades needed to be dressed using a carborundum stone. If they were dressed, two types of stone needed to be studied. In the case when the blades were not dressed, an abrasive slurry mix was used; two types of slurry mix would then need to be studied distinguishing on whether or not the slurry was warm at the start of the cutting.

The grit size of the wheel used, was also a factor of interest. Thus, the factors and levels of interest were as follows:

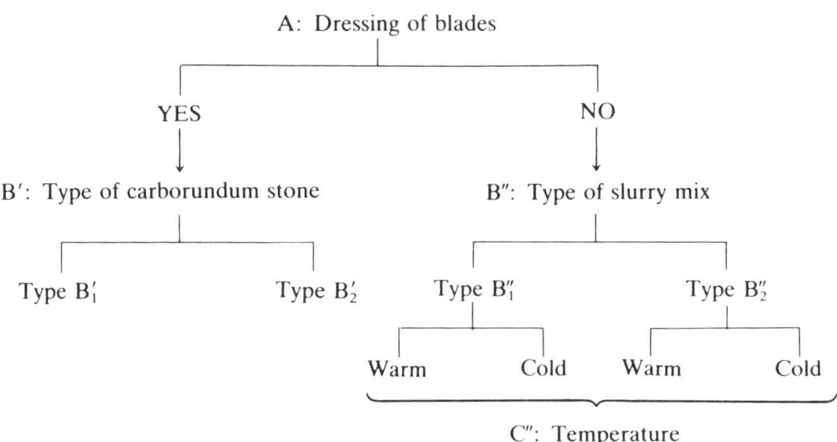

The above factors were assigned to $OA_8(2^7)$ as shown in Table 3.8a.

Note that B can be considered as a pseudo-factor which becomes B′ (type of dressing stone) when A is at level 1 (i.e. when the blades are to

TABLE 3.8a

Factors:	A	B	C	D	e	Data
Columns:	1	2 3	4 5	6	7	(seconds)
Trial		B′	e = C′			
1	1	1	1	1	1	55
2	1	1	2	2	2	22
3	1	2	1	2	2	25
4	1	2	2	1	1	45/147
		B″	C″			
5	2	1	1	1	2	50
6	2	1	2	2	1	35
7	2	2	1	2	1	10
8	2	2	2	1	2	20/115
						262

be dressed), and becomes B″ (type of slurry mix) when A is at level 2 (i.e. when the blades are not dressed). Also, C is another pseudo-factor which becomes C″ (temperature of the slurry mix) when A is at level 2, and is not defined when A is at level 1.

The target value for the orientation was an angle of 14′00″. The response results represented the average absolute deviation from the target (in seconds) and are also shown in Table 3.8a.

Analysis

For the factors A and D, the usual procedure is as follows:

$$CF = \frac{(262)^2}{8}, \qquad TSS = (55^2 + \cdots + 20^2) - CF = 1803.5$$

$$S_A = \frac{(147)^2 + (115)^2}{4} - CF = 128 \qquad (\text{df} = 1)$$

$$S_D = \frac{(170)^2 + (92)^2}{4} - CF = 760.5 \qquad (\text{df} = 1).$$

The factors B′ and B″ are analysed separately within each level of A:

Within level A(1),

$$S_{B'} = \frac{(B_1')^2 + (B_2')^2}{2} - \frac{(B_1' + B_2')^2}{4} = \frac{(77)^2 + (70)^2}{2} - \frac{(147)^2}{4} = 12.25$$

$$(\text{df} = 1).$$

Within level $A(2)$,

$$S_{B''} = \frac{(B_1'')^2 + (B_2'')^2}{2} - \frac{(B_1'' + B_2'')^2}{4} = \frac{(85)^2 + (30)^2}{2} - \frac{(115)^2}{4} = 756.25$$

(df = 1).

The factor C'' is analysed within level A(2),

$$S_{C''} = \frac{(C_1'')^2 + (C_2'')^2}{2} - \frac{(115)^2}{4} = \frac{60^2 + 55^2}{2} - \frac{(115)^2}{4} = 6.25 \qquad (\text{df} = 1)$$

Finally, for the residual

$$RSS = TSS - (S_A + S_D + S_{B'} + S_{B''} + S_{C''}) = 140.25 \qquad (\text{df} = 2).$$

(A pooled error sum of squares is obtained by pooling together RSS, $S_{B'}$ and $S_{C''}''$.)

ANOVA

Source	df	S	F-ratio
A	1	128	3.2
B { B′	1	12.25p	
B { B″	1	756.25	19.1*
C″	1	6.25p	
D	1	760.5	19.2*
RSS	2	140.25p	
(*e*-pooled)	(4)	(158.75)	
Total	7	1803.5	

The above analysis suggests that, although there is no evidence for or against the dressing of blades, there is, however, evidence for the existence of a significant effect due to the type of slurry mix used in the absence of blade-dressing. There is also evidence for a significant effect due to the large size of the wheel used, with the 240 grit wheel resulting in better performance.

3.3.5.4 Split-unit designs

These designs are otherwise known as 'split-plot' designs and used when there are certain restrictions on randomization in the order of experimentation, e.g. when there are factors with levels which are difficult to change. Use of split-unit designs in these cases can lower the cost of the experiment and speed its completion.

The arrays used are basically the same orthogonal designs mentioned so far, but the factors are assigned to certain columns according to their difficulty in changing their levels, and appropriate data analysis procedures are followed.

3.3.5.4a Assignment of factors On the basis of their changing ability, factors are divided into groups of order 1, 2, etc. (order groups) with the highest order group including the factors whose levels are the easiest to change. At the bottom of each of Taguchi's orthogonal arrays (Appendix D) there are brackets indicating to which group each column belongs. This can help assign factors with different degree of difficulty in level-changing. Of course, such grouping is meaningless when there is no difference in the difficulty of changing the levels of the factors.

When interactions between factors need to be studied, the following rules apply:

(a) The interaction of two factors which belong to different groups is assigned to a column belonging to the higher-order group. For example, let us consider the $OA_8(2^7)$.

Column 2 belongs to the second-order group, column 7 belongs to the third-order group and their interaction, which is column 6, belongs to the third-order group (the higher of the two orders).

(b1) The interaction between two 2-level factors of the same group, is assigned to a column belonging to a group of lower order. For example, the interaction between columns 5 and 6 of $OA_8(2^7)$, which belong to a third-order group, is column 3 which appears in a group of lower order (order 2).

(b2) When the interaction (of two factors of the same group) corresponds to more than one column (e.g. the interaction between two 3-level factors), only one column among those appears in a group of lower order.

For example, the interaction between columns 5 and 7 (which belong to the same group of order 3) of the $OA_{27}(3^{13})$, corresponds to columns 1 and 6, only the first of which (i.e. column 1) belongs to a group of lower order (order 1).

These rules are simply the consequence of the rules governing interaction columns in an orthogonal array (see the triangular interaction matrices in Appendix D).

3.3.5.4b Analysis of effects For analysing the effects of the factors in a split-unit design the following basic rule applies.

Basic analysis rule: the variability of a source (factor or interaction) of a particular order group is tested against the error variation of *this* group, provided this error variation is significant when tested against the error variation of the next higher order group. If a group's error variation is not retrievable, the sources of this group are tested against the error variation of the next higher-order group.

To help retrieve a group's error variation, successive groups may be taken together to form a mixed group. For example, group 1 and group 2 are usually mixed to form the first-order group (primary group) as Table 3.9 shows for the case when one needs to assign two difficult to change factors A, B and two easy to change factors C, D, on $OA_8(2^7)$.

TABLE 3.9

Factors:	A	B	e	C	A × C	D	e
Columns:	1	2	3	4	5	6	7
Trial							
1	1	1	1	1	1	1	1
2	1	1	1	2	2	2	2
3	1	2	2	1	1	2	2
4	1	2	2	2	2	1	1
5	2	1	2	1	2	1	2
6	2	1	2	2	1	2	1
7	2	2	1	1	2	2	1
8	2	2	1	2	1	1	2
Group	1	2		3			
	Primary group			Secondary group			

The four different combinations $(1, 1, 1)$, $(1, 2, 2)$, $(2, 1, 2)$, and $(2, 2, 1)$ within the primary group are called primary units (or 'plots'). Experimentally, the levels of the factors assigned to any of the columns of the primary group will change (at most) 4 times. In other words, the levels of up to three factors which are difficult to change, will remain the same during four runs.

When it is very difficult to change factors, a mixed group of groups 1, 2, and 3 may be constructed. This allows for some of the columns of the (primary) group to be left unassigned and hence for the group's error variation to be retrieved.

A case study

Returning to the case study on the orientation of crystal resonators (see Section 3.3.5.3), an experiment took place to assess the effect on the process performance of 9 factors, one of which was the newness of the cutting blades. The 'newness factor' was difficult to vary, as it required replacement of all the cutting blades with new ones for half of the experimental trials.

Two other factors, 'blade tension' and 'machine parallelism', were also difficult to change although to a lesser degree. The levels of all the other factors were easy to change. The factors of interest, and their levels to be studied were as follows

		Levels	
Groups	Factors	1	2
Primary	A: Blade newness	old	new
Secondary	B: Blade tension	1.86 kN per blade	2.23 kN per blade
	C: Machine parallelism	4 μ	12 μ
Third-order	D: Slurry mix	80 μ	20 μ
	E: Temperature of mix	Cold	Warm
	F: Initial strength of hydraulic pressure	12 bar	$12\frac{1}{4}$ bar
	G: Final strength of hydraulic pressure	$12\frac{1}{4}$ bar	$12\frac{1}{2}$ bar
	H: Wax type	Yellow	Shellac
	I: Blade parallelism	2 μ	6 μ

The interactions $A \times B$, $D \times E$, and $C \times I$ were also to be studied. A split-unit design was utilized based on $OA_{16}(2^{15})$ as Table 3.10 shows.

For each trial, two sticks were cut, and their orientation was measured. The response data in Table 3.10 represents the absolute deviation (in seconds) from the target value.

Analysis

The sum of squares for all the effects can be found in the usual way; for example

$$CF = \frac{(1010)^2}{32} = 31878.125$$

$$S_A = \frac{(616)^2 + (394)^2}{16} - CF = 1540.125, \qquad (\text{df} = 1)$$

etc.

The source variations for each group will have to be tested against the error variation of the group, which can usually be found by calculating

TABLE 3.10

Factors:	A	e	e	B	A × B	C	D × E	H	D	I	F	C × I	e	E	G	Data
Columns:	1	2	3	4	5	6	7	8	9	10	11	12	13	14	15	
Trial																
1	1	1	1	1	1	1	1	1	1	1	1	1	1	1	1	50,45
2	1	1	1	1	1	1	1	2	2	2	2	2	2	2	2	43,40
3	1	1	1	2	2	2	2	1	1	1	1	2	2	2	2	35,31
4	1	1	1	2	2	2	2	2	2	2	2	1	1	1	1	30,28
5	1	2	2	1	1	2	2	1	1	2	2	1	1	2	2	55,48
6	1	2	2	1	1	2	2	2	2	1	1	2	2	1	1	48,49
7	1	2	2	2	2	1	1	1	1	2	2	2	2	1	1	30,33
8	1	2	2	2	2	1	1	2	2	1	1	1	1	2	2	25,26
9	2	1	2	1	2	1	2	1	2	1	2	1	2	1	2	25,21
10	2	1	2	1	2	1	2	2	1	2	1	2	1	2	1	29,31
11	2	1	2	2	1	2	1	1	2	1	2	2	1	2	1	15,18
12	2	1	2	2	1	2	1	2	1	2	1	1	2	1	2	21,23
13	2	2	1	1	2	2	1	1	2	2	1	1	2	2	1	18,16
14	2	2	1	1	2	2	1	2	1	1	2	2	1	1	2	32,31
15	2	2	1	2	1	1	2	1	2	2	1	2	1	1	2	24,28
16	2	2	1	2	1	1	2	2	1	1	2	1	2	2	1	30,32
Order groups		1			2						3					1010

the variation corresponding to the unassigned (free) columns of the group.

For example, the error variation for the primary group can be found through the calculation of the sum of squares for columns 2 and 3 (as though there were factors assigned to them), and then

$$S_{e_1} = S_{\text{col.1}} + S_{\text{col.2}}$$

$$= \left\{ \frac{(\text{sum of 1 of col. 1})^2 + (\text{sum of 2 of col. 1})^2}{16} - CF \right\}$$

$$+ \left\{ \frac{(\text{sum of 1 col. 2})^2 + (\text{sum of 2 of col. 2})^2}{16} - CF \right\} = 50 + 8 = 58$$

$$(\text{df} = 2).$$

Another way of calculating S_{e_1}, is by subtracting S_A from the total variation between primary units i.e.

$$S_{e_1} = \frac{(A_{111})^2 + (A_{122})^2 + (A_{212})^2 + (A_{221})^2}{8} - CF - S_A = 58 \qquad (\text{df} = 2).$$

The secondary error variance (error for group 2) is not retrievable (no free column available), whereas the third-order variance is found through

$$S_{e_3} = S_{\text{col. 13}} = 12.5 \qquad (\text{df} = 1).$$

Note that, if the 16 trials had been performed in a completely random sequence, the inter-experimental error sum of squares (error between trials) would be given by

$$S_{e_1} + S_{e_2} + S_{e_3}.$$

Finally, the replication error variance (error within trials), which now represents a fourth-order variance, can be obtained by, either subtracting from the total sum of squares all the sum of squares calculated so far, or from the usual formula [see (2.52) in Chapter 2]

$$S_{e_4} = \left[50^2 + 45^2 - \frac{(50+45)^2}{2}\right] + \cdots + \left[30^2 + 32^2 - \frac{(30+32)^2}{2}\right] = 86$$

(df = 16).

The ANOVA table for this split-unit design is as shown in Table 3.10a.

TABLE 3.10a. ANOVA (split-unit design)

	Source	df	S	MS	F-ratio
Primary source	A	1	1540.125	1540.125	53.1*
First-order error	e_1	2	58.0	29.0	5.15*
Secondary sources	B	1	722.0	772.0	128.2**
	A × B	1	512.0	512.0	90.9**
	C	1	6.125 p		
	D × E	1	190.125	190.125	33.8**
Second-order error	e_2	—	—	—	
Third-order sources	H	1	21.125	21.125	3.75
	D	1	325.125	325.125	57.7**
	I	1	8.0 p		
	F	1	4.5 p		
	C × I	1	18.0	18.0	3.2
	E	1	21.125	21.125	3.75
	G	1	1.125 p		
Third-order error	e_3	1	12.5 p		
Fourth-order error	e_4	16	86	5.375	
(pooled error)		(21)	(118.25)	(5.631)	
Total		31	3525.875		

Note that since e_3 is not significant when tested against the replication error e_4, the existence of a third-order error variance cannot be asserted. Note also that since e_2 is not retrievable the existence of a second-order error variance cannot be asserted either. Pooling together (with S_{e_4}) the sum of squares of the sources which are not significant (to at least 10% level) when tested against the replication error (i.e. of e_3, G, F, I, and C), we obtain a pooled error sum of squares (with 21 degrees of freedom) against which all other sources, except A, are tested. Since e_1 is significant, A is tested against e_1.

The principle behind the factor-assignment and data analysis for split-unit designs covered in this section apply not only for cases of orthogonal arrays, but also for cases of non-orthogonal arrays to be described next.

3.3.6 Taguchi's non-orthogonal arrays

Occasionally, it might be necessary, perhaps due to no further experimentation capability, that 3-level factors will have to be studied using only 2-level arrays; or, that 2-level factors will have to be assigned on 3-level columns. Taguchi suggests certain procedures (described below) for dealing with these cases, at the expense of the loss of part of the orthogonality in the designs employed. Consequently, certain deviations from the usual data-analysis are also suggested.

3.3.6.1 Assigning 2-level factors to 3-level arrays

3.3.6.1a 'Dummy-level' technique We can assign a 2-level factor A onto a 3-level column, by turning A into a 3-level factor by repeating one of the levels. So the third level of A is a repetition of one of the two levels. Usually the level to be repeated is the one considered the more important of the two, and for which more information is needed.

For example, we can assign a 2-level factor on to the first column of the $OA_9(3^4)$ array by repeating level 1 in place of level 3 as Table 3.11 shows. Of course, column 1 is now non-orthogonal to the other columns and this makes the design partly unbalanced.

The sum of squares for the 2-level factor A can be calculated bearing in mind that level 1 is repeated twice as many times as level 2. Then

$$S_A = \frac{(\text{total sum in level 1})^2}{(\text{No. of observations in level 1})} + \frac{(\text{total sum in level 2})^2}{(\text{No. of observations in level 2})} - CF$$

$$= \frac{(A_1 + A_1')^2}{6r} + \frac{(A_2)^2}{3r} - \frac{(A_1 + A_2 + A_1')^2}{9r},$$

TABLE 3.11

Columns:	1	2	3	4
Factors:	A	B	C	D
Trial				
1	1	1	1	1
2	1	2	2	2
3	1	3	3	3
4	2	1	2	3
5	2	2	3	1
6	2	3	1	2
7	1′(=3)	1	3	2
8	1′(=3)	2	1	3
9	1′(=3)	3	2	1

where r is the number of replications for each trial, and A_i = sum of data-values in level i, $i = 1, 2, 1'$.

Recall that

$$CF = \text{Correction factor} = \frac{(\text{grand total})^2}{\text{total no. of observations}} = \frac{(A_1 + A_2 + A_1')^2}{9r}.$$

This is called the 'dummy level' (or 'pseudo-level') technique.

3.3.6.1b Combining technique We can use a single 3-level column for the study of two 2-level factors A, B, under the assumption that there is no significant interaction between these factors. This is achieved by forming a 3-level 'combination' factor [AB] whose three levels are defined as

[AB](1) = A(1)B(1) (=Level 1 for A and Level 1 for B)

[AB](2) = A(2)B(1) (=Level 2 for A and Level 1 for B)

[AB](3) = A(1)B(2) (=Level 1 for A and Level 2 for B).

By assigning this combination factor [AB] on to a 3-level column, it is possible to obtain information about the effects of both A and B. In fact, the main effect of A can be obtained from the difference between [AB](1) and [AB](2), whereas the main effect of B can be obtained from the difference between [AB](1) and [AB](3).

For example, suppose a 3-level column of $OA_9(3^4)$ is used. The total effect of the combination factor [AB] can be found in the usual way through the sum of squares for the 3-level factor [AB] as

$$S_{[AB]} = \frac{[(AB)_1^2 + (AB)_2^2 + (AB)_3^2]}{3r} - CF$$

where r is the number of replications per trial, with the parenthesis-terms representing level-totals. Now

$$S_A = \frac{[(AB)_1 - (AB)_2]^2}{6r} = \frac{(A_1B_1 - A_2B_1)^2}{6r}$$

and

$$S_B = \frac{[(AB)_1 - (AB)_3]^2}{6r} = \frac{(A_1B_1 - A_1B_2)^2}{6r}.$$

Thus, finding the main effects of two 2-level factors A and B, is the same as comparing the three levels of factors [AB] which is a combination of A and B. Note that, since A and B are not orthogonal, S_A and S_B *do not* sum up to $S_{[AB]}$.

A case-study (ceramic experiment)

An experiment was carried out to determine the effect of 5 variables on the hardness of ceramic tiles:

C: Catalyst – two types: C(1) and C(2)
H: Hardener – two types: H(1) and H(2)
M: Type of clay material used – two types: M(1) and M(2)
T: Firing temperature – three levels: T(1) = 300, T(2) = 400 and T(3) = 500 °C
D: Firing duration – three levels: D(1) = 50, D(2) = 70 and D(3) = 90 min.

The above 5 factors were assigned to the $OA_9(3^4)$ design making use of both the 'dummy level' and the 'combining' technique.

First the two 3-level factors T and D were assigned to columns 3 and 4 of $OA_9(3^4)$ respectively. The 2-level factor M was assigned to column 2, applying the 'dummy level' technique. The dummy level is a repetition of M(1) (which is the clay material currently used) inserted in level 3 of the second column. For the two 2-level factors C and H, we create the 3-level combination factor [CH] with levels [CH](1) = C(1)H(1), [CH](2) = C(2)H(1) and [CH](3) = C(1)H(2).

The combination factor [CH] was assigned to column 1. Two ceramic tiles were used in each trial. The hardness of the tile was measured by the maximum pressure (in kilograms) needed to be applied before the tile broke. The design used and the resulting data can be seen in Table 3.12.

TABLE 3.12

Factors:	[CH]	M	T	D	
Columns:	1	2	3	4	Data
Trial					
1	$C_{(1)}H_{(1)}$	$M_{(1)}$	$T_{(1)}$	$D_{(1)}$	23,21
2	$C_{(1)}H_{(1)}$	$M_{(2)}$	$T_{(2)}$	$D_{(2)}$	28,29
3	$C_{(1)}H_{(1)}$	$M_{(1')}$	$T_{(3)}$	$D_{(3)}$	20,20
4	$C_{(2)}H_{(1)}$	$M_{(1)}$	$T_{(2)}$	$D_{(3)}$	33,31
5	$C_{(2)}H_{(1)}$	$M_{(2)}$	$T_{(3)}$	$D_{(1)}$	15,12
6	$C_{(2)}H_{(1)}$	$M_{(1')}$	$T_{(1)}$	$D_{(2)}$	22,24
7	$C_{(1)}H_{(2)}$	$M_{(1)}$	$T_{(3)}$	$D_{(2)}$	29,30
8	$C_{(1)}H_{(2)}$	$M_{(2)}$	$T_{(1)}$	$D_{(3)}$	6,11
9	$C_{(1)}H_{(2)}$	$M_{(1')}$	$T_{(2)}$	$D_{(1)}$	44,40

Analysis

We have

$$CF = \frac{(\text{grand total})^2}{\text{no. of obs.}} = \frac{(23 + \cdots + 40)^2}{9 \times 2} = 10\,658.$$

For the total sum of squares

$$TSS = (23^2 + \cdots + 40^2) - CF = 1650.$$

The sum of squares for the 3-level factors D and T can be calculated in the usual way

$$S_D = \frac{(D_1)^2 + (D_2)^2 + (D_3)^2}{3 \times r} - CF$$

$$= \frac{(23 + 21 + 15 + 12 + 44 + 40)^2 + (28 + \cdots + 30)^2 + (20 + \cdots + 11)^2}{3 \times 2}$$

$$- 10\,658 = 160.33.$$

$$S_T = \frac{(T_1)^2 + (T_2)^2 + (T_3)^2}{3 \times r} - CF$$

$$= \frac{(23 + \cdots + 11)^2 + (28 + \cdots + 40)^2 + (20 + \cdots + 30)^2}{3 \times 2} - 10\,658$$

$$= 900.33.$$

Since the first level of M is repeated twice as many times as the second level we have

$$S_{\mathrm{M}} = \frac{(M_1 + M_1')^2}{6 \times r} + \frac{(M_2)^2}{3 \times r} - CF$$
$$= \frac{(23 + \cdots + 40)^2}{6 \times 2} + \frac{(28 + \cdots + 11)^2}{3 \times 2} - 10\,658$$
$$= \frac{(337)^2}{12} + \frac{(101)^2}{6} - 10\,658 = 506.25.$$

For the combination factor [CH], in the usual way,

$$S_{[\mathrm{CH}]} = \frac{(\mathrm{CH})_1^2 + (\mathrm{CH})_2^2 + (\mathrm{CH})_3^2}{3 \times r} - CF$$
$$= \frac{(23 + \cdots + 20)^2 + (33 + \cdots + 24)^2 + (29 + \cdots + 40)^2}{3 \times 2} - 10\,658$$
$$= \frac{141^2 + 137^2 + 160^2}{6} - 10\,658 = 50.33 \qquad (\mathrm{df} = 2).$$

To resolve the effects of C and H we have

$$S_{\mathrm{C}} = \frac{[(\mathrm{CH})_1 - (\mathrm{CH})_2]^2}{6 \times r} = \frac{(\mathrm{C_1H_1} - \mathrm{C_2H_1})^2}{6 \times r} = \frac{(141 - 137)^2}{12} = 1.33 \qquad (\mathrm{df} = 1).$$

$$S_{\mathrm{H}} = \frac{[(\mathrm{CH})_1 - (\mathrm{CH})_3]^2}{6 \times r} = \frac{(\mathrm{C_1H_1} - \mathrm{C_1H_2})^2}{6 \times r} = \frac{(141 - 160)^2}{12} = 30.08 \quad (\mathrm{df} = 1).$$

Since C and H are not orthogonal $S_{\mathrm{C}} + S_{\mathrm{H}} \neq S_{[\mathrm{CH}]}$.

The variation between trials (first-order variance) is found through the 'between trials' sum of squares.

$$S_{e_1} = \tfrac{1}{2}[(23 + 21)^2 + \cdots + (44 + 40)^2] - CF - [S_{\mathrm{D}} + S_{\mathrm{T}} + S_{\mathrm{M}} + S_{[\mathrm{CH}]}]$$
$$= 0.76.$$

The 'within trials' sum of squares S_{e_2} will provide the means of estimating the variation within trials (second-order variance).

$$S_{e_2} = TSS - S_{e_1} - [S_{\mathrm{D}} + S_{\mathrm{T}} + S_{\mathrm{M}} + S_{[\mathrm{CH}]}] = 32.$$

The analysis of variance summary can now be seen in Table 3.12a

It is clear from Table 3.12a that the first-order variation is small (MS_{e_1} is insignificant when compared to MS_{e_2}). So, we can pool S_{e_1} with S_{e_2} to create the residual sum of squares, compared with which the effects of D, T, M, and [CH] will turn out to be significant. This residual variation cannot be used for elucidating the separate effects of C and H. However, it is clear from the values of S_{C} and S_{H} that the significant effect of [CH]

TABLE 3.12a

Source	df	S	MS
D	2	160.33	80.165
T	2	900.33	450.165
M	1	506.25	506.25
[CH]	2	50.33	25.165
C	(1)	(1.33)	
H	(1)	(30.08)	
e_1	1	0.76	0.76
e_2	9	32	3.56
Total	17	1650	

is due to H rather than C. Therefore, although [CH] was assigned to column 1, it can be assumed that the factor H alone appeared in column 1 with its first level as a dummy level (i.e. repeated twice as many times).

In this case we can recalculate the sum of squares due to H once again from column 1 on this basis (as we calculated S_M) i.e.:

$$S'_H = \frac{(H_1 + H'_1)^2}{6 \times r} + \frac{(H_2)^2}{3 \times r} - CF$$

$$= \frac{(23 + \cdots + 24)^2}{12} + \frac{(29 + \cdots + 40)^2}{6} - 10\,658$$

$$= \frac{(278)^2}{12} + \frac{(160)^2}{6} - 10\,658 = 49.$$

The final ANOVA table is Table 3.12b.

TABLE 3.12b. ANOVA for ceramic experiment

Source	df	S	MS	F-ratio
D	2	160.33	80.165	25.9** sign to 1% level
T	2	900.33	450.165	145.2**
M	1	506.25	506.25	163.3**
H	1	49.00	49.00	15.8**
Residual	11	34.09	3.1	
Total	17	1650		

Remark: Note that $S'_H = S_{[CH]} - S_C$
Therefore, although $S_H + S_C \neq S_{[CH]}$, we nevertheless have

$$S'_H + S_C = S_{[CH]} \text{ (or } S_H + S'_C = S_{[CH]}),$$

where $S'_H(S'_C)$ is the sum of squares due to H(C) when H(C) is assumed to be a dummy-level factor assigned to the column to which the combination factor [CH] was initially assigned.

3.3.6.2 Assigning multi-level factors to 2/3-level arrays

3.3.6.2a Multi-level formation with dummy-level technique We can assign a 3-level factor on a 2-level array by first creating a 4-level column (see Section 3.3.5.1) and then applying the dummy-level technique by repeating one of the three actual levels. For example, having created a 4-level column combining the first three columns of $OA_8(2^7)$, we can assign to it a 3-level factor A by repeating the most important level (A_1 say) as follows.

	Columns				
Trial	(123) A	4	5	6	7
1	A(1)	1	1	1	1
2	A(1)	2	2	2	2
3	A(2)	1	1	2	2
4	A(2)	2	2	1	1
5	A(3)	1	2	1	2
6	A(3)	2	1	2	1
7	A(4) = A(1′)	1	2	2	1
8	A(4) = A(1′)	2	1	1	2

Similarly, we can assign a 7-level factor onto $OA_{16}(2^{15})$, by first creating the $OA_{16}(8 \times 2^8)$ (see Table 3.7b) and then applying the dummy-level technique on the 8-level column with one of the 7 actual levels of the factor as a dummy. This can be expanded to 3-level arrays. For example, an 8-level factor can be studied with the $OA_{27}(3^{13})$ array by first creating a 9-level column (see Table 3.7c) and then utilizing the dummy-level technique.

3.3.6.2b 'Idle-column' technique If one uses the technique of the previous section, at least 1 degree of freedom is usually wasted. Indeed, if we assign a 3-level factor on the 4-level column of $OA_8(4 \times 2^4)$, we 'make use' of three columns of 1 degree of freedom each (the three 2-level columns used to form the 4-level column), i.e. 3 degrees of freedom, for the study of a factor of only 2 degrees of freedom. Moreover, using $OA_8(2^7)$, only one 3-level factor can be studied with this technique, because only one *independent* 4-level column can be created in the $OA_8(2^7)$ design (see Section 3.3.5.1). So, if one wishes to study three, say, 3-level factors with this method, the OA_8 cannot be used. A technique by which one uses only as many 2-level columns as the number of degrees of freedom involved, is the 'idle-column' technique:

3-*level creation*: We first choose one of the columns to be the 'idle-column', I, with levels I(1) and I(2).

Suppose A is a 3-level factor with levels A(1), A(2) and A(3). We assign A to a column in such a way that levels A(1) and A(2) are within level I(1), and levels A(2) and A(3) are within level I(2), with level A(2) repeated in the column twice as many times as levels A(1) or A(3) (as in the dummy-level case).

If the interaction column for the two columns of I and A is erased, then it is possible to obtain the main effect of A through levels I(1) and I(2). Indeed, from I(1) we can obtain the difference between A(1) and A(2), and from I(2) we can obtain the difference between A(2) and A(3). For example, the first three columns of $OA_8(2^7)$ could be used to obtain the following:

Factors					I	A
Columns	1	2	3		1	2 3
	1	1	1		1	1
	1	1	1		1	1
	1	2	2	→	1	2
	1	2	2		1	2
	2	1	2		2	2
	2	1	2		2	2
	2	2	1		2	3
	2	2	1		2	3

So far, the technique seems equivalent to the 'multi-level formation and dummy level' technique of Section 3.3.6.2a. However, the difference here is that the idle column 'I' can be used more than once. In fact, it can be used up to 3 times in the $OA_8(2^7)$, taking care of three 3-level factors. This would be as follows

Factors									I	A	B	C
Columns	1	2	3	4	5	6	7		1	2 3	4 5	6 7
	1	1	1	1	1	1	1		1	1	1	1
	1	1	1	2	2	2	2		1	1	2	2
	1	2	2	1	1	2	2		1	2	1	2
	1	2	2	2	2	1	1	→	1	2	2	1
	2	1	2	1	2	1	2		2	2	2	2
	2	1	2	2	1	2	1		2	2	3	3
	2	2	1	1	2	2	1		2	3	2	3
	2	2	1	2	1	1	2		2	3	3	2

Note that 'I' (the first column) is combined with columns 2, 4, and 6 to create 3-level columns with the following level-correspondence

$$(1,1) \rightarrow 1$$

$$(1,2) \rightarrow 2$$

$$(2,1) \rightarrow 2$$

$$(2,2) \rightarrow 3$$

Then the interaction columns 3 $(=1 \times 2)$, 5$(=1 \times 4)$ and 7 $(=1 \times 6)$ are erased from the array and a design is obtained for the study of three 3-level factors.

The above level-correspondence can be used for the creation of 3-level columns in any 2-level orthogonal array, e.g. $OA_{16}(2^{15})$. If possible, no factor should be assigned to the idle-column (hence the name 'idle'). This is because, one-half of the difference between the first and the third level (of the newly created 3-level column) is confounded with the difference between I(1) and I(2), which corresponds to the main effect of the factor assigned to the idle-column. Of course, since, for example, the difference between A(1) and A(3) can be obtained, it is possible to correct the effect of A from the effect of I. But this has to be repeated for every 3-level factor which was assigned using I. In order to avoid tedious correction calculation, it is best to leave column I unassigned, or assign to it a factor of secondary interest such as a 'block' factor.

Multi-level creation

The idle-column technique can be applied to the study of any n-level factors $(n \geq 3)$, using any 2-level orthogonal array $OA_K(2^{K-1})$ where K is not a multiple of n.

The 'multi-level formation' technique (Section 3.3.5.1) can help in the creation of the *n*-level columns as long as

(a) The idle-column appears in the design.

(b) One of the *n* levels appears twice as many times as the other levels; half of the time in the part of the column corresponding to level I(1), and half the time in the part of the column corresponding to level I(2). This level is called the 'control' level and helps in the comparison of all the levels. Usually one selects the most important level to be the control level. One can have more than one control level. This is, for example, when a 6-level factor needs to be studied and two of those levels are repeated twice as many times as the others. Half of these times corresponding to I(1) and the other half corresponding to I(2).

The 'idle-column' technique can be seen as a method of creating a pseudo-factor which has different levels depending on the levels of the idle-column. These levels can be considered as being nested within the levels of the idle-column, and so the analysis for the pseudo-factor is similar to the analysis of a nested design obtained through the transfactor technique (see Section 3.3.5.3). All the above are demonstrated by the following example.

A case-study

To determine operating life duration with respect to mechanical and electrical parameters of silicon devices (of various metallurgies) assembled on a substrate using Tape Automated Bonding (TAB) techniques, destructive pull test data was analysed for the study of five controllable factors:

ST:	Substrate temperature (°C);	3 levels
UP:	Ultrasonic power (mwatt);	3 levels
BT:	Bonding time (msec);	2 levels
P:	Pressure (grams);	2 levels
M:	Metallurgy type;	7 levels

For constructing an experimental design, the idle-column technique was used to create one 7-level column and two 3-level columns on the $OA_{16}(2^{15})$ design. The resulting design and the average pull strength data (in grams) for each trial is shown in Table 3.13.

Note that column 1 of $OA_{16}(2^{15})$ was used as the idle-column I. A 7-level column was constructed using columns 1–7 with level 1 as the

TABLE 3.13. TAB experimental design

Factors:	I	M	ST	UP	BT	P	Pull-strength
Columns:	1	(2–7)	(8 9)	(10 11)	14	15	data
Trial							
1	1	1	1	1	1	1	23
2	1	1	2	2	2	2	55
3	1	2	1	1	2	2	30
4	1	2	2	2	1	1	60
5	1	3	1	2	2	2	56
6	1	3	2	1	1	1	20
7	1	4	1	2	1	1	40
8	1	4	2	1	2	2	35
9	2	1	1	1	1	2	18
10	2	1	3	3	2	1	48
11	2	5	1	1	2	1	30
12	2	5	3	3	1	2	32
13	2	6	1	3	2	1	59
14	2	6	3	1	1	2	25
15	2	7	1	3	1	2	45
16	2	7	3	1	2	1	42
							618

'control' level; M can now be considered as a pseudo-factor with four levels, which differ according to the levels of the idle-column. These levels are:

Idle-col.	Pseudo-factor M			
I(1)	1	2	3	4
I(2)	1	5	6	7

The idle-column I was also used for the construction of two 3-level columns to accommodate the 3-level factors ST and UP. We have:

	Pseudo-factor			
Idle-col.	ST		UP	
I(1)	1	2	1	2
I(2)	1	3	1	3

Analysis

$$CF = \frac{(618)^2}{16} = 23\,870.25$$

$$TSS = (23^2 + \cdots + 42^2) - CF = 2991.75 \qquad (\text{df} = 15)$$

$$S_{\text{p}} = \frac{(P_1)^2 + (P_2)^2}{8} - CF = \frac{(322)^2 + (296)^2}{8} - 23\,870.25 = 42.25 \qquad (\text{df} = 1)$$

$$S_{\text{BT}} = \frac{(BT_1)^2 + (BT_2)^2}{8} - CF = \frac{(263)^2 + (355)^2}{8} - 23\,870.25 = 529 \qquad (\text{df} = 1).$$

For the analysis of the 'pseudo-factors' UP, ST, and M we follow a similar analysis to the analysis of 'nested' factors (see Section 3.3.5.3). For example, the levels of UP differ depending on the levels of the idle-column I; so we perform a different analysis of those levels for each level of column 1: The sum of squares due to UP when levels 1 and 2 are used within level I(1) is given by

$$S_{\text{UP}(1,2)} = \frac{(\text{UP}_1)^2 + (\text{UP}_2)^2}{4} - CF_{\text{I}(1)} = \frac{(108)^2 + (211)^2}{4} - \frac{(319)^2}{8} = 1326.125 \qquad (\text{df} = 1).$$

Moreover, the sum of squares due to UP when levels 1 and 3 are used within level I(2) is given by

$$S_{\text{UP}(1,3)} = \frac{(\text{UP}_1)^2 + (\text{UP}_3)^2}{4} - CF_{\text{I}(2)} = \frac{(115)^2 + (184)^2}{4} - \frac{(299)^2}{8} = 595.125 \qquad (\text{df} = 1).$$

Similarly, for ST

$$S_{\text{ST}(1,2)} = \frac{(149)^2 + (170)^2}{4} - \frac{(319)^2}{8} = 55.125 \qquad (\text{df} = 1).$$

and

$$S_{\text{ST}(1,3)} = \frac{(152)^2 + (147)^2}{4} - \frac{(299)^2}{8} = 3.125 \qquad (\text{df} = 1).$$

and for M

$$S_{\text{M}(1,2,3,4)} = \frac{(M_1)^2 + (M_2)^2 + (M_3)^2 + (M_4)^2}{2} - CF_{\text{I}(1)}$$

$$= \frac{(78)^2 + (90)^2 + (76)^2 + (75)^2}{2} - \frac{(319)^2}{8} = 72.375 \qquad (\text{df} = 3).$$

$$S_{M(1,5,6,7)} = \frac{(M_1)^2 + (M_5)^2 + (M_6)^2 + (M_7)^2}{2} - CF_{I(2)}$$

$$= \frac{(66)^2 + (62)^2 + (84)^2 + (87)^2}{2} - \frac{(299)^2}{8} = 237.375 \qquad (\text{df} = 3).$$

The ANOVA results are shown in Table 3.13a.

TABLE 3.13a. ANOVA: TAB experiment

Source		df	*S*	*MS*	*F*-ratio
P		1	42.25	42.25	0.97
BT		1	529	529	12.1* sign at 5% level
UP	$UP_{(1,2)}$	1	1326.125	1326.125	30.3* sign at 5% level
	$UP_{(1,3)}$	1	595.125	595.125	13.6* sign at 5% level
ST	$ST_{(1,2)}$	1	55.125	55.125	1.26
	$ST_{(1,3)}$	1	3.125	3.125	0.07
M	$M_{(1,2,3,4)}$	3	72.375	24.125	0.55
	$M_{(1,5,6,7)}$	3	237.375	79.125	1.81
Residual		3	131.25	43.75	
Total		15	2991.75		

The above analysis shows evidence of a significant effect due to Bonding Time (BT) and Ultrasonic Power (UP).

Remark: Using the principles of the idle-column technique we can assign multi-level factors to 3-level orthogonal arrays. For example, up to four 7-level factors can be studied independently using the $OA_{27}(3^{13})$. Indeed, using the first column as the idle-column and assuming level 1 to be the control level for all the factors, we can obtain the design of Table 3.14.

3.3.6.3 Partially supplemented designs

When a factor has numerous levels and assignment using designs mentioned so far is difficult, the number of rows of a particular orthogonal array can be expanded to accommodate such a factor; the analysis of the data resulting from such a supplemented design is then done in a way to ensure that orthogonality for the rest of the factors is still valid.

For example, although the assignment of a 4-level factor to a 2-level orthogonal array is a simple matter of creating a 4-level column at the

TABLE 3.14

Factors:	Idle	A	B	C	D
Columns:	1	(2 3 4)	(5 6 7)	(8 9 10)	(11 12 13)
Trial					
1	1	1	1	1	1
2	1	1	2	2	2
3	1	1	3	3	3
4	1	2	1	2	3
5	1	2	2	3	1
6	1	2	3	1	2
7	1	3	1	3	2
8	1	3	2	1	3
9	1	3	3	2	1
10	2	1	1	1	1
11	2	1	4	4	4
12	2	1	5	5	5
13	2	4	1	4	5
14	2	4	4	5	1
15	2	4	5	1	4
16	2	5	1	5	4
17	2	5	4	1	5
18	2	5	5	4	1
19	3	1	1	1	1
20	3	1	6	6	6
21	3	1	7	7	7
22	3	6	1	6	7
23	3	6	6	7	1
24	3	6	7	1	6
25	3	7	1	7	6
26	3	7	6	1	7
27	3	7	7	6	1

expense of only three 2-level columns (see Section 3.3.5.1), the assignment of a 5-level factor would require the creation of an 8-level column with three dummy levels resulting in a waste of columns (and degrees of freedom).

For these cases, Taguchi suggests the creation, first, of an orthogonal array accommodating as many levels (of this factor) as possible, and then the partial supplementing of this array by an appropriate number of rows to accommodate the rest of the levels.

For example, suppose we want to study a 5-level factor A, and four 2-level factors B, C, D, and E. We first create a 4-level column in $OA_8(2^7)$ using the technique of Section 3.3.5.1. The orthogonal array $OA_8(4 \times 2^4)$ is then obtained. For the fifth level of A, we supplement this array with two additional trial runs under the same conditions as level A(4). The resulting design is shown in Table 3.15

TABLE 3.15. A partially supplemented design

Trial	Factors A	B	C	D	E		Response (hypothetical)
1	1	1	1	1	1	$OA_8(4 \times 2^4)$	Y_1
2	1	2	2	2	2		Y_2
3	2	1	1	2	2		Y_3
4	2	2	2	1	1		Y_4
5	3	1	2	1	2		Y_5
6	3	2	1	2	1		Y_6
7	4	1	2	2	1		Y_7
8	4	2	1	1	2		Y_8
9	5	1	2	2	1	Supplement	Y_9
10	5	2	1	1	2		Y_{10}

Note that the level combination B(1)C(2)D(2)E(1) appears twice in the design (in trials 7 and 9). Similarly B(2)C(1)D(1)E(2) appears twice, in trials 8 and 10. The other combinations in trials 1 to 6 [e.g. B(1)C(1)D(1)E(1) etc.] appear only once. This makes the effects of B, C, D, and E non-orthogonal. To make these effects orthogonal for the purpose of analysis, we assume that combinations 1 to 6 have been used twice, both with the same result. This, of course, introduces an additional part of error variance of $6 \times \sigma_e^2$, where σ_e^2 is the actual experimental error (between and within trials) variance. This excess will have to be taken into consideration when we test for significance of the effects.

In general, when we have partially supplemented the orthogonal array OA_N and have conducted experiments under $(N + M)$ different conditions (i.e. N conditions due to OA_N plus M additional—supplemented – ones) where $M < N$, then by assuming (during the analysis) that $N - M$ trials (conditions) have been made twice each, we introduce an excess part of error variance which is

$$(N - M)\sigma_e^2,$$

where σ_e^2 is the actual experimental error variance. This excess part of error, is distributed equally, $[(N-M)/N] \times \sigma_e^2$ each, among the N degrees of freedom of the sources (including the correction factor) of the original orthogonal array OA_N.

For example, for the case of Table 3.15, we have $N=8$ and $M=2$. Then the excess part of error that enters the sum of squares of the sources of $OA_8(4 \times 2^4)$ are as follows

$$\left(\frac{6}{8}\right)\sigma_e^2 \text{ for the correction factor } CF,$$

$$\left(\frac{6}{8}\right)\sigma_e^2 \text{ for each of the (1-degree-of-freedom) 2-level factors B, C, D and E}$$

and

$$\left(3 \times \frac{6}{8}\right)\sigma_e^2 \text{ for the (3-degree-of-freedom) 4-level factor A.}$$

Therefore the excess part of error variance that enters the mean sum of squares for the *sources of the supplemented design* of Table 3.15, is found by dividing the above by the associated degrees of freedom (1 df for B, C, D, E and 4 df for A). Finally, the total (old plus excess) amount of error variance expected to enter the mean sum of squares of the sources of the supplemented design is

$$\sigma_e^2 + \frac{6}{8}\sigma_e^2 = \left(\frac{7}{4}\right) \times \sigma_e^2 \text{ for each of } CF\text{, B, C, D, and E,}$$

and

$$\sigma_e^2 + \frac{3 \times \frac{6}{8}}{4}\sigma_e^2 = \left(\frac{25}{16}\right) \times \sigma_e^2 \text{ for the 5-level factor A.}$$

If W is the coefficient involved in the product with σ_e^2 above, i.e.

$$W = \begin{cases} \frac{7}{4} \text{ for B, C, D, and E} \\ \frac{25}{16} \text{ for A,} \end{cases}$$

then W can be used to correct the mean sum of squares for each source in the supplemented design. Once we calculate the MSs using the available data under the assumption that trials 1 to 6 have taken place twice, we can correct for this assumption by dividing MS by the associated W.

3.3.6.3a A theoretical example (no-replication) As a demonstration let us use the design of Table 3.15 for which we assume only one observation per trial (for cases of replicated trials, see case-study below). We have

$$CF = [2(Y_1 + \cdots + Y_6) + Y_7 + \cdots + Y_{10}]^2/16$$

$$TSS = \{2(Y_1^2 + \cdots + Y_6^2) + Y_7^2 + \cdots + Y_{10}^2\} - CF$$

$$S_A = \frac{A_1^2 + A_2^2 + A_3^2}{4} + \frac{A_4^2 + A_5^2}{2} - CF$$

$$= \frac{[2(Y_1 + Y_2)]^2 + [2(Y_3 + Y_4)]^2 + [2(Y_5 + Y_6)]^2}{4}$$

$$+ \frac{(Y_7 + Y_8)^2 + (Y_9 + Y_{10})^2}{2} - CF$$

$$S_B = \frac{B_1^2 + B_2^2}{8} - CF$$

$$= \frac{(2Y_1 + 2Y_3 + 2Y_5 + Y_7 + Y_9)^2 + (2Y_2 + 2Y_4 + 2Y_6 + Y_8 + Y_{10})^2}{8} - CF$$

Similarly for S_C, S_D, and S_E.

Finally

$$S_e = TSS - (S_A + S_B + S_C + S_D + D_E).$$

From these, the ANOVA Table 3.15a is constructed

TABLE 3.15a. ANOVA (hypothetical data)

Source	df	S	MS	Corrected MS MS/w	F-ratio
A	4	S_A	$S_A/4$	$MS_A \times \frac{16}{25}$	
B	1	S_B	S_B	$MS_B \times \frac{4}{7}$	
C	1	S_C	S_C	$MS_C \times \frac{4}{7}$	CMS/MS_e
D	1	S_D	S_D	$MS_D \times \frac{4}{7}$	
E	1	S_E	S_E	$MS_D \times \frac{4}{7}$	
Error	1	S_e	S_e	MS_e	
Total	10	TSS			

3.3.6.3b A case-study (with replication) An experiment took place to investigate the effects of three factors of interest on the yield of a new process for the separation of the main components (isomers) of an important dyestuff intermediate. The factors and the levels needed to be studied were as follows:

	Levels		
A: Temperature	20	30	40
B: Dilution factor	500	1000	
C: Amount of chloride	1.0	2.0	

There was an experimental capability of only 6 trials with two yield responses for each.

The two levels of factors B and C and two out of the 3 levels of A are assigned to the $OA_4(2^3)$ design. This orthogonal array is then supplemented by two additional trials to correspond to the third level of A; these trials are under the same conditions as the trials corresponding to level 2 of A. The supplemented design is shown in Table 3.16.

TABLE 3.16

Trial	Factors A	B	C		Yield	
1	1	1	1	$OA_4(2^3)$	26.6 27.3	To be repeated during analysis
2	1	2	2		30.1 32.9	
3	2	1	2		20.8 21.0	
4	2	2	1		33.5 35.2	
5	3	1	2	Supplement	25.5 26.0	
6	3	2	1		38.1 37	

Analysis

To make the effects of B and C orthogonal, we assume that trials 1 and 2 have been repeated twice with the same yield results. Therefore, in order to find the main effects of A, B, C we analyse the data of Table 3.16a.

TABLE 3.16a

Trial	Factors A	B	C	Data	Sub-total	Trial	Factors A	B	C	Data	Sub-total	Total
1	1	1	1	26.6 27.3	53.9	1′	1	1	1	26.6 27.3	53.9	107.8
2	1	2	2	30.1 32.9	63	2′	1	2	2	30.1 32.9	63	126.0
3	2	1	2	20.8 21.0	41.8	5	3	1	2	25.5 26.0	51.5	93.3
4	2	2	1	33.5 35.2	68.7	6	3	2	1	38.1 37	75.1	143.8

As usual

$$CF = \frac{(470.9)^2}{16} = 13\,859.18$$

$$S_A = \frac{(A_1)^2}{8} + \frac{(A_2)^2}{4} + \frac{(A_3)^2}{4} - CF$$

$$= \frac{[2(53.9) + 2(63)]^2}{8} + \frac{(41.8 + 68.7)^2}{4} + \frac{(51.5 + 75.1)^2}{4} - CF = 33.08$$

$$(\text{df} = 2).$$

$$S_B = \frac{(B_1)^2 + (B_2)^2}{8} - CF = \frac{(107.8 + 93.3)^2 + (126.0 + 143.8)^2}{8} - CF$$

$$= 294.98 \qquad (\text{df} = 1)$$

$$S_C = \frac{(C_1)^2 + (C_2)^2}{8} - CF = \frac{(107.8 + 143.8)^2 + (126.0 + 93.3)^2}{8} - CF = 65.2$$

$$(\text{df} = 1).$$

Since there is a replication for each trial we first find the inter-experimental (inter-trial) error sum of squares

$$S_{e_1} = \tfrac{1}{2}\{2(53.9^2 + 63^2) + (41.8)^2 + (68.7)^2 + (51.5)^2 + (75.1)^2\}$$
$$- CF - \{S_A + S_B + S_C\} = 1.37 \qquad (\text{df} = 1).$$

The replication error (within-trial-error) is found only from the part where experimentation was actually conducted. In other words, the data from trials 1 and 2 *are not* calculated doubly. Following the proper procedure for assessing replication errors (see Chapter 2) we have for the within-trial sum of squares

$$S_{e_2} = \left[26.6^2 + 27.3^2 - \frac{(53.9)^2}{2}\right] + \left[30.1^2 + 32.9^2 - \frac{(63)^2}{2}\right] + \cdots$$
$$+ \left[38.1^2 + 37^2 - \frac{(75.1)^2}{2}\right] = 6.36 \qquad (\text{df} = 6).$$

Finally, for the total sum of squares we have

$$TSS = S_A + S_B + S_C + S_{e_1} + S_{e_2} = 400.99 \quad (\text{df} = 11).$$

Note that the total degrees of freedom is one less than the number of *actual* data values.

For correcting the mean sum of squares for the effects of A, B, and C we note that, as we used the $OA_4(2^3)$ design and supplemented it with 2 additional trials, we have $N = 4$ and $M = 2$. So, we introduced (through the repetition of trials 1 and 2) an excess part of error variance which is

$$(4-2)\sigma_e^2 = 2\sigma_e^2$$

where σ_e^2 represents the inter-trial and within-trial (actual) error variance. This is distributed equally, $(2\sigma_e^2)/4$ each, among the 4 degrees of freedom of the sources of the original OA_4 (including the correction factor). Since everyone of these sources (*CF*, A, B, and C) had 1 degree of freedom [in the $OA_4(2^3)$], there is an $(\frac{1}{2})\sigma_e^2$ excess part of error that enters the sum of squares of the sources of $OA_4(2^3)$. Consequently, the excess part of error variance that enters the mean sum of squares for the sources of the supplemented design of Table 3.16, is

$$(\tfrac{1}{2})\sigma_e^2 \text{ for CF, B, and C} \quad \text{(with 1 degree of freedom)}$$

and

$$\frac{(\frac{1}{2})\sigma_e^2}{2} = \frac{\sigma_e^2}{4} \text{ for A} \quad \text{(with 2 degrees of freedom).}$$

The total amount of error variance entering *MS* for each factor is

$$\sigma_e^2 + \tfrac{1}{2}\sigma_e^2 = (\tfrac{3}{2})\sigma_e^2 \text{ for } CF\text{, B, and C}$$

and

$$\sigma_e^2 + \frac{\sigma_e^2}{4} = (\tfrac{5}{4})\sigma_e^2 \text{ for A.}$$

The mean sum of squares can be corrected by dividing them with

$$W = \begin{cases} \frac{3}{2} \text{ for B and C.} \\ \frac{5}{4} \text{ for A.} \end{cases}$$

From these, the ANOVA Table 3.16b can be constructed.

TABLE 3.16b. ANOVA for yield data

Source	df	*S*	*MS*	Corrected *MS*	*F*-ratio
A	2	33.08	16.54	$16.5 \times \frac{4}{5} = 13.23$	11.98*
B	1	294.88	294.98	$294.98 \times \frac{2}{3} = 196.65$	178.13**
C	1	65.2	65.2	$65.2 \times \frac{2}{3} = 43.47$	39.38**
e_1	1	1.37p	1.37	1.37	
e_2	6	6.36p	1.06	1.06	
(e_{pooled})	(7)	(7.73)	(1.104)	(1.104)	
Total	11	400.99			

Note that as e_1 was not significant in comparison with e_2, a pooled residual was constructed by pooling together S_{e_1} and S_{e_2} (if e_1 were significant, the factor effects *MS*'s would have had to be compared to MS_{e_1}). We conclude that all factors B, C, and (to a lesser degree) A significantly affect the yield.

3.3.6.4 Interaction partially omitted designs

With continuous factors at more than 2-levels, and interactions between only low-order terms (e.g. only the interaction between the linear effects) the designs mentioned so far can be used by ignoring all higher-order interactions with only a small loss of efficiency.

We will briefly refer to some representative cases which indicate the appropriate columns to be used for the study of the interaction only between linear effects.

3.3.6.4a Study of low-order interaction between a 4-level and 2-level factor Suppose A is a 4-level continuous factor and B is a 2-level (not necessarily continuous) factor. The OA_8 (4×2^4) design can be used (see Table 3.7a) for the study of A, B, and also $A_L \times B$ as Table 3.17 shows.

TABLE 3.17

Factors:	A	B	$A_L \times B$	C	D
Columns:	1 2 3	4	5	6	7
Trial					
1	1	1	1	1	1
2	1	2	2	2	2
3	2	1	1	2	2
4	2	2	2	1	1
5	3	1	2	1	2
6	3	2	1	2	1
7	4	1	2	2	1
8	4	2	1	1	2

The design of Table 3.17 is an interaction partially omitted design because only $A_L \times B$ is assigned to one column whereas the interaction components $A_Q \times B$ and $A_C \times B$ are 'omitted'. Taguchi has shown that column 5 can be used to obtain information about $A_L \times B$ with 80% efficiency, meaning that only 20% of the effect of $A_L \times B$ 'enters' column 6. So column 6 can be used to study another factor, C say, provided we 'correct for the effect' of $A_L \times B$ by subtracting one-fourth of the $A_L \times B$

effect from that of C. Sometimes even correction is unnecessary especially if the $A_L \times B$ is less than 1% significant (i.e. significant to a level $\alpha > 1\%$).

The sum of squares for $A_L \times B$ can be obtained by viewing $A_L \times B$ as the factor assigned to the 2-level column 5, i.e.

$$S_{A_L \times B} = \frac{[(\text{sum of 1 of col. 5})^2 + (\text{sum of 2 of col. 5})^2]}{4} - CF$$

or

$$S_{A_L \times B} = \frac{[(\text{sum of 2 of col. 5}) - (\text{sum of 1 of col. 5})]^2}{8}.$$

This formula can be used for any column used for the study of the interaction between the linear effects of two factors.

3.3.6.4b Study of low-order interaction between 4-level factors Two 4-level continuous factors A, B, two 2-level factors C, D, and the interactions $A_L \times B_L$, $A_L \times C$, $B_L \times C$, $A_L \times D$, and $C \times D$ can be assigned to $OA_{16}(2^{15})$, but using the 'multi-level formation' technique (Section 3.3.5.1) and using the information of the triangular interaction matrix (see Appendix D) as Table 3.18 shows. Column 5 possesses an efficiency of 64%. One-fourth of the effect of $A_L \times B_L$ has entered columns 6 and 9.

TABLE 3.18

Factors:	A	B	$A_L \times B_L$	C	$A_L \times C$	$B_L \times C$	D	$A_L \times D$	$C \times D$
Columns:	1 2 3	4 8 12	5	15	14	11	6	7	9
Trial									
1	1	1	1	1	1	1	1	1	1
2	1	2	1	2	2	2	1	1	2
3	1	3	2	2	2	1	2	2	1
4	1	4	2	1	1	2	2	2	2
5	2	1	1	2	2	2	2	2	1
6	2	2	1	1	1	1	2	2	2
7	2	3	2	1	1	2	1	1	1
8	2	4	2	2	2	1	1	1	2
9	3	1	2	2	1	2	1	2	2
10	3	2	2	1	2	1	1	2	1
11	3	3	1	1	2	2	2	1	2
12	3	4	1	2	1	1	2	1	1
13	4	1	2	1	2	1	2	1	2
14	4	2	2	2	1	2	2	1	1
15	4	3	1	2	1	1	1	2	2
16	4	4	1	1	2	2	1	2	1

3.3.6.4c Study of low-order interaction between 3-level factors Suppose that the idle-column technique was used to create two 3-level columns on the $OA_{16}(2^{15})$ design (see Section 3.3.6.2b). Two 3-level factors A, B, two 2-level factors C, D and the interactions $A_L \times B_L$, $C \times D$ can be assigned as Table 3.19 shows. Column 6 possesses an efficiency of 100% which makes the independent evaluation of $C \times D$ under column 7 possible.

TABLE 3.19

Factors:	Idle	A	B	$A_L \times B_L$	$C \times D$	C	D
Columns:	1	2,3	4,5	6	7	10	13
Trial							
1	1	1	1	1	1	1	1
2	1	1	1	1	1	2	2
3	1	1	2	2	2	1	2
4	1	1	2	2	2	2	1
5	1	2	1	2	2	2	1
6	1	2	1	2	2	1	2
7	1	2	2	1	1	2	2
8	1	2	2	1	1	1	1
9	2	2	2	1	2	1	2
10	2	2	2	1	2	2	1
11	2	2	3	2	1	1	1
12	2	2	3	2	1	2	2
13	2	3	2	2	1	2	2
14	2	3	2	2	1	1	1
15	2	3	3	1	2	2	1
16	2	3	3	1	2	1	2

3.4 Bibliography

Some of the material in this chapter is classical and is available in well-known texts on experimental design. A modern example is Box, Hunter, and Hunter (1978). One of the first comprehensive books was Cochran and Cox (1957). Other texts are listed in the references. Two excellent texts which are less technical in approach but can be considered as classics are the original book of R. A. Fisher (1966, 8th edn) and Cox (1958). These are particularly good for their discussions of blocking and more intuitive aspects of experimental layouts. For highly technical material on combinatorial methods in experimental design the reader should consult Raghavarao (1971) or more recent texts of which Street and Street (1987) is an excellent example.

4
FURTHER DESIGN AND ANALYSIS TECHNIQUES

In this chapter we outline some further methods, strongly recommended by Taguchi for promoting the efficiency of an experimental study.

Simple procedures are described for studying numerous factors using only a small experimental design, and to ensure the lack of interaction effects before the experiment takes place, through a proper choice of factor levels. It is proposed that proper prior consideration of interrelationships between the factors can help in the proper choice of factor levels which, apart from cancelling interactions, can also help in some special situations, such as in mixture experiments.

The concept of 'contribution ratios' plays a vital role in 'tolerance analysis' to be described in later chapters. Techniques for predicting the short-term and long-term process performance are described with a description of the 'discount' (beta) coefficients' which could be used to improve estimation. The common problem of 'missing values' is considered, and analysis techniques for such cases are suggested.

Analysis techniques for binary data are outlined, and these include Taguchi's 'minute' analysis for life-test data and 'accumulating' analysis for categorical and ranked-type data. Finally, a method of fitting models on small sets of data is described and demonstrated.

4.1 Assignment of numerous factors on small designs

When one needs to study many factors, and, due to a restrictive experimentation capability, none of the techniques covered in Chapter 3 can produce a satisfactory design, then a last resort is the 'confounding technique' described below.

To optimize numerous characteristics during one set of experimental trials, it is possible to use only a small design and simultaneously perform many types of study, using the 'compounding technique' described below.

4.1.1 Confounding technique

There might be occasions when one needs to study the effects of numerous multi-level sources totalling n, say, degrees of freedom, but the only 'affordable' designs are those allowing for m degrees of freedom, where $m < n$. Prior experience and engineering judgement will then have to play a role in identifying those factors which are not 'expected' to have a highly significant effect. The effects of these factors will be allowed to be 'confounded'. For example, two of these factors, A and B say, with 2 levels each, will be assigned to a single 2-level column by forming the two-level factor [AB] with levels

$$[AB](1) = A(1)B(1) \quad \text{and} \quad [AB](2) = A(2)B(2)$$

If the effect of [AB] is found to be insignificant, it may be assumed that neither A nor B is significant. Of course, if [AB] is found to have a significant effect, it is always possible that, either A, or B, or both A and B, could be significant, and thus, an additional experiment might be necessary in order to separate their effects.

Nevertheless, in the presence of numerous factors, especially during screening experimentation, this method can be followed (by even combining more than 2 factors together) in order to reduce a prohibitively large number of sources into a manageable set of 'effects' which may include some confounded ones. Once this reduction is achieved, some subsequent experimentation might be possible for a more detailed study and proper assessment.

4.1.2 Compounding technique

There are many occasions when the number of characteristics of interest is large. It is possible to perform many types of study (as many as the number of responses) using the same design and performing only one experiment, by assigning many factors to the same columns of the design. During the analysis of a particular response, it is assumed that the sources of the design are related to this response.

For example, suppose there are three responses of interest R_1, R_2, and R_3, with which the nine 2-level factors A, B, . . . , I are related in the manner shown in Table 4.1.

TABLE 4.1. Factors and related responses

	A	B	C	D	E	F	G	H	I
R_1	×	×	×	×	×				
R_2	×	×		×			×	×	
R_3	×			×		×		×	×

The '×' mark signifies that the associated factor might have an influence on the corresponding response. Table 4.2 shows how one could study 15 main effects and 3 interaction effects at the expense of only 8 experimental trials using the $OA_8(2^7)$ design. The number of replications per trial could differ according to the response under consideration. Different ANOVA analysis, of course, will have to be performed, one for each response.

TABLE 4.2. Assignment by compounding

R_1	A	B	C	D	e	B×D	E			
R_2	A	B	A×B	D	A×D	G	H			
R_3	A	e	e	D	F	I	H			
								Data:		
Column	1	2	3	4	5	6	7	R_1	R_2	R_3
Trial										
1	1	1	1	1	1	1	1	·	·	·
2	1	1	1	2	2	2	2	·	·	·
⋮	⋮	⋮	⋮	⋮	⋮	⋮	⋮			
8	2	2	1	2	1	1	2			

4.2 Choosing the factor levels

One of the most important procedures that have to be followed prior to experimentation is the proper selection of factor levels.

4.2.1 Avoiding useless information

Careful consideration of the experimental range for each factor and of the interrelationship between factors, can help an appropriate choice of levels. It must not include any range which, from prior experience, is known to be unusable or worse than the others in terms of cost, acceptable yield, etc. This eliminates the possibility of failed trials or useless information.

4.2.2 Cancelling interactions

One should always carefully consider whether or not the range of a particular factor depends on the levels of another factor. If such an interrelationship exists, the levels of one of the factors should be defined in a different way for every level of the other factors. For example, in chemical reaction experiments it is common that the range of the reaction time (A), in which one wishes to experiment, varies according to the levels of temperature (B) used. One could then define the levels of the two-level factor A, in the manner indicated by the following numerical (hypothetical) example.

	A	A(1)	A(2)
B	B_1: 50 °C	5 hours	7 hours
	B_2: 60 °C	4 hours	6 hours
	B_3: 70 °C	3 hours	5 hours

Data analysis is conducted in the usual manner.

Forming the levels of A as above, apart from avoiding experiments in an unnecessary or unusable range, has most importantly the advantage of cancelling the interaction between A and B. Cancellation of interactions can heighten the precision of data analysis and can improve the choice of an affordable experimental design. It rarely happens that there is no interaction between the factors under study. The effect of every factor A usually differs according to other conditions B, C, ... However, as long as one considers interrelationships among the factors prior to experimentation, it is often possible to ensure that there will be monotonicity of the factorial effects and additivity in the response model. In Taguchi's opinion, when monotonicity or additivity of factorial effects do not hold, in other words when there is interaction, it is because insufficient research has been done on the characteristics and the interrelationships between the different factors. In such cases the reliability of the experimental results is lower. In fact, he strongly believes that 'as long as it is not possible to minimize interactions between controllable factors, it is impossible to render an experiment efficient'!

4.2.3 Dealing with mixture experiments

When mixing various substances in percentages that have to add up to 100, levels can be formed by one of the two following methods:

4.2.3.1 Expanding–shrinking method

After the range for the various substances has been decided and initial levels have been formed and assigned on the appropriate experimental design, the initial levels are allowed to 'expand' or 'shrink' so that they add up to 100 for each experimental trial.

For example, suppose that four 3-level substances are assigned to the four columns of $OA_9(3^4)$ in their following initial levels (in percentages):

Factors	Levels 1	2	3
A	30	40	50
B	20	35	40
C	5	10	15
D	15	20	25

Since the first trial of $OA_9(3^4)$ is an experiment, in which all factors (substances) have to be in their first level, the total of the initial levels for the first trial will amount to $30+20+5+15=70\%$ and not 100%. To perform this experimental trial, we first expand the levels used in such a way that they add up to 100; to do this, we multiply each of the original quantities (for the first trial) by 100/70. So, for the first trial of $OA_9(3^4)$, the following quantities will be used.

A(1)′	B(1)′
$30 \times \frac{100}{70} = 42.9\%$	$20 \times \frac{100}{70} = 28.6\%$

C(1)′	D(1)′
$5 \times \frac{100}{70} = 7.1\%$	$15 \times \frac{100}{70} = 21.4\%$

In general, before we perform a trial every level is multiplied by $100/k$ where k is the total sum of the initial levels corresponding to this trial. In the above example, for the first trial k was 70, and the multiplication by 100/70 has as an effect on the expansion of the initial levels for this trial.

For the last trial of $OA_9(3^4)$ we initially have

$$A(3)+B(3)+C(2)+D(1)=50+40+10+15=115$$

and the multiplication by 100/115 will have as an effect the shrinking of the initial values to

$$A(3')=50\times\frac{100}{115}=43.5\%, \qquad B(3)'=34.8\%,$$

$$C(2)'=8.7\% \quad \text{and} \quad D(1)'=13\%$$

so that they add up to 100%.

4.2.3.2 Planning out method

One of the substances is chosen not to be considered as a factor, so that it can be used as a 'remainder' to bring the sum total of the levels of the other factors up to 100%.

For example, in the case considered in Section 4.2.3.1, substance A can be used as a remainder, and enter the experimental trial with a proportion of 60% when B, C, and D are all at level 1, or enter with a proportion of 20% when B, C, and D are all at level 3, etc.

In order to ensure that the level total of the other factors does not exceed 100%, usually the substance used in the greatest quantity is chosen as the remainder. Of course, this should not constitute a rule. For example, if a substance is to be studied in terms of quantity as well as in terms of type, or if it is considered in combination with another substance for purposes of cancelling their interaction (see Section 4.2.2), then it should not be used as a remainder despite being the substance of the greatest quantity.

4.3 Net variation and contribution ratio

According to Taguchi, the F-test, which is only able to evaluate qualitatively whether factorial effects exist or not, is not as important as the 'contribution ratios' of the sources which can be used for a quantitative evaluation. So, quantitative judgement is considered far more important than qualitative judgement.

The 'contribution ratio' for each source represents the variability contribution of this source (in percentage terms) to the total variability of the experimental results. A resolution of the total dispersion in such a way is similar to 'spectral resolution' used in engineering, where one resolves the power of the whole frequency spectrum into powers of the frequency components; the proportion of power of each component relative to the total power is its contribution ratio.

One of the main uses of the contribution ratios lies in the field of 'tolerance-design' to be explained in a later chapter. In defining the notion of 'net variation' and 'contribution ratio' we will distinguish the case of a single error-variation component and the case of many error-variation components.

4.3.1 Single error-variation component

As explained in Chapter 2, in an analysis of variance table, the mean sum of squares for each effect is defined as the ratio of the sum of squares of the effect to its degrees of freedom. So, for the (main or interaction) effect A:

$$MS_A = S_A/\mathrm{df}_A$$

and for the residual

$$MS_e = S_e/\mathrm{df}_e$$

where S_e is the error sum of squares. MS_e is otherwise known as the error variance. In testing for significance for the effect of the source A, the following F-ratio is calculated

$$F_A = \frac{MS_A}{MS_e} = \frac{S_A}{\mathrm{df}_A \times MS_e}.$$

If the effect of the source A were negligible, S_A should be about $\mathrm{df}_A \times MS_e$. Therefore, the net effect of A can be estimated from

$$S'_A = S_A - \mathrm{df}_A \times MS_e.$$

S'_A is called the net variation (or 'pure variation') of source A. The net variation for the error, S'_e, is obtained by subtracting the net variations of all the sources from the total sum of squares TSS. To obtain the portion of the whole dispersion for which a source is responsible we divide the source's net variation by TSS. Therefore, the extent to which the source A accounts for the variation, in percentage terms, is given by

$$\rho_A = \frac{S'_A}{TSS} \times 100 = \frac{S_A - \mathrm{df}_A \times MS_e}{TSS} \times 100$$

which is called the contribution ratio of A.

Of course, we similarly have

$$\rho_e = \frac{S'_e}{TSS} \times 100.$$

The above also covers the case when e represents a (single) pooled error variation component. It is sufficient to calculate only the contribution ratios of significant sources.

Example: If we recall the ceramic experiment—case-study of Section 3.3.6.1b, the final ANOVA results can be extended as Table 4.3 shows (see also Table 3.12b) by incorporating the net variations and contribution ratios. Looking at the last column of Table 4.3, the highest contributors to the variability of the experimental results of Table 3.12 (see Section 3.3.6.1b), is the firing temperature (T), whose differences among its three levels (300, 400, and 500 °C) account for 54.2% of the total variation in the hardness of the ceramic tiles used in the experiment.

TABLE 4.3. Extended ANOVA results for ceramic experiment

Source	df	S	S'	MS	F-ratio	ρ (in %)
D	2	160.33	154.13	80.165	25.9	9.3
T	2	900.33	894.13	450.165	145.2	54.2
M	1	506.15	503.15	506.25	163.3	30.5
H	1	49.00	45.90	49.00	15.8	2.8
Residual	11	34.09	52.69	3.1		3.2
Total	17	1650.00	1650.00			100

Similarly, 30.5% of the total variation is attributable to the difference between the two types [M(1) and M(2)] of the clay material (M) used. Only 3.2% of the total variation is attributable to individual tile differences.

Contribution ratios can be similarly calculated, and interpreted for interaction effects (of any order).

4.3.2 Many error-variation components

When there are many error-variation components, a similar procedure is followed for the calculation of ρ for a source. The only difference is that the MS_e used in the basic formula for ρ is the particular variation component of error against which the source is tested, assuming this error-component is significant.

Examples:

(a) When there are replications, and the inter-experiment error e_1 (error between trials) is found to be significant when tested against the replication error e_2 (error within the trials), then for any source A

$$\rho_A = \frac{S_A - \text{df}_A \times MS_{e_1}}{TSS} \times 100$$

Of course, if e_1 is not significant, it is pooled with e_2 and the principles of Section 4.3.1 apply. Even if e_1 is significant, it is sufficient to calculate one contribution ratio ρ_e for the total error $(e_1 + e_2)$ by subtracting the contribution ratios of all the other sources from 100%.

(b) In the case of split-unit designs (see Section 3.3.5.4) a similar principle applies for each 'order-group'. The MS_e used in the basic formula of ρ for a source belonging to a particular order-group, is the error-variation component of *this* order-group, assuming it is significant when tested against the error of a higher order-group. As before, it is sufficient to calculate only one ρ_e for the total error, by subtraction. However, if the contribution ratio of a particular error-component, say e_1, needs to be calculated the following simple procedure can be followed.

(i) Calculate the (total) contribution ratio ρ_{T1} of the first-order group using MS_{e_2} (if significant), i.e. from

$$\rho_{T1} = \frac{S_{T1} - df_{T1} \times MS_{e_2}}{TSS} \times 100$$

where S_{T1} and df_{T1} are respectively the (total) sum of squares and degrees of freedom corresponding to the first-order group.

(ii) subtract from ρ_{T1} the contribution ratios of all the other (significant) sources of the first-order group; ρ_{e_1} will then be obtained.

The above are demonstrated on the following ANOVA table

Source	df	S	MS	F-ratio	S'	$\rho\%$
A	1	480	480	20*	456	64.41
e_1	2	48	24	6*	60	8.475
B	1	60	60	15**	56	7.91
C	2	56	28	7*	48	6.78
A × B	1	32	32	8*	28	3.95
e_2	8	32	4		60	8.4
Total	15	708			708	100

Note that

$$S'_{e_1} = \underbrace{[S_{T1} - df_{T1} \times MS_{e_2}]}_{\text{net variation of first-order group}} - [S'_A] = [(480 + 48) - 3 \times 4] - [456] = 60.$$

4.4 Estimation of process performance

After an experiment has taken place and the optimal settings of the studied parameters have been determined, there is often a need to estimate the process average performance under the optimal conditions. A method predicting the long-run average performance would be useful for ‘what if?’ situations, e.g. in cases when cost considerations prohibit the setting of a particular factor at its ‘optimal’ level, and a different suboptimal setting needs to be considered and its effect on the long-run process performance to be evaluated.

Of course, every estimation procedure will have to be based on the experimental results, and the estimated process average can be considered valid only under the assumption that the state of experimentation continues. For example, since the overall mean of the available data enters the estimation calculations, and since it is considered as the effect of the *fixed* causes at the time of experimentation, it will have to be assumed that these causes will remain fixed. Nevertheless, since changes in the state of experimentation (and so of the general mean) cannot be helped, it is still possible to assume that the differences between the optimal process average (average estimated under the optimal levels) and those estimated under other combinations of levels will remain approximately the same.

Taguchi recommends an estimation formula, simply based on the individual differences between the average of the preferred factor levels and the overall mean (Section 4.4.1). A confidence interval for the estimated process average can easily be constructed on the basis of the 'efficient number of replications' to be described in Section 4.4.2. Taguchi also recommends a refinement of his estimation formulae through the use of the 'discount (Beta) coefficients' to be described in Section 4.4.3.

4.4.1 A simple estimation formula

Suppose one is interested in estimating the long-run average yield $\hat{\mu}$ of a process on the basis of the data obtained from an experiment which studied the effects of only two factors A and B, whose analysis indicated the selection of the third level of A and the second level of B as optimal for high yield. Under the optimal levels A(3) and B(2) the process average of the yield $\hat{\mu}$ (from now on) is estimated by

$$\begin{aligned}\hat{\mu} &= (\text{mean of all experiments, } \bar{T}) \\ &\quad + (\text{increase in yield above } \bar{T} \text{ when A(3) is selected}) \\ &\quad + (\text{increase in yield above } \bar{T} \text{ when B(2) is selected}) \\ &= \bar{T} + (\bar{A}_3 - \bar{T}) + (\bar{B}_2 - \bar{T}).\end{aligned}$$

where $\bar{A}_3$ and $\bar{B}_2$ are respectively the average yield in levels A(3) and B(2) ($\bar{A}_3$ can be easily calculated by averaging all the data values obtained when factor A was at its level 3; similarly for $\bar{B}_2$. $\bar{T}$, of course, is the grand average of all the available data values). The above formula can include not just main effects but also interaction effects. Averages for the interaction levels (e.g. $\overline{A_1B_1}$, $\overline{A_1B_2}$, etc.) will then have to be calculated and a selection to be made for comparison with $\bar{T}$. In general,

$$\hat{\mu} = \bar{T} + \sum (\text{average of selected source level} - \bar{T}). \qquad (4.1)$$

Sources which are not important (for example, sources whose sum of squares are being pooled to construct a pooled error sum of squares) can be disregarded, since their inclusion in eqn (4.1) will have only a minor effect on the estimate.

Clearly, $\hat{\mu}$ is only an estimate for the real process average μ under the selected conditions. However, confidence limits for μ can be constructed on the basis of $\hat{\mu}$ and of the error variance obtained through ANOVA. We distinguish between the case of a single error-variation component and the case of many error variation components.

4.4.1.1 Confidence limits on the basis of a single error-variation component

Suppose that during ANOVA a single error-variation component e is obtained (this generally covers the case when e is either retrievable or is obtained by pooling). Then, confidence limits for μ, with (100-α)% confidence, can be obtained by using the formula

$$\hat{\mu} \pm \sqrt{F(1, \text{df}_e; \alpha) \times MS_e \times (1/n_e)}, \qquad (4.2)$$

where MS_e is the mean sum of squares for e (error variance) in the ANOVA table, df_e is the residual degrees of freedom (df associated with the error sum of squares S_e), $F(1, \text{df}_e; \alpha)$ is the critical value from the F-table depending on 1 and df_e of freedom at level of significance α, and n_e is the 'effective number of replications' to be described in Section 4.4.2.

Equation (4.2) provides confidence limits (within a certain level of confidence) for the real process average expected *in the long run* under selected conditions, assuming of course that the 'state of experimentation' continues.

To predict the process average during r runs of confirmatory trials (conducted for the purpose of ascertaining whether or not the conclusions drawn from the experiment were correct), the confidence limits are given from

$$\hat{\mu} \pm \sqrt{F(1, \text{df}_e; \alpha) \times MS_e \times \left(\frac{1}{n_e} + \frac{1}{r}\right)}.$$

Note that

$$MS_e \times \left(\frac{1}{n_e}\right) \quad \text{or} \quad MS_e \times \left(\frac{1}{n_e} + \frac{1}{r}\right)$$

represents the variance of the estimate $\hat{\mu}$, $V(\hat{\mu})$.

4.4.1.2 Confidence limits on the basis of many error–variation components

When the experimental design used requires the resolution of the error-variation into different components, such as in the case of split-unit designs (see Section 3.3.5.4), a similar resolution takes place in the square-root part of (4.2).

For example, if e_1, e_2, and e_3 represent the error-components of the first, second, and third order-group respectively, in an experiment using a split-unit design, then the confidence limits for the long-term process average are given by

$$\hat{\mu} \pm \sqrt{F(1, \mathrm{df}_{e_1}; \alpha) \times \frac{MS_{e_1}}{n_{e_1}} + F(1, \mathrm{df}_{e_2}; \alpha) \times \frac{MS_{e_2}}{n_{e_2}} + F(1, \mathrm{df}_{e_3}; \alpha) \times \frac{MS_{e_3}}{n_{e_3}}} \tag{4.3}$$

where n_{e_i} is the effective number of replications corresponding to the ith-order-group, $i = 1, 2, 3$, to be described in Section 4.4.2.3.

A special case of (4.3) is when there is a retrievable inter-experiment error e_1 which is significant in comparison with the replication error e_2. If e_1 and e_2 are the only error-components, (4.3) can still be applied. However, in such a case it can be shown (see Remark of Section 4.4.2.3) that $n_{e_2} = \infty$ and so (4.3) reduces to (4.2) with $e \equiv e_1$.

4.4.2 *Effective number of replications*

To determine of confidence limits for the real process average (Section 4.4.1) the 'effective number of replications' n_e is used. Taguchi's general equation for n_e is as follows

$$n_e = \frac{TND}{DCF + \Sigma\,(ENDF)}, \tag{4.4}$$

where TND is the total number of data values (i.e. the size of the experiment) or the total number of degrees of freedom (including those for the CF), $\Sigma\,(ENDF)$ is the total 'effective' number of degrees of freedom attributable to the sources that have not been disregarded in estimating the process average, and DCF represents the 'effective' number of degrees of freedom for the Correction factor (usually with the value of 1, except during accumulating analysis—see Section 4.7).

The 'effective' number of degrees of freedom ($ENDF$) of the (non-disregarded) sources, depends on the experimental design used. Some representative cases are as follows:

4.4.2.1 The case for orthogonal designs

Suppose that an experiment took place using a $OA_N(2^P \times 3^Q)$ orthogonal design for the study of P 2-level factors and Q 3-level factors in the expense of N experimental trials with r replications per trial. Suppose also that p 2-level and q 3-level factors were found to be significant along with three 2-way interactions, one between two 2-level factors, one between two 3-levels factors and one between a 2-level and a 3-level factor.

The 'effective' number of degrees of freedom in this case is simply the usual degrees of freedom for each non-disregarded source i.e.

$$\begin{aligned}\sum (ENDF) &= p \times (\text{df for a 2-level factor}) \\ &\quad + q \times (\text{df for a 3-level factor}) \\ &\quad + \underbrace{(2-1)(2-1) + (3-1)(3-1) + (2-1)(3-1)}_{\text{(df for the interactions)}} \\ &= p \times (2-1) + q \times (3-1) + 1 + 4 + 2.\end{aligned}$$

So, for the effective number of replications using (4.4)

$$n_e = \frac{N \times r}{1 + p + 2q + 7}.$$

A similar procedure is followed for the calculation of n_e for any number of multi-level factors assigned to orthogonal arrays. The *ENDF* for each source considered, is equivalent to the usual degrees of freedom.

4.4.2.2 The case for dummy-level designs

The dummy-level technique is a convenient method of assigning an m-level factor on an n-level comumn when $m < n$ (see Section 3.3.6.1a). Generally, one of the m actual levels is repeated $(n - m + 1)$ times. So, $(n - m)$ dummy levels are created for the factor, which, together with the m actual levels are assigned to the n-level column. The effective number of degrees of freedom of such a factor is

$$\frac{n}{n - m + 1} - 1 = \frac{\text{number of levels before creating dummy}}{\text{number of duplications of dummy level}} - 1.$$

If the factor does not require dummy levels, i.e. $n = m$, then the degrees of freedom are the usual ones, i.e. $(n - 1)$.

For the interaction effects we multiply the individual *ENDF*s for each factor in the interaction. For example, suppose we conducted an experiment having assigned one 2-level factor A, two 3-level factors B and C, and two 4-level factors D and E on the $OA_{16}(4^5)$ design. Using the dummy level technique we repeated one of the 2 levels of A three times, and one of the 3 levels of B and C twice. Assuming r replications for each of the 16 trials and assuming that the effects of A, B, C, and E were not disregarded, we have

$$n_e = \frac{16r}{\underset{\text{A}}{1 + (\tfrac{4}{3} - 1)} + \underset{\text{B}}{(\tfrac{4}{2} - 1)} + \underset{\text{C}}{(\tfrac{4}{2} - 1)} + \underset{\text{E}}{(4 - 1)}} = \frac{16r}{(5 + \tfrac{4}{3})} \approx 2.5r.$$

The above procedure for calculating n_e also covers the case when one uses the 'combining' technique (see Section 3.3.6.1b). For example, if [AB] is a 3-level combination factor with levels [AB](1) = A(1)B(1), [AB](2) = A(1)B(2) and [AB](3) = A(2)B(1), and, through the appropriate analysis, B is found to be insignificant, then it can be assumed that A alone appears on the column of the design, with its first level as a dummy. The effective number of the degrees of freedom for A will be $(\tfrac{3}{2} - 1)$ with B being disregarded in the calculation of n_e.

4.4.2.3 The case for 'split-unit' designs

When there is a difficulty in changing the levels of the factors, the factors are divided into different classes (order groups) according to their changing-difficulty, and a 'split-unit' design is used (see Section 3.3.5.4).

In such cases there is a resolution of the total error variation into components associated with each order-group and a different value of n_e associated with each variation component and each order group.

For the first-order group eqn (4.4) can be used, i.e.

$$n_{e_1} = \frac{TND}{1 + (TENDF)_1}, \tag{4.4a}$$

where $(TENDF)_1$ is the sum of the efficient number of degrees of freedom of the sources that have not been disregarded when tested against the *first-order* error variance.

For the other order-groups,

$$n_{e_2} = \frac{TND}{(TENDF)_2} \quad \text{and} \quad n_{e_3} = \frac{TND}{(TENDF)_3}, \tag{4.4b}$$

where $(TENDF)_i$ $i = 2, 3$ are defined similarly to $(TENDF)_1$ considering, of course, the error variance of the ith order group.

If the sources include factors with dummy levels, the procedure of Section 4.4.2.2 can be followed. One may also pool non-significant sources with corresponding errors, with appropriate changes in the residual degrees of freedom df_{e_i} $i = 1, 2, 3$, needed in eqn (4.3).

Remark: When there is a retrievable inter-experiment error e_1 which is significant compared with the replication error e_2, and e_1, e_2 are the only error components, (4.4a) and (4.4b) can again be applied. Of course, all the degrees of freedom of all the sources involved will be used in (4.4a). Consequently if one applies (4.4b) a value of ∞ will be obtained for n_{e_2}. This explains why formula (4.3) reduces to (4.2) in the calculation for confidence limits of the long-term process average (see Section 4.4.1).

4.4.2.4 The case for 'pseudo-factor' designs

When the experimental design is constructed using relatively complicated techniques such as the 'trans-factor' technique (see Section 3.3.5.3) or the 'idle-column' technique (see Section 3.3.6.2b), the calculation of the effective number of replications needed in eqn (4.2), is troublesome. For these cases, Taguchi suggests the following simple approximate formula instead of (4.2):

$$\hat{\mu} \pm 3\sqrt{MS_e} \tag{4.5}$$

where MS_e is the error variance (mean sum of squares of error).

4.4.2.5 The case for 'partially-supplemented' designs

Recall that in the analysis of the data from a partially-supplemented design (see Section 3.3.6.3) we need to 'correct' for the excess part of error introduced when we assumed that some of the trials (conditions) have each been experimented upon twice. A 'correction coefficient' W is used to correct the mean sum of squares for each source in the supplemented design. These coefficients W will again have to be used to 'adjust' n_e, whose formula is the same as that defined by the procedures of the previous sections, except that the ENDF for each source (including the '1' for the correction factor CF) is multiplied by the corresponding coefficient W for the source.

For example, for the case-study of Section 3.3.6.3b we had 3 significant sources A, B, and C and we found

$$W = \begin{cases} \frac{3}{2} & \text{for } CF, \text{ B and C} \\ \frac{5}{4} & \text{for A.} \end{cases}$$

Then

$$\begin{aligned} n_e &= \frac{TND}{1 \times W_{CF} + \mathrm{df}_A W_A + \mathrm{df}_B W_B + \mathrm{df}_C W_C} \\ &= \frac{12}{1 \times \frac{3}{2} + 2 \times \frac{5}{4} + 1 \times \frac{3}{2} + 1 \times \frac{3}{2}} = \frac{12}{7}. \end{aligned}$$

In general, we have

$$n_e = \frac{TND}{DCF \times W_{CF} + \Sigma\,(ENDF \text{ of source}) \times W_{\text{source}}}. \qquad (4.6)$$

Equation (4.6) is a simple extension of (4.4) applicable to partially supplemented designs.

4.4.2.6 The case for a contrast of factor levels

Suppose A is a factor with k levels, each corresponding to r data values. Let A_i represent the sum of all r values corresponding to the ith level of A.

A linear combination of A_i's, $i = 1, \ldots, k$ of the form

$$L = c_1A_1 + \cdots + c_kA_k$$

so that

$$c_1 + \cdots + c_k = 0$$

is termed a 'contrast'. The sum of squares for the contrast L is given by

$$S_A(L) = \frac{L^2}{(c_1^2 + \cdots + c_k^2)r} = \frac{(c_1A_1 + \cdots + c_kA_k)^2}{(c_1^2 + \cdots + c_k^2)r}, \qquad (4.7)$$

and corresponds to 1 degree of freedom. Then $S_A(L)$ represents one variation component of S_A, the sum of squares for factor A (see Section 2.6.3.1).

If $L_1 = c_1A_1 + \cdots + c_kA_k$ and $L_2 = c_1'A_1 + \cdots + c_k'A_k$ are two contrasts, for which the inner product of their coefficients is zero, i.e.

$$c_1c_1' + c_2c_2' + \cdots + c_kc_k' = 0,$$

then $S(L_1)$ and $S(L_2)$ are mutually independent variations of 1 degree of freedom each. L_1 and L_2 are then 'orthogonal'. S_A can be resolved to $k-1$ mutually independent variation components with 1 degree of freedom each. Recall that

$$S_A = \frac{A_1^2 + \cdots + A_k^2}{r} - \frac{(A_1 + \cdots + A_k)^2}{k \cdot r},$$

corresponding to $(k-1)$ degrees of freedom.

For example suppose $k = 3$. The following contrasts

$$L_1 = \frac{A_3}{r} - \frac{A_1}{r} \quad \text{and} \quad L_2 = \frac{A_1 + A_3}{2r} - \frac{A_2}{r}$$

are clearly two mutually orthogonal contrasts, considering that

$$c_1 = -\frac{1}{r}, \qquad c_2 = 0, \qquad c_3 = \frac{1}{r}$$

and

$$c_1' = \frac{1}{2r}, \qquad c_2' = -\frac{1}{r}, \qquad c_3' = \frac{1}{2r}.$$

Note that, applying (4.7)

$$S_A(L_1) = \frac{\left(\frac{A_3}{r} - \frac{A_1}{r}\right)^2}{\left\{\left(\frac{1}{r}\right)^2 + \left(-\frac{1}{r}\right)^2\right\} r} = \frac{(A_3 - A_1)^2}{2r}$$

and

$$S_A(L_2) = \frac{\left(\frac{A_1 + A_3}{2r} - \frac{A_2}{r}\right)^2}{\left\{\left(\frac{1}{2r}\right)^2 + \left(\frac{1}{2r}\right)^2 + \left(-\frac{1}{r}\right)^2\right\} r} = \frac{(A_1 - 2A_2 + A_3)^2}{6r}$$

with

$$S_A(L_1) + S_A(L_2) \equiv S_A.$$

If A is a quantitative factor with three equidistant levels, then the above two variation-components $S_A(L_1)$ and $S_A(L_2)$ represent the resolution of S_A into its linear and quadratic components respectively, each with 1 degree of freedom (see Chapter 2).

Confidence limits for a contrast $L = c_1A_1 + \cdots + c_kA_k$ can be found in the usual way by

$$L \pm \sqrt{F(1, \mathrm{df_e}; \alpha) \frac{MS_e}{\eta_e(L)}},$$

where, in this case, the effective number of replications $\eta_e(L)$ is given by

$$\eta_e(L) = \frac{1}{[c_1^2 + \cdots + c_k^2] r}. \tag{4.8}$$

When the number of replications r is different for each of $A_1, \ldots, A_k$, say, $r_1, \ldots, r_k$, then

$$n_e(L) = \frac{1}{c_1^2 r_1 + \cdots + c_k^2 r_k}. \tag{4.9}$$

4.4.2.7 The case for a linear combination of data values

A special case of (4.9) is when $r_1 = \cdots = r_k = 1$. Therefore (4.9) (or (4.8) with $r = 1$) can be applied for the case of a linear combination of n

observed single values $Y_1, \ldots, Y_n$ of the form

$$L = c_1 Y_1 + \cdots + c_n Y_n. \tag{4.10}$$

Irrespective of the fact that (4.10) may or may not be a 'contrast', we have

$$n_e(L) = \frac{1}{c_1^2 + \cdots + c_n^2}. \tag{4.11}$$

The estimate $\hat{\mu}$ of the process average as expressed by (4.1), can be considered as a linear combination of individual experimental data values, appropriately weighted depending on the sources involved in (4.1). So, theoretically (4.11) can be applied for the calculation of n_e needed in (4.2). However, when numerous sources are involved, finding the coefficients $c_1, c_2, \ldots,$ for the respective data values is rather troublesome. This is why eqn (4.4) is strongly recommended.

Since the average of a factor level or the total average of all experimental data values is a linear combination of data values, (4.11) can be applied. For example, if

$$\bar{\mathrm{A}}_1 = \frac{Y_1 + Y_2 + Y_3 + Y_4}{4},$$

then by (4.11)

$$n_e[\mathrm{A}(1)] = \frac{1}{(\frac{1}{4})^2 + (\frac{1}{4})^2 + (\frac{1}{4})^2 + (\frac{1}{4})^2} = 4.$$

Therefore, if one is interested in finding confidence limits for the mean of each factor level, in order to determine how much representation the effect of that factor possesses, the 'effective number of replications' is simply the number of data values used to calculate the average for that level.

For example, limits for the real mean of level 1 of A are given by

$$\bar{\mathrm{A}}_1 \pm \sqrt{F(1, \mathrm{df}_e; \alpha) MS_e \frac{1}{n_e[\mathrm{A}(1)]}}. \tag{4.12}$$

Since

$$\widehat{\mathrm{var}}(\bar{\mathrm{A}}_1) = \frac{MS_e}{n_e[\mathrm{A}(1)]} \tag{4.13}$$

is an estimate of the variance of the sample mean $\bar{A}_1$, (4.12) is equivalent to

$$\bar{\mathrm{A}}_1 \pm t(\mathrm{df}_e; \alpha)\sqrt{\widehat{\mathrm{var}}(\bar{\mathrm{A}}_1)}$$

or

$$\bar{\mathrm{A}}_1 \pm t(\mathrm{df}_e; \alpha) \cdot S\hat{D}(\bar{\mathrm{A}}_1).$$

Equation (4.11) can also be applied for the case of the grand average $\bar{T}$, yielding

$$n_e(T) = TND = \text{total number of data values.}$$

For the linear combination (4.10), the sum of squares is given by [as in (4.7)]

$$S(L) = \frac{L^2}{c_1^2 + \cdots + c_n^2}. \tag{4.14}$$

Therefore, if $L = \bar{T} = (Y_1 + \cdots + Y_n)/n$, then by (4.14)

$$S_{\bar{T}} = S(L) = \frac{L^2}{\left(\frac{1}{n}\right)^2 \times n} = \left(\frac{Y_1 + \cdots + Y_n}{n}\right)^2 \times n = \frac{(T)^2}{n} \equiv CF. \tag{4.15}$$

The above demonstrates that the correction factor CF (used in ANOVA) represents the sum of squares due to the general mean. This corresponds to 1 degree of freedom and is subtracted from the sum total of squared data values

$$Y_1^2 + \cdots + Y_n^2 \quad \text{('uncorrected' sum of squares)}$$

in order to obtain the ('corrected') total sum of squares TSS (see Chapter 2).

4.4.3 Discount (beta) coefficients

Taguchi strongly believes that 'significance testing' should be considered *only* as a 'preparatory stage for performing a better estimation'. He advocates that to accept the significance of an effect with total confidence just because its F-ratio happens to be slightly greater than a critical value (from the F-tables), and to completely disregard this effect if its ratio happens to be slightly less than this critical value, constitutes 'a total departure from common sense!'. He also suggests that the usefulness of F-ratios lie only on their role as a 'substitute' for the 'discount (or beta) coefficients' which can be used to improve eqns (4.1) and (4.4) and therefore to improve estimation and long-run prediction of the process performance.

4.4.3.1 Definition

The use of the discount coefficients is based on the principle that every estimate (e.g. $\hat{\mu}$, $\bar{T}$, $\bar{A}_2 - \bar{T}$, etc.) is usually an overestimate of the true value of the effect (or of the parameter) which is estimated. There is, therefore, a need for some 'discounting'.

For example, if one is interested in the difference between two levels of factor A whose true mean values respectively are μ_1^A and μ_2^A, one estimates this difference by

$$\bar{A}_1 - \bar{A}_2,$$

which is an overestimate of the true difference $\mu_1^A - \mu_2^A$.

One would therefore improve on the estimate, if one could calculate a 'discount coefficient', β, which when applied to the initial estimate, maximizes the precision. Equivalently, one should find β which minimizes the mean square of

$$D = \beta(\bar{A}_1 - \bar{A}_2) - (\mu_1^A - \mu_2^A).$$

In general, least-squares methods are employed in order to estimate the discount coefficient $\beta(\hat{L})$. This, when multiplied by $\hat{L}$, gives a better estimate for the real L. Here, $\hat{L}$ could be either a linear combination of data values, such as the grand average $\bar{T}$, or a contrast such as $\bar{A}_1 - \bar{A}_2$, $\bar{A}_1 - \bar{T}$, etc., and $\beta(\hat{L})$ is chosen so that the mean square of the error

$$D = \beta(\hat{L}) \times \hat{L} - L$$

is minimized.

It can be shown (see Section 4.4.3.2 below) that

$$\beta(\hat{L}) = 1 - \frac{1}{F_{\hat{L}}} \tag{4.16}$$

when $F_{\hat{L}}$ is the variance ratio between the variance of $\hat{L}$ and the error variance MS_e. Since the variance of $\hat{L}$ will normally be a variation component corresponding to one degree of freedom, we have

$$F_{\hat{L}} = \frac{S(\hat{L})}{MS_e}$$

where $S(\hat{L})$ represents the sum of squares of $\hat{L}$ as given by (4.7) or (4.14). When the variance ratio $F_{\hat{L}}$ approaches ∞, the value of $\beta(\hat{L})$ approaches 1; conversely, when $F_{\hat{L}}$ approaches 1, $\beta(\hat{L})$ approaches zero. When $F_{\hat{L}}$ is less than 1, $\beta(\hat{L})$ is regarded as zero.

So in general

$$\beta(\hat{L}) = \begin{cases} 1 - \dfrac{1}{F_{\hat{L}}} & \text{if} \quad F_{\hat{L}} = \dfrac{MS_{\hat{L}}}{MS_e} > 1. \\ 0 & \text{if} \quad F_{\hat{L}} \leqslant 1. \end{cases}$$

The above demonstrates a willingness not to disregard any effect unless its variance ratio is less than 1. On the other hand, Taguchi is also reluctant to accept any estimation 'at its full value', even if there is a high degree of significance. For example, even if the variance ratio is as high as 10, he advises to take 0.9 times the effect 'observed' during experimentation as a more reliable estimate of the real effect. This is because he believes that it is very rare that a larger effect (greater than the one based on the experimental results) will be obtained in the long run. In most cases, less than the 'expected' effect is obtained, and this justifies the 'discounting' during estimation.

It is clear that the main use of the discount coefficients is for improving the reliability of the estimating eqns (4.1) and (4.4) [and therefore (4.2)]. These now take the form

$$\hat{\mu}_\beta = \bar{T} \times \beta(\bar{T}) + \sum [\hat{L} \times \beta(\hat{L})] \tag{4.17}$$

and

$$n_{e,\beta} = \frac{TND}{\beta(\bar{T}) + \Sigma\,[(ENDF) \times \beta(\hat{L})]}, \tag{4.18}$$

where by $\hat{L}$ we now mean the contrast

$$\hat{L} = (\text{average of selected source level} - \bar{T})$$

and the sum Σ is over the sources that have not been disregarded in estimating μ, or equivalently over sources for which $\beta(\hat{L}) > 0$.

When $\hat{L}$ is a contrast between the two levels of a 2-level factor with an equal number of m data values in each level, $F_{\hat{L}}$ (needed in (4.16)), is simply the F-ratio used in the ANOVA table. This is because

$$\hat{L} = \bar{A}_1 - \bar{T}$$

implies

$$\hat{L} = \bar{A}_1 - \frac{\bar{A}_1 + \bar{A}_2}{2} = \frac{\bar{A}_1 - \bar{A}_2}{2} = \frac{A_1 - A_2}{2m},$$

and

$$S_{\hat{L}} = \frac{(A_1 - A_2)^2}{2m} = \frac{A_1^2 + A_2^2}{m} - \frac{(A_1 + A_2)^2}{2m} = S_{\mathrm{A}} = MS_{\mathrm{A}},$$

giving

$$F_{\hat{L}} = \frac{S_{\hat{L}}}{MS_{\mathrm{e}}} = \frac{MS_{\mathrm{A}}}{MS_{\mathrm{e}}} = \mathrm{F}_{\mathrm{A}}.$$

For other cases, such as when A is a multi-level factor or a factor with dummy levels etc, (4.7) has to be used for determining $S_{\hat{L}}$ and hence of $F_{\hat{L}}$. Such cases will be demonstrated in the case-study of Section 4.4.3.3.

4.4.3.2 Estimation of discount coefficients

We will concentrate on the case when $\hat{L}$ is a contrast between the data-average of a selected level, $\bar{A}_s$ say, and the grand average $\bar{T}$, that is,

$$\hat{L} = \bar{A}_s - \bar{T}.$$

Other cases for L can be similarly dealt with. Assuming the source A has k levels,

$$\hat{L} = \bar{A}_s - \frac{\bar{A}_1 + \bar{A}_2 + \cdots + \bar{A}_s + \cdots + \bar{A}_k}{k},$$

and assuming that each level corresponds to the same effective number of replications, m say,

$$\hat{L} = \frac{(k-1)A_s - (A_1 + \cdots + A_{s-1} + A_{s+1} \cdots + A_k)}{k \cdot m} \tag{4.19}$$

where A_i is the sum total of the m data values for the ith level $i = 1, \ldots, k$.

If the mean values for T (grand total) and A_i $i = 1, \ldots, k$ are respectively μ_T and $\mu_T + a_i$ $i = 1, \ldots, k$, with

$$a_1 + a_2 + \cdots + a_k = 0,$$

then $\bar{T}$ and $\bar{A}_i$ $i = 1, \ldots, k$ are their unbiased estimators; that is,

$$E(\bar{T}) = \mu_T \quad \text{and} \quad E(\bar{A}_i) = \mu_T + a_i \quad i = 1, \ldots, k.$$

We are interested in estimating the discount coefficient $\beta = \beta(\hat{L})$ so that

$$\beta(\bar{A}_s - \bar{T}) = (\mu_T + a_s) - \mu_T = a_s.$$

We can estimate β by finding the value of β which minimizes the mean square of the error

$$D = \beta\hat{L} - L = \beta(\bar{A}_s - \bar{T}) - a_s$$

which is given by

$$E(D^2) = E\{\beta(\bar{A}_s - \bar{T}) - a_s\}^2.$$

But we know that (see Appendix B)

$$\begin{aligned} E(D^2) &= [(E(D)]^2 + \text{var}(D) \\ &= \{\beta \cdot E(\bar{A}_s - \bar{T}) - E(a_s)\}^2 + \text{var}\{\beta(\bar{A}_s - \bar{T}) - a_s\}, \end{aligned}$$

giving

$$E(D^2) = \{\beta E(\hat{L}) - a_s\}^2 + \beta^2 \,\text{var}(\hat{L}). \tag{4.20}$$

Now

$$E(\hat{L}) = E(\bar{A}_s - \bar{T}) = E(\bar{A}_s) - E(\bar{T}) = \mu_T + a_s - \mu_T = a_s,$$

and because of (4.19)

$$\operatorname{var}(\hat{L}) = \operatorname{var}\left[\frac{(k-1)\bar{A}_s - (\bar{A}_1 + \cdots + \bar{A}_{s-1} + \bar{A}_{s+1} + \cdots + \bar{A}_k)}{k}\right].$$

Since $\operatorname{var}(\bar{A}_i) = \sigma_e^2/m$, $i = 1, \ldots, k$ where σ_e^2 is the error variance,

$$\operatorname{var}(\hat{L}) = \frac{1}{k^2}\left\{(k-1)^2\frac{\sigma_e^2}{m} + (k-1)\frac{\sigma_e^2}{m}\right\} = \frac{(k-1)}{k}\frac{\sigma_e^2}{m}.$$

So, (4.20) becomes

$$E(D^2) = (\beta-1)^2 a_s^2 + \frac{\beta^2(k-1)}{km}\sigma_e^2.$$

Differentiating now $E(D^2)$ with respect to β, and putting the result equal to zero, we can obtain the value of β which minimizes $E(D^2)$ as

$$\beta = \frac{a_s^2}{a_s^2 + \dfrac{(k-1)}{km}\sigma_e^2}. \tag{4.21}$$

Now, as in (4.20) we can find

$$E(\hat{L}^2) = [E(\hat{L})]^2 + \operatorname{var}(\hat{L}) = a_s^2 + \frac{(k-1)}{km}\sigma_e^2. \tag{4.22}$$

Also, using (4.7) the sum of squares of the contrast (4.19) can be found as

$$S(\hat{L}) = \frac{\hat{L}^2}{\left\{\left(\dfrac{k-1}{km}\right)^2 + (k-1)\left(\dfrac{-1}{km}\right)^2\right\}m} = \frac{\hat{L}^2}{(k-1)/km}.$$

Thus, from (4.22)

$$E(S(\hat{L})) = \frac{E(\hat{L}^2)}{(k-1)/km} = \frac{a_s^2 + \dfrac{k-1}{km}\sigma_e^2}{\dfrac{(k-1)}{km}},$$

from which we obtain

$$a_s^2 = \frac{(k-1)}{km}\{E[S(\hat{L})] - \sigma_e^2\}.$$

Hence, an estimate for a_s^2 is

$$\hat{a}_s^2 = \frac{(k-1)}{km}[S(\hat{L}) - MS_e]. \tag{4.23}$$

Using (4.21) and (4.23) an estimate of $\beta = \beta(\hat{L})$ can be found as

$$\hat{\beta} = \frac{\frac{(k-1)}{km}[S(\hat{L}) - MS_e]}{\frac{(k-1)}{km}[S(\hat{L}) - MS_e] + \frac{(k-1)}{km}MS_e} = \frac{S(\hat{L}) - MS_e}{S(\hat{L})} = 1 - \frac{1}{F_{\hat{L}}},$$

as required in (4.16). When $\hat{L} = \bar{T}$, by a similar procedure we can show that

$$\beta(\bar{T}) = 1 - \frac{1}{F_{\bar{T}}}, \tag{4.24}$$

and using (4.15),

$$F_{\bar{T}} = \frac{S(\bar{T})}{MS_e} = \frac{CF}{MS_e},$$

with CF being the 'correction factor'.

4.4.3.3 A case-study

Let us consider again the case-study of Section 3.3.6.1b (ceramic experiment), in which three 2-level factors and two 3-level factors were assigned to the $OA_9(3^4)$ design using the 'combining' and 'dummy-level' techniques (see Section 3.3.6.1).

The two 2-level factors 'C' and 'H' were combined to create a 3-level combination factor [CH]. During the analysis, after separating the effects C and H, only factor H was found to be significant. Although [CH] was assigned the 3-level column 1 of $OA_9(3^4)$, it was eventually assumed that, only H alone appeared in column 1 with its first level as a dummy level (i.e. repeated twice as many times as level 2). The 2-level factor 'M' was assigned to the 3-level column 2 with its first level as a dummy level. The 3-level factors 'T' and 'D' were assigned to columns 3 and 4 (see Table 3.12). Factors D, T, M, and H were all found to be significant.

The level-totals for each significant factor were as follows:

For factor D: $D_1 = 155, \quad D_2 = 162, \quad D_3 = 121$

For factor T: $T_1 = 107, \quad T_2 = 205, \quad T_3 = 126$

For factor M: $M_1^d = M_1 + M_1' = 167 + 170 = 337, \quad M_2 = 101$

For factor H: $H_1^d = H_1 + H_1' = 141 + 137 = 278, \quad H_2 = 160.$

Since the ideal response (breaking strength) should be as high as possible it is clear that the 'optimum' level-combination of the above factors is

$$\text{D(2)T(2)M(1)H(2).}$$

We are interested in estimating the long-run process average under the 'optimal' conditions using the discount coefficients. Using formula (4.17) we have

$$\hat{\mu}_\beta = (GM) \times \beta(GM) + (\bar{D}_2 - GM) \times \beta(\bar{D}_2) + (\bar{T}_2 - GM) \times \beta(\bar{T}_2) + (\bar{M}_1^d - GM) \times \beta(\bar{M}_1^d) + (\bar{H}_2 - GM) \times \beta(\bar{H}_2), \quad (4.25)$$

where GM represents the grand mean of all the experimental results and $\bar{M}_1^d$ indicates that level 1 of M was also used as a dummy level. We know that (see Table 3.12b)

$$MS_e = 3.1,$$
$$GM = 24.33,$$

and

$$CF = S_{GM} = MS_{GM} = 10\,658.$$

So by (4.24)

$$\beta(GM) = 1 - \frac{1}{F_{GM}} = 1 - \left(\frac{MS_{GM}}{MS_e}\right)^{-1} = 1 - \left(\frac{10\,658}{3.1}\right)^{-1} = 0.9997.$$

To calculate $\beta(\bar{D}_2)$ we calculate the 1 degree of freedom sum of squares of the contrast

$$L(D_2) = \bar{D}_2 - GM = \bar{D}_2 - (\bar{D}_1 + \bar{D}_2 + \bar{D}_3)/3,$$

which in terms of level totals can be expressed as

$$L(D_2) = \frac{2D_2 - D_1 - D_3}{3m} = \left(\frac{2}{3m}\right) D_2 + \left(\frac{-1}{3m}\right) D_1 + \left(\frac{-1}{3m}\right) D_3$$

where m is the effective number of replications in each level; in this case, $m = 6$, because there are six experimental data values for each level of D. By (4.7) we have

$$S_{L(D_2)} = \frac{[L(D_2)]^2}{\left[\left(\frac{2}{3 \times m}\right)^2 + \left(\frac{-1}{3 \times m}\right)^2 + \left(\frac{-1}{3 \times m}\right)^2\right] m} = \frac{\left[\frac{2 \times 162 - 155 - 121}{3 \times 6}\right]^2}{\frac{6}{(3 \times 6)^2} \times 6}$$

$$= \frac{(48)^2}{36} = 64.$$

Therefore

$$F_{L(D_2)} = \frac{MS_{L(D_2)}}{MS_e} = \frac{S_{L(D_2)}}{MS_e} = \frac{64}{31} = 20.65$$

and so by (4.16)

$$\beta[L(D_2)] = 1 - \frac{1}{F_{L(D_2)}} = 1 - \frac{1}{20.65} = 0.9516.$$

Similarly we find that

$$\beta[L(T_2)] = 1 - \frac{1}{F_{L(T_2)}} = 1 - \frac{1}{280.73} = 0.9964.$$

For the dummy-level factor M, the optimal level is the one which was used as a dummy. So we are dealing with the contrast

$$\begin{aligned} L(M_1^d) &= \bar{M}_1^d - GM = \frac{(\bar{M}_1 + \bar{M}_1')}{2} - \frac{(\bar{M}_1 + \bar{M}_1' + \bar{M}_2)}{3} \\ &= \frac{(\bar{M}_1 + \bar{M}_1') - 2\bar{M}_2}{6} = \frac{(M_1 + M_1') - 2M_2}{6 \times 6} \\ &= \left(\frac{1}{36}\right) M_1 + \left(\frac{1}{36}\right) M_1' + \left(\frac{-2}{36}\right) M_2. \end{aligned}$$

We now proceed as before by (4.7)

$$\begin{aligned} S_{L(M_1^d)} &= \frac{[L(M_1^d)]^2}{\left[\left(\frac{1}{36}\right)^2 + \left(\frac{1}{36}\right)^2 + \left(\frac{-2}{36}\right)^2\right] \times 6} = \frac{[(M_1 + M_1') - 2M_2]^2}{36} \\ &= \frac{[337 - 2 \times 101]^2}{36} = 506.25, \end{aligned}$$

and so

$$\beta(L(M_1^d)) = 1 - \left(\frac{506.25}{3.1}\right)^{-1} = 0.9939.$$

For the factor H with level 1 as a dummy, the optimal level is level 2. So the contrast is as follows

$$L(H_2) = \bar{H}_2 - GM = \bar{H}_2 - \frac{(\bar{H}_1 + \bar{H}_1' + \bar{H}_2)}{3} = \frac{2\bar{H}_2 - (\bar{H}_1 + \bar{H}_1')}{3},$$

and so

$$L(H_2) = \left(\frac{2}{18}\right) H_2 + \left(-\frac{1}{18}\right) H_1 + \left(-\frac{1}{18}\right) H_1'.$$

Therefore

$$S_{L(H_2)} = \frac{[L(H_2)]^2}{\left[\left(\frac{2}{18}\right)^2 + \left(-\frac{1}{18}\right)^2 + \left(-\frac{1}{18}\right)^2\right]6} = \frac{[2H_2 - (H_1 + H_1')]^2}{36}$$

$$= \frac{[2 \times 160 - 278]^2}{36} = 49$$

and

$$\beta(L(H_2)) = 1 - \left(\frac{49}{3.1}\right)^{-1} = 0.9367.$$

From the above, (4.25) becomes

$$\hat{\mu}_\beta = (24.33) \times (0.9997) + \left(\frac{162}{6} - 24.33\right) \times (0.9516) + \left(\frac{205}{6} - 24.33\right) \times (0.9964) + \left(\frac{337}{12} - 24.33\right) \times (0.9939) + \left(\frac{160}{6} - 24.33\right) \times (0.9367),$$

and so

$$\hat{\mu}_\beta = 42.58.$$

Confidence limits for the real long-run process average can be found from (4.2)

$$\hat{\mu}_\beta \pm \sqrt{F(1, \mathrm{df_e}; \alpha) MS_\mathrm{e} \frac{1}{n_{\mathrm{e},\beta}}} \tag{4.26}$$

where $n_{e,\beta}$ is calculated using (4.18). Indeed,

$$n_{\mathrm{e},\beta} = \frac{TND}{\beta(GM) + \Sigma_L (ENDF) \times \beta(L)},$$

where the sum is over all the significant sources D, T, M, and H.

For the 3-level factors D and T we have $ENDF = 3 - 1 = 2$. For the dummy-level factors M and H we have (see Section 4.4.2.2) for each $ENDF = \frac{3}{2} - 1 = 0.5$. Therefore

$$n_{\mathrm{e},\beta} = \frac{9 \times 2}{\underset{GM}{(0.9997)} + \underset{\mathrm{D}}{(3-1) \times (0.9516)} + \underset{\mathrm{T}}{(3-1) \times (0.9964)} + \underset{\mathrm{M}}{(\frac{3}{2}-1) \times (0.9939)} + \underset{\mathrm{H}}{(\frac{3}{2}-1) \times (0.9367)}}$$

$$= \frac{18}{5.861} = 3.07.$$

Since (see Section 3.3.6.1b) $df_e = 11$, $MS_e = 3.1$, and at $\alpha = 5\%$ level of significance $F(1, 11; 5) = 4.84$, by (4.26) we have

$$42.58 \pm \sqrt{(4.84) \times (3.1) \times \frac{1}{3.07}} = 42.58 \pm 2.21,$$

which means that a 95% confidence interval for the real long-run process average (assuming the state of experimentation continues) ranges from 40.37 to 44.79.

If we are interested in a 95% confidence interval for the real process average during the next 10 confirmatory trials we have

$$42.58 \pm \sqrt{(4.84) \times (3.1) \times \left(\frac{1}{3.07} + \frac{1}{10}\right)} = 42.58 \pm 2.53,$$

i.e. from 40.05 to 45.11.

4.5 Dealing with missing values

There are occasions when part of the data has been lost as a result of carelessness or accident. This could result in an unequal number of replications per trial, or even to the loss of the whole set of experimental results for one or many experimental trials, i.e. for one or many combinations of factor levels.

If a lost data value *is not* the result of the effects of the factors involved, then this is rendered a 'missing value'. However, if a particular combination of levels led to the destruction of a product unit, and as a result there are no response values available, these cannot be considered as missing values. This is because some important information has been acquired regarding something the experimenter had not known in advance (for example, that a particular combination of factor levels is 'unworkable').

For the case of 'proper' missing values, if the experimental trial cannot be re-done, certain procedures have to be followed for the analysis of the available 'incomplete' data. Some special cases for simple randomized block designs have already been considered in Section 3.2.2. In this section we will consider the general case for any factorial experiments (which covers the special cases considered before). We will distinguish between the case of non-zero but unequal number of replications (which occurs frequently), and the case when not even one data value exists for certain trials (factor-level-combinations).

4.5.1 *The case for non-zero but unequal number of replications*

If every trial in an experimental design has been successful in terms of yielding at least one response value, but the number of replications per trial differs due to accidental loss of data (accidental damage of an experimental unit during the course of an experimental trial, carelessness, etc.), a certain analysis method can be followed using only the mean responses per trial (as though they were the only response values for each trial) and then adjusting the 'within-trial' variability by the 'harmonic mean' (reciprocal of mean of reciprocals) of the numbers of replications (of the respective trials).

TABLE 4.4

Trial	A	B	Data	No. of replications	Mean per trial
1	1	1	6	1	6.0
2	1	2	7, 8	2	7.5
3	1	3	3, 5, 5	3	4.3
4	2	1	6, 7	2	6.5
5	2	2	7, 9, 8	3	8.0
6	2	3	4	1	4.0
			75		36.3

This can be demonstrated by the following example. One 2-level factor A and one 3-level factor B were assigned to a full factorial design requiring 6 trials. The resulting data is shown in Table 4.4. The last column of Table 4.4 shows the mean values per trial, which we analyse as though they were representing a single value per trial:

$$(CF)_1 = \frac{(36.3)^2}{6} = 219.62$$

$$S_A = \frac{(6+7.5+4.3)^2 + (6.5+8+4)^2}{3} - (CF)_1 = 0.08 \qquad (\text{df}=1)$$

$$S_B = \frac{(6+6.5)^2 + (7.5+8)^2 + (4.3+4)^2}{2} - (CF)_1 = 13.08 \qquad (\text{df}=2)$$

and

$$(TSS)_1 = (6)^2 + (7.5)^2 + \cdots + (4)^2 - (CF)_1 = 13.3, \qquad (\text{df}=5)$$

from which

$$S_{A\times B} = S_{e_1} = (TSS)_1 - S_A - S_B = 0.22. \qquad (\text{df}=2)$$

Now the variation between replications (within trials) is found from

$$S_{e_2} = \left(\frac{1}{H}\right) S'_{e_2},$$

where H is the harmonic mean (reciprocal of mean of reciprocals) of the number of replications, i.e. (from Table 4.4),

$$H = \{\tfrac{1}{6}((1)^{-1} + (2)^{-1} + (3)^{-1} + (2)^{-1} + (3)^{-1} + (1)^{-1})\}^{-1} = 1.64$$

and S'_{e_2} represents the sum of the variation between replications for the respective experimental conditions, i.e. [as in (2.52)]

$$\begin{aligned} S'_{e_2} = &\left[6^2 - \frac{6^2}{1}\right] + \left[7^2 + 8^2 - \frac{(7+8)^2}{2}\right] \\ &+ \left[3^2 + 5^2 + 5^2 - \frac{13^2}{3}\right] + \left[6^2 + 7^2 - \frac{13^2}{2}\right] \\ &+ \left[7^2 + 9^2 + 8^2 - \frac{24^2}{3}\right] + \left[\left(4^2 - \frac{4^2}{1}\right)\right] = 5.67. \qquad (\text{df} = 6). \end{aligned}$$

So, finally

$$S_{e_2} = \frac{1}{H} S'_{e_2} = 5.6/1.64 = 3.46. \qquad (\text{df} = 6)$$

We can now construct the ANOVA Table 4.5.

TABLE 4.5

Source	df	S	MS	F-ratio	Corrected S	ρ (%)
A	1	0.08 p	0.08			
B	2	13.08	6.54	15.6**	12.24	73%
(A × B)e_1	2	0.22 p	0.11			
e_2	6	3.46 p	0.58			
(e_{pooled})	(9)	(3.76)	(0.42)		4.6	27%
Total	11	16.84			16.84	100%

For these cases, confidence limits for the long-run process average can be obtained from

$$\hat{\mu} \pm \sqrt{F \times MS_{\text{e}}\left(\frac{1}{n_{\text{e}}} + H\right)}$$

with n_{e} being calculated as though the mean-responses per trial (last column of Table 4.4) were the only values available. For example, in the case considered above, where only the factor B was not disregarded, we have (see Section 4.4.2.1)

$$n_e = \frac{6}{1 + (3 - 1)} = 2.$$

4.5.2 *The case of zero replications in some trials*

For dealing with the cases when no data values are available for some of the experimental trials, we distinguish between the case when error degrees of freedom remain (i.e. when there is a retrievable error variance, *not* by pooling) and the case when there are no degrees of freedom of error or when df_e is very small.

4.5.2.1 When error degrees of freedom exist

For the cases when there is a retrievable error sum of squares (not by pooling insignificant sources) one could estimate the missing values by following the method already described in Section 3.2.2.1, otherwise known as the 'Fisher–Yates' method:

4.5.2.1a Single error variation component The estimated values are the ones which minimize the (single) residual (error) sum of squares S_e. Differentiating S_e with respect to x, with x representing a missing value, and putting the result equal to zero, we obtain an equation to be solved with respect to x (method of least squares). When there is more than one missing value, the same method is followed, and a set of simultaneous equations is obtained which are as many as the number of missing values. Solving these equations, estimates of the missing values can be obtained, to be inserted in place of the lost measurements. Analysis of variance then proceeds as usual, except that, there is a subtraction of 1 degree of freedom from the residual degrees of freedom for every missing value. This is because, despite the estimation, the missing values have not been reproduced, and the lost information has not been regained.

For a demonstration of this method the reader is referred to Section 3.2.2.1.

4.5.2.1b Many error variation components When there are at least two error variation components as in the case of split-unit designs (see Section 3.3.5.4), the estimates of the missing values are usually obtained by minimizing the highest order error sum of squares except in one special case when a low-order error sum of squares is minimized. This special case is when one of two level combinations of the low-order factors (factors more difficult to change) consist entirely of missing values.

For example, when $OA_{27}(3^{13})$ is used as a split-unit design, the first 4 columns are usually reserved for the first-order group. One (or both) of columns 3 and 4 are reserved for the calculation of first-order variation; the remaining first-order columns are such that nine level-combinations of the first-order factors can be studied [e.g. for two first-order factors these are the combination $(1, 1), (1, 2), \ldots, (3, 3)$]. If *all* three trials for any one or two of those combinations contain no results, then the missing value for each trial is estimated by minimizing S_{e_1}. For any other case S_{e_2} is minimized. Note that in such a design, S_{e_2} is the highest-order error sum of squares.

4.5.2.2 When no error degrees of freedom exist

When there is no retrievable error sum of squares S_e, or when a retrievable S_e corresponds to a small number of degrees of freedom, one could pool together the sum of squares of the smaller effects to construct a pooled error sum of squares with adequate degrees of freedom, and then apply the 'Fisher–Yates' method for estimating missing values.

However, the above method of minimizing the error variation, is one whose precision is highest only when none of the factorial effects is disregarded, through pooling. Taguchi recommends a sequential approximation method which can deal with any number of missing values, and can also be used even when sources are multi-level and numerous, even when $df_e = 0$, and can also allow pooling.

The method requires the sequential application of the ANOVA and regression techniques.

The basic steps to follow are briefly as follows:

1. In place of the missing values substitute a suitable value, such as the mean of all the available actual data-values.
2. Perform the usual ANOVA for determination of non-negligible factorial effects (pooling is allowed).
3. Using the non-negligible effects as independent variables, fit a regression model for the response, using all the data (actual and estimated). Using the model, re-estimate the missing values.
4. With the new estimated values in place of the missing value, repeat the procedure from step (2).

Usually, it is adequate to repeat the procedure three times. After the final estimation of the missing values has taken place, the final ANOVA table is constructed, appropriately 'adjusted' for the missing values, i.e.

by subtracting from the total and from the error degrees of freedom as many df's as the number of missing values. This method is particularly useful when missing values are very numerous, and can be used instead of the 'Fisher–Yates' method, which in such cases requires more calculations.

4.6 Analysis of binary data

Situations involving the division of certain phenomenon into two states of categories (win or lose, live or die, defective or non-defective) require analysis of 'binary data' usually expressed by a series of the numbers 0 and 1 representing the two categories.

To analyse binary data, it is suggested that one follows exactly the same techniques usually employed for continuous data values, by simply considering the numbers 0 and 1 as the resulting data of a continuous variable. We can demonstrate this through a simple example, distinguishing between the case of an equal number of replications per trial, and the case of an unequal number of replications per trial.

4.6.1 When the number of replications per trial is the same

In the injection moulding industry, defective items are usually produced when the plastic being forced into the cavity of a mould does not fill the entire cavity leaving a part empty or a section missing. In order to assess the effect of the injection pressure and of the injection speed on the moulding performance, five tests took place for all level combinations between the 3-level pressure factor and the 4-level speed factor. For each of the tests a value of 0 was assigned if the unit was found to be defective and a value 1 if found to be non-defective. The results can be seen in Table 4.6.

From Table 4.6 we note that from the whole experiment of 60 tests, moulding was successful in 36 tests. Therefore the total number of experimental data values is 60 and the grand total of the values is 36. So, in the usual way

$$CF = \frac{(\text{grand total})^2}{\text{no. of data values}} = \frac{(36)^2}{60} = 21.6$$

$$TSS = \sum Y^2 - CF = (1^2 + 0^2 + \cdots + 1^2) - 21.6 = 36 - 21.6 = 14.4 \qquad (\text{df} = 59)$$

$$S_A = \frac{(P_1)^2 + (P_2)^2 + (P_3)^2}{20} - CF = \frac{(16)^2 + (12)^2 + (8)^2}{20} - 21.6 = 1.6 \qquad (\text{df} = 2)$$

TABLE 4.6

Trial	Pressure (A)	Speed (B)	Binary data	Total non-defectives
1	1	1	1, 0, 1, 1, 0	3
2	1	2	1, 1, 1, 1, 1	5
3	1	3	1, 1, 1, 1, 0	4
4	1	4	1, 0, 1, 1, 1	4
5	2	1	0, 1, 1, 0, 1	3
6	2	2	1, 1, 1, 0, 0	3
7	2	3	1, 1, 1, 0, 1	4
8	2	4	1, 1, 0, 0, 0	2
9	3	1	1, 1, 0, 0, 0	2
10	3	2	0, 0, 1, 0, 1	2
11	3	3	1, 0, 0, 0, 0	1
12	3	4	1, 0, 1, 0, 1	3
				36

$$S_B = \frac{(B_1)^2 + \cdots + (B_4)^2}{15} - CF = \frac{8^2 + 10^2 + 9^2 + 9^2}{15} - 21.6 = 0.13 \quad (\text{df} = 3)$$

$$S_{A\times B} = \frac{3^2 + 5^2 + \cdots + 3^2}{5} - CF - S_A - S_B = 1.07 \quad (\text{df} = 6)$$

and

$$S_e = TSS - S_A - S_B - S_{A\times B} = 14.4 - 1.6 - 0.13 - 1.07 = 11.6 \quad (\text{df} = 48).$$

Therefore

ANOVA

Source	df	S	MS	F-ratio
A	2	1.6	0.8	3.33*
B	3	0.13	0.043	0.18
A × B	6	1.07	0.178	0.24
e	48	11.6	0.24	
Total	59	14.4		

Only the pressure (A) is found to be affecting the performance of the moulding with an optimal setting at level 1.

4.6.2 *When the number of replications per trial varies*

The techniques described in Section 4.5.1 apply. The mean value per trial is found and ANOVA first proceeds on the basis of only those mean values. Then the error sum of squares is found as the sum of the deviations within each trial, and adjusted by the harmonic mean.

Of course, the mean value for each trial will now represent the proportion of non-defectives per trial. The analysis proceeds in exactly the same way as in the example of Section 4.5.1 by considering the numbers 1 and 0 as response data for a continuous variable.

4.6.3 *Estimation through logit transformation*

When dealing with proportions of defectives or non-defectives as in Section 4.6.2, estimating the process performance under the optimal conditions [using (4.1)] may sometimes produce unrealistic results. For example, suppose the experimental analysis of the proportion of defectives for each level-combination of two 2-level factors A and B has produced the following results:

	A(1)	A(2)	Total
B(1)	0.03	0.62	0.65
B(2)	0.01	0.22	0.23
	0.04	0.84	0.88

The process proportion of defects under the optimal settings $A(1)$ and $B(2)$ is given by [using (4.1)]:

$$\hat{\mu} = \bar{T} + (\bar{A}_1 - \bar{T}) + (\bar{B}_2 - \bar{T}) = \bar{A}_1 + \bar{B}_2 - \bar{T}$$

$$= \frac{0.04}{2} + \frac{0.23}{2} - \frac{0.88}{4} = 0.02 + 0.115 - 0.22 = -0.085$$

which is an unrealistic estimate.

For these cases Taguchi recommends a transformation of the proportions p into decibels (dB) using the 'logit' transformation (or 'omega method'):

$$-10\log_{10}\left\{\frac{1}{p} - 1\right\}.$$

After the decibel value of $\hat{\mu}$ is found, the inverse transformation is applied so that a result in terms of proportions is obtained.

For example, for the above case

$$\begin{aligned}(\text{decibel value of } \hat{\mu}) &= [(\text{decibel value of } \bar{A}_1) + (\text{decibel values of } \bar{B}_2)] \\ &\quad - (\text{decibel value of } \bar{T}) \\ &= -10\log\left(\frac{1}{0.02}-1\right) - 10\log\left(\frac{1}{0.115}-1\right) \\ &\quad - \left[-10\log\left(\frac{1}{0.22}-1\right)\right] \\ &= -16.902 - 8.8962 + 5.497 = -20.267\ \text{dB},\end{aligned}$$

and transforming back to the original scale

$$\hat{\mu} \approx 0.0093.$$

So the expected proportion of defects of the process under the optimal conditions is 0.93%.

For confidence limits for $\hat{\mu}$, recall that, in general, if

$$\varepsilon = \sqrt{F(1, \mathrm{df}_e; \alpha)\frac{MS_e}{n_e}},$$

one can use the equation (see 4.2)

$$\hat{\mu} \pm \varepsilon.$$

When the logit transformation is used, one should first use the equation:

$$(\text{decibel value of } \hat{\mu}) \pm (\text{decibel value of } \varepsilon)$$

and then transform back to the original scale.

Taguchi has shown that

$$\text{decibel value of } \varepsilon = \frac{k \times \varepsilon}{\bar{T}(1-\bar{T})}, \tag{4.27}$$

where

$$k = \frac{10}{\log_e 10} = 4.343.$$

4.6.4 *Life-test data and 'Minute' analysis*

In performing life-tests, the experimenter is interested in estimating the survival period of a particular product unit under certain conditions which are simulated in an experimental environment. Life-tests are not necessarily 'destruction' tests. If the duration of the testing period is predetermined, some of the test units could survive the duration of the experiment. When the resulting data is expressed in the continuous mode of the 'length of life' for each unit, with the additional specification on whether it was a 'failure' or a 'survival' for the whole of the testing period, 'Weibull analysis' is usually performed to assess the reliability of the product unit. However, such an analysis requires the assumption that the failure distribution of the unit follows a Weibull type probability distribution.

A technique for life-test analysis not requiring such an assumption, is the method of 'minute' analysis (or 'minimum unit' analysis or 'minute accumulating' analysis) recommended by Taguchi, whereby, not only the effects of the factors affecting the unit's survival can be assessed, but the survival distribution function can also be estimated.

The method is based on the breakdown of the testing period into regular intervals of a specific duration called 'minimum units' ('minutes') or 'cycles'. The duration of a minimum unit depends on the problem under investigation. For example, in discussing the life of small light-bulbs, probably one or two hours will be used as the minimum unit. In other cases, it could be anything from a second, one day, etc., to a year and more. In cases of 'accelerated ageing' a minimum unit of one hour could represent the length of a year, etc. Essentially, the method constitutes an investigation of *which* minimum unit the end of the life took place in, with respect to the test piece used in the experiment. The data is expressed in binary form with '1' ('0') signifying that the test unit was still functioning (not functioning) by the end of a particular cycle.

The division of the test period into minimum units, $\omega(1), \ldots, \omega(n)$ say, introduces an additional factor, the 'minimum unit factor', or cycle factor or ('omega') ω-factor with n levels. It also introduces an additional error component, the between 'minutes' error or ω-order error. This will have to be considered during analysis.

The ANOVA is performed in the usual way appropriate for binary data, but bearing in mind that:

(a) There are now three components of error variance to be considered:
  (i) the inter-experiment error ('first-order' or 'between-experiments')
  (ii) the replication error ('second-order' or 'within-experiments' or 'between-replications' error) in case there are replications per trial, and
  (iii) the (omega) ω-order error ('between-minimum-units' error)

(b) The replication error (if it exists), the ω-factor and any ω-related effect (such as interactions between the ω-factor and the other factors) are tested for significance against the ω-order error.

The following case-study will demonstrate the above.

A test on the durability of fluorescent lamps

In order to improve the durability of fluorescent lamps, the following factors were selected for study:

A: Additive (2 types), B: Fluorescent agents (2 types).

C: Coating method (2 types), D: Tube-washing method (2 types).

The four 2-level factors were assigned on $OA_8(2^7)$. For each of the 8 level-combinations the durability of two produced lamps were tested. The whole testing period was 10 days. Every two days how many of the lamps were still functioning was recorded. The number '1' was recorded for those lamps still 'alive', and a '0' otherwise. The minimum unit (cycle) for the above case is '2 days'. Since the test-duration was 10 days, we are dealing with a cycle-factor (ω-factor) of five levels $\omega(1), \ldots, \omega(5)$. The resulting binary data for the test appears in Table 4.7.

TABLE 4.7

Trial	Design A	B	C	D	Data (ω) $\omega(1)$	$\omega(2)$	$\omega(3)$	$\omega(4)$	$\omega(5)$	Total for replications	Total for trial
1	1	1	1	1	1	1	0	0	0	2	5
					1	1	1	0	0	3	
2	1	1	2	2	1	0	0	0	0	1	3
					1	1	0	0	0	2	
3	1	2	1	2	1	1	1	0	0	3	4
					1	0	0	0	0	1	
4	1	2	2	1	1	1	0	0	0	2	6
					1	1	1	1	0	4	
5	2	1	1	2	1	1	1	1	0	4	9
					1	1	1	1	1	5	
6	2	1	2	1	1	1	1	0	0	3	8
					1	1	1	1	1	5	
7	2	2	1	1	1	1	1	0	0	3	6
					1	1	1	0	0	3	
8	2	2	2	2	1	1	1	1	1	5	10
					1	1	1	1	1	5	
Totals					16	14	11	6	4	51	51

Analysis

The factor-effects are analysed in the usual way: The total number of observations is $8 \times 2 \times 5 = 80$. So

$$CF = \frac{(51)^2}{80} = 32.51$$

$$TSS = 1^2 + \cdots + 1^2 - CF = 51 - 32.51 = 18.49 \qquad (\mathrm{df} = 79),$$

$$S_A = \frac{(A_1)^2 + (A_2)^2}{40} - CF = \frac{(18)^2 + (33)^2}{40} - 32.51 = 2.81 \qquad (\mathrm{df} = 1).$$

Similarly

$$S_B = 0.01, \qquad S_C = 0.11, \qquad S_D = 0.01 \qquad (\mathrm{df} = 1).$$

The effects of the main sources (factors A, B, C, and D) will have to be tested against the inter-experimental error (first-order variance) to be found through (see (2.50), Section 2.6.3.3)

$$S_{e_1} = \frac{5^2 + 3^2 + \cdots + 10^2}{10} - CF - (S_A + S_B + S_C + S_D) = 1.25$$

$$(\mathrm{df} = 7 - 4 = 3).$$

Since there are 2 replications per trial, the replication error (second-order variance) will also have to be assessed through [see (2.52), Section 2.6.3.3]

$$S_{e_2} = \left[\frac{2^2 + 3^2}{5} - \frac{(2+3)^2}{10}\right] + \left[\frac{1^2 + 2^2}{5} - \frac{(1+2)^2}{10}\right] + \cdots$$
$$+ \left[\frac{5^2 + 5^2}{5} - \frac{(5+5)^2}{10}\right].$$

So

$$S_{e_2} = 1.5 \qquad (\mathrm{df} = 8).$$

(The inter-experiment error will be tested against the replication error.) Now, the effects of the ω-related sources (main effect of ω-factor and its interactions with the other factors) can be assessed in the usual way:

$$S_\omega = \frac{(\omega_1)^2 + \cdots + (\omega_5)^2}{16} - CF = \frac{(16)^2 + (14)^2 + \cdots + (4)^2}{16} - 32.51$$
$$= 6.55 \qquad (\mathrm{df} = 4).$$

Also

$$S_{A \times \omega} = \frac{(A_1\omega_1)^2 + (A_1\omega_2)^2 + \cdots + (A_1\omega_5)^2 + (A_2\omega_1)^2 + \cdots + (A_2\omega_5)^2}{8}$$
$$- CF - S_A - S_\omega,$$

i.e.

$$S_{A\times\omega} = \frac{8^2 + 6^2 + \cdots + 0^2 + 8^2 + \cdots + 4^2}{8} - 32.51 - 2.81 - 6.55 = 1.00$$

(df = 4).

Similarly

$$S_{B\times\omega} = 0.05, \qquad S_{C\times\omega} = 0.45 \quad \text{and} \quad S_{D\times\omega} = 0.8 \qquad (\text{df} = 4).$$

The replication error and the ω-related sources will have to be tested against the ω-order error (third-order variance), the sum of squares of which can be calculated by subtraction as

$$S_{e_3} = TSS - [S_A + S_B + S_C + S_D + S_{e_1} + S_{e_2} + S_\omega + S_{A\times\omega} + S_{B\times\omega} + S_{C\times\omega} + S_{D\times\omega}]$$

from which

$$S_{e_3} = 3.95 \qquad (\text{df} = 44).$$

The ANOVA Table 4.7a can be constructed. Testing is narrowed down for only the important sources.

In Table 4.7a, e_1 (when tested against e_2) is not significant, but when pooled with e_2 and tested against e_3 is found to be significant. The only other sources found to be significant are factor A [when tested against a pooled $(e_{1,2})$], and sources ω and $A \times \omega$, when tested against a pooled (e_3).

TABLE 4.7a. ANOVA

Source	df	S	MS	F-ratio	S'	$\rho\%$
A	1	2.81	2.81	13.4**	2.60	14.1
B	1	0.01 p_1	0.01			
C	1	0.11 p_1	0.11			
D	1	0.01 p_1	0.01			
e_1	3	1.25 p_1	0.42			
e_2	8	1.5 p_1	0.19			
$(e_{1,2})$	(14)	(2.88)	(0.21)	(2.44*)	1.80	9.7
ω	4	6.55	1.64	19.1**	6.21	33.6
$A\times\omega$	4	1.00	0.25	2.9*	0.66	3.6
$B\times\omega$	4	0.05 p_2	0.01			
$C\times\omega$	4	0.45 p_2	0.11			
$D\times\omega$	4	0.80	0.20	2.3	0.46	2.5
e_3	44	3.95 p_2	0.09			
(e_3)	(52)	(4.45)	(0.086)		6.76	36.5
Total	79	18.49			18.49	100.0

The sum of squares of sources indicated by 'p_1' are pooled with $e_{1,2}$, whereas those indicated by 'p_2' are pooled with e_3. The contribution ratios of the important sources are calculated following the procedure described in Section 4.3.2.

Estimation of survival-rate distribution

The survival rate distribution of the lamps can be estimated by first estimating the process survival average at each cycle. Since the only significant sources are A, ω, and A $\times$ ω, the process average will be calculated under the optimal level of A (which is A(2) as having the highest survival rate) for each ω_j, $j = 1, \ldots, 5$.

Using (4.1)

$$\mu_j = \bar{T} + (\bar{A}_2 - \bar{T}) + (\bar{\omega}_j - \bar{T}) + (\overline{A_2\omega_j} - \bar{T}),$$

i.e.

$$\mu_j = A_2 + \bar{\omega}_j + \overline{A_2\omega_j} - 2\bar{T} \qquad j = 1, \ldots, 5, \tag{4.28}$$

where

$$\bar{T} = \frac{51}{80} = 0.6375,$$

from which

$$(\text{decibel value of } \bar{T}) = -10 \log_{10}\left(\frac{1}{\bar{T}} - 1\right) = 2.45$$

and similarly

$$\bar{A}_2 = \frac{33}{40} = 0.825 \approx 6.73 \text{ dB}$$

$$\bar{\omega}_1 = \frac{16}{16} = 1 \approx \infty \text{ dB}, \qquad \overline{A_2\omega_1} = \frac{8}{8} = 1 \approx \infty \text{ dB}$$

$$\bar{\omega}_2 = \frac{14}{16} = 0.875 \approx -8.45 \text{ dB}, \qquad \overline{A_2\omega_2} = 1 \approx \infty \text{ dB}$$

$$\bar{\omega}_3 = 0.6875 \approx 3.42 \text{ dB}, \qquad \overline{A_2\omega_3} \approx \infty \text{ dB}$$

$$\bar{\omega}_4 = 0.375 \approx -2.22 \text{ dB}, \qquad \overline{A_2\omega_4} = \tfrac{5}{8} = 0.625 \approx 2.22 \text{ dB}$$

and

$$\bar{\omega}_5 = 0.25 \approx -4.77 \text{ dB}, \qquad \overline{A_2\omega_5} = 0.5 \approx 0.0 \text{ dB}.$$

So, using (4.28), the process survival rate for each cycle is given by

	$\omega(1)$	$\omega(2)$	$\omega(3)$	$\omega(4)$	$\omega(5)$
Decibel value	∞	∞	∞	1.83	−2.94
Process average	100%	100%	100%	60.4%	33.7%

Confidence limits can also be calculated using

$$(\text{decibel value of } \hat{\mu}_j) \pm (\text{decibel value of } \varepsilon)$$

where, by (4.27) and (4.3)

$$\text{decibel value of } \varepsilon = \frac{4.343}{\bar{T}(1-\bar{T})}\sqrt{F(1, 14; 5\%)\frac{MS_{(e_{1,2})}}{n_1} + F(1, 52; 5\%)\frac{MS_{(e_3)}}{n_2}}$$

where

$$\bar{T} = 0.6375$$

$$F(1, 14; 5\%) = 4.6, \qquad MS_{(e_{1,2})} = 0.21$$

$$F(1, 52; 5\%) = 4.03, \qquad MS_{(e_3)} = 0.086$$

and (see Section 4.4.2.3)

$$n_1 = \frac{80}{1 + \text{df}_A} = \frac{80}{2} = 40$$

and

$$n_2 = \frac{80}{\text{df}_\omega + \text{df}_{A\times\omega}} = \frac{80}{4+4} = 10.$$

Therefore

$$\text{decibel value for } \varepsilon = 4.56.$$

So, for example, a 95% confidence interval for the process average in $\omega(5)$ is given in decibels by

$$(\text{decibel value of } \hat{\mu}_5) \pm 4.56 = -2.94 \pm 4.56 \text{ dB},$$

i.e. from -7.5 to 1.62 dB or from 15.1% to 59.2%.

Only lower limits are calculated for the cycles $\omega(1)$, $\omega(2)$, and $\omega(3)$. The lower limit in decibels for such a case is given by

$$-\left\{\left[\text{decibel value of } \frac{1}{2}\left(\frac{1}{n_1} + \frac{1}{n_2}\right)\right] + (\text{decibel value of } \varepsilon)\right\}, \tag{4.29}$$

i.e.

$$-\left\{\left[\text{decibel value of } \frac{1}{2}\left(\frac{1}{40} + \frac{1}{10}\right)\right] + 4.56\right\}$$

or

$$-[-11.76 + 4.56] = 7.2 \text{ dB},$$

which is equivalent to 84%. (The negative of eqn (4.29) is applied as an upper limit of the case when $\hat{\mu} = 0\%$.)

The final table expressing the survival distribution of the lamps over the 5 cycles is as shown in Table 4.7b. The values of Table 4.7b can now be graphically illustrated for a visual representation of survival rate curves.

TABLE 4.7b. Estimation of process survival rate %

	ω_1	ω_2	ω_3	ω_4	ω_5
$\hat{\mu}$	100	100	100	60.4	33.7
Upper 95% limit	100	100	100	81.3	59.2
Lower 95% limit	84	84	84	34.8	15.1

4.7 'Accumulating' analysis

For the analysis of ranked data, categorized, say, into bad, fair, good, and excellent, or for the cases when the characteristic values, although fundamentally continuous, have been divided into a number of classes depending on their magnitude, Taguchi recommends 'accumulating' analysis. The method can also be used to analyse data which is mixed categorical-continuous in nature. The steps to follow for such an analysis can be best described through an actual case-study which follows.

4.7.1 A case study: The duplicator experiment

An experiment took place in order to determine the best operating conditions for the paper-feeding phase at high speeds (208 r.p.m.) of an offset duplicator. Twelve factors at two levels were selected, as given in Table 4.8.

TABLE 4.8

Factor	Description	Level 1	Level 2
A	Vacuum header type	Normal	Lightweight
B	Feed cam type	Normal	Smoothed
C	Master cylinder cam	Smoothed	Normal
D	Air rifle setting	Normal	High
E	Chain gripper release cam	Normal	Advanced
F	Paper weight bar spring	Without	With
G	Release blowdown spray	OFF	ON
H	Buckle setting	Normal (7)	High (15)
I	Paperweight bar	Light	Heavy
J	Paperweight bar position	Normal	Back
K	Impression roller setting	Normal (7)	High (15)
L	Vacuum setting	Normal	High

The 12 factors A, . . . , L and one interaction between factors F and I were assigned on the OA_{16} (2^{15}) design; four tests were carried out in each trial. The results are shown in Table 4.9. The data values represent the number of paper-sheets successfully fed through the duplicator at each test. The letter 'f' indicates a complete failure in the feeding performance for that test. The asterisks indicate that some tests were censored; for example, '*377' means that the test was stopped after 377 paper sheets successfully passed through the machine, in the expectation that the paper-feeding would continue beyond 377 sheets.

TABLE 4.9

	Design															Data			
																Tests			
Factors:	A	B	C	D	E	F		G	H	F×I		I	J	K	L				
Columns:	1	2	3	4	5	6	7	8	9	10	11	12	13	14	15	1	2	3	4
Trial																			
1	1	1	1	1	1	1	1	1	1	1	1	1	1	1	1	2	9	f	3
2	1	1	1	1	1	1	1	2	2	2	2	2	2	2	2	124	46	f	3
3	1	1	1	2	2	2	2	1	1	1	1	2	2	2	2	21	7	f	2
4	1	1	1	2	2	2	2	2	2	2	2	1	1	1	1	3	9	f	6
5	1	2	2	1	1	2	2	1	1	2	2	1	1	2	2	*377	13	7	7
6	1	2	2	1	1	2	2	2	2	1	1	2	2	1	1	*379	*359	f	*341
7	1	2	2	2	2	1	1	1	1	2	2	2	2	1	1	*372	43	f	184
8	1	2	2	2	2	1	1	2	2	1	1	1	1	2	2	330	5	143	*337
9	2	1	2	1	2	1	2	1	2	1	2	1	2	1	2	2	3	f	4
10	2	1	2	1	2	1	2	2	1	2	1	2	1	2	1	1	3	f	f
11	2	1	2	2	1	2	1	1	2	1	2	2	1	2	1	4	34	3	f
12	2	1	2	2	1	2	1	2	1	2	1	1	2	1	2	1	3	3	3
13	2	2	1	1	2	2	1	1	2	2	1	1	2	2	1	*500	*500	219	77
14	2	2	1	1	2	2	1	2	1	1	2	2	1	1	2	*500	*500	*500	*500
15	2	2	1	2	1	1	2	1	2	2	1	2	1	1	2	*500	489	9	8
16	2	2	1	2	1	1	2	2	1	1	2	1	2	2	1	45	46	f	218

Analysis

To perform accumulating analysis, the test-data are divided into the following 4 categories

Category	Description
I	f(paper-feeding failed)
II	[1, 168]
III	[169, 336]
IV	$[337, \infty]$

Clearly, Category IV is the most desired category, while Category I is the least desired one.

The frequency distribution per trial with respect to the four categories is shown in Table 4.10a. We are basically dealing with binary data; indeed, the data of a category 'c', can be regarded as having value 1 if in category 'c', and 0 if not. Table 4.10b also shows the cumulative frequencies for all trials. The cumulative categories are denoted with parentheses; thus (III) means sum of Categories I, II, and III. We can now calculate the frequencies and cumulative frequencies for all the factor levels (see Table 4.11).

TABLE 4.10

	a: Frequencies				b: Cumulative frequencies			
Category:	I	II	III	IV	(I)	(II)	(III)	(IV)
Trial								
1	1	3	0	0	1	4	4	4
2	1	3	0	0	1	4	4	4
3	1	3	0	0	1	4	4	4
4	1	3	0	0	1	4	4	4
5	0	3	0	1	0	3	3	4
6	1	0	0	3	1	1	1	4
7	1	1	1	1	1	2	3	4
8	0	2	1	1	0	2	3	4
9	1	3	0	0	1	4	4	4
10	2	2	0	0	2	4	4	4
11	1	3	0	0	1	4	4	4
12	0	4	0	0	0	4	4	4
13	0	1	1	2	0	1	2	4
14	0	0	0	4	0	0	0	4
15	0	2	0	2	0	2	2	4
16	1	2	1	0	1	3	4	4
					11	46	50	64

For example, the cumulative frequency of the first level of A under category (II) is obtained by summing all the cumulative frequencies under (II) (from Table 4.10b) of the first 8 trials. The cumulative frequency of the second level of B under Category (III) is obtained by summing the cumulative frequencies under (III) (from Table 4.10b) of trials 5–8 and 13–16.

TABLE 4.11

Categories:	a: Frequencies I	II	III	IV	b: Cumulative frequencies (I)	(II)	(III)	(IV)
A(1)	6	18	2	6	6	24	26	32
A(2)	5	17	2	8	5	22	24	32
B(1)	8	24	0	0	8	32	32	32
B(2)	3	11	4	14	3	14	18	32
C(1)	5	17	2	8	5	22	24	32
C(2)	6	18	2	6	6	24	26	32
D(1)	6	15	1	10	6	21	22	32
D(2)	5	20	3	4	5	25	28	32
E(1)	5	20	1	6	5	25	26	32
E(2)	6	15	3	8	6	21	24	32
F(1)	7	18	3	4	7	25	28	32
F(2)	4	17	1	10	4	21	22	32
G(1)	5	19	2	6	5	24	26	32
G(2)	6	16	2	8	6	22	24	32
H(1)	6	18	2	6	6	24	26	32
H(2)	5	17	2	8	5	22	24	32
F(1)I(1)	3	10	2	1	3	13	15	16
F(1)I(2)	4	8	1	3	4	12	13	16
F(2)I(1)	1	11	1	3	1	12	13	16
F(2)I(2)	3	6	0	7	3	9	9	16
I(1)	4	21	3	4	4	25	28	32
I(2)	7	14	1	10	7	21	22	32
J(1)	5	18	1	8	5	23	24	32
J(2)	6	17	3	6	6	23	26	32
K(1)	5	16	1	10	5	21	22	32
K(2)	6	19	3	4	6	25	28	32
L(1)	7	16	3	6	7	23	26	32
L(2)	4	19	1	8	4	23	24	32
Totals	11	35	4	14	11	46	50	64

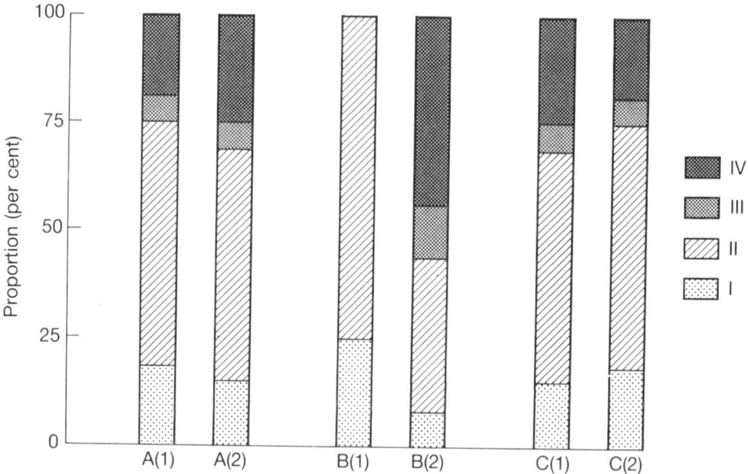

FIG. 4.1. 'Accumulating' analysis: Duplicator experiment.

The frequencies of Table 4.11a can be used for the construction of graphs such as, for example, the one depicted in Fig. 4.1 of the levels of factors A, B, and C. Note that the cumulative frequency in Category (IV) for each level (Table 4.11b) represents the total number of tests performed for that level.

The techniques appropriate for the analysis of binary data (Section 4.6) can now be applied to the calculation of the source-variation effects separately for each of the Categories (I), (II), and (III). It is clear that such a calculation is not necessary for Category (IV).

A weighted variation component for each source, covering all Categories (I), (II) and (III), can then be calculated using the weights $W_{(\mathrm{I})}$, $W_{(\mathrm{II})}$, and $W_{(\mathrm{III})}$, which are the reciprocals of the binomial error variances $P_c(1-P_c)$, where P_c is the proportion of data in cumulative category (c). So, from the sum totals of Table 4.11b.

$$W_{(\mathrm{I})} = \frac{1}{P_{(\mathrm{I})}(1-P_{(\mathrm{I})})} = \frac{1}{\frac{11}{64}\left(1-\frac{11}{64}\right)} = \frac{64^2}{11(64-11)} = 7.026.$$

Similarly

$$W_{(\mathrm{II})} = \frac{64^2}{46(64-46)} = 4.947$$

and

$$W_{(\mathrm{III})} = \frac{64^2}{50(64-50)} = 5.851.$$

By using these weights, the analysis of variance proceeds as follows: The correction factor of each Category (I), (II), and (III) is weighed by the corresponding weights and an overall CF is calculated as

$$CF = CF_{(I)} \times W_{(I)} + CF_{(II)} \times W_{(II)} + CF_{(III)} \times W_{(III)}$$
$$= \frac{11^2}{64} \times 7.026 + \frac{46^2}{64} \times 4.947 + \frac{50^2}{64} \times 5.851$$
$$= 405.398 \qquad (\text{df} = 3).$$

The number of degrees of freedom corresponding to the overall CF is 3, which is the same as the number of cumulative categories being analysed.

On the same basis

$$TSS = (TSS \text{ for (I)}) \times W_{(I)} + (TSS \text{ for (II)}) \times W_{(II)} + (TSS \text{ for (III)}) \times W_{(III)}.$$

But

$$TSS \text{ for (I)} = 11 - \frac{11^2}{64} = (64) \times P_{(I)} - (11) \times P_{(I)}$$
$$= (64) \times P_{(I)}(1 - P_{(I)}) = \frac{64}{W_{(I)}}.$$

So

$$(TSS \text{ for (I)}) \times W_{(I)} = 64.$$

The same applies for the other categories (II) and (II).

Therefore

$$TSS = 64 + 64 + 64 = 3 \times 64 \qquad (\text{df} = 3 \times 63).$$

So in general

$$TSS = [(\text{number of classes being analysed}) \times (\text{total number of measured values})].$$

Now, for the sum of squares of the sources, similar principles apply. For example, the sum of squares for factor A is given by

$$S_A = [S_A \text{ for (I)}] \times W_{(I)} + [S_A \text{ for (II)}] \times W_{(II)} + [S_A \text{ for (III)}] \times W_{(III)}$$
$$(\text{df} = 3 \times 1 = 3).$$

So

$$S_A = \left\{\frac{6^2 + 5^2}{32} - \frac{11^2}{64}\right\} W_{(I)} + \left\{\frac{24^2 + 22^2}{32} - \frac{46^2}{64}\right\} W_{(II)} + \left\{\frac{26^2 + 24^2}{32} - \frac{50^2}{64}\right\} W_{(III)}$$
$$= \frac{(6^2 + 5^2) \times W_{(I)} + (24^2 + 22^2) \times W_{(II)} + (26^2 + 24^2) \times W_{(III)}}{32} - CF$$
$$= \frac{61 \times 7.026 + 1060 \times 4.947 + 1252 \times 5.851}{32} - 405.398 = 0.785$$
$$(\text{df} = 3).$$

Similarly

$$S_B = \frac{(8^2+3^2)\times 7.026 + (32^2+14^2)\times 4.947 + (32^2+18^2)\times 5.851}{32} - 405.398$$

$$= 45.708 \qquad (\text{df} = 3\times 1 = 3),$$

$$S_C = 0.785, \quad S_D = 4.638, \quad S_E = 1.713, \quad S_F = 5.516,$$
$$S_G = 0.785, \quad S_H = 0.785, \quad S_I = 5.516,$$

and

$$S_{F\times I} = \frac{(3^2+4^2+1^2+3^2)\times W_{(I)} + (13^2+12^2+12^2+9^2)\times W_{(II)} + (15^2+13^2+13^2+9^2)\times W_{(III)}}{16}$$
$$- CF - S_F - S_I = 0.785 \qquad (\text{df} = 3\times 1 = 3),$$
$$S_J = 0.476, \quad S_K = 4.638, \quad S_L = 1.354 \qquad (\text{df} = 3).$$

Now, using the data of Table 4.10b, the inter-experiment (between trials) error variance can be found from

$$S_{e_1} = \frac{(1^2+1^2+\cdots+0^2+1^2)\times W_{(I)} + (4^2+4^2+\cdots+2^2+3^2)\times W_{(II)} + (4^2+4^2+\cdots+2^2+4^2)\times W_{(III)}}{4}$$
$$- CF - (S_A + \cdots + S_L) = 5.128 \qquad (\text{df} = 3\times 2 = 6).$$

Finally, the replication error sum of squares can be found by subtraction

$$S_{e_2} = TSS - (S_{e_1} + S_A + \cdots + S_L) = 113.388 \qquad (\text{df} = 144).$$

The ANOVA results can now be presented as shown in Table 4.12.

Note that e_1 is significant when compared with e_2. The sum of squares for e_1 is then pooled with the sum of squares of small magnitude (denoted by 'p'), yielding a ($e_{1,\text{pooled}}$) variance against which all the other sources are tested.

Estimation

Table 4.12 indicates that factors B, D, F, I, and K cannot be disregarded. Since the most desired category is Category IV, using Table 4.11a, we can see that the optimal settings for the above factors is

$$\text{B(2)D(1)F(2)I(2)K(1)}.$$

In order to estimate the long-run performance of the duplicator under the optimal settings with respect to each category we follow the techniques of Section 4.6.3 using the logit transformation and the cumulative frequencies of Table 4.11b. From Table 4.11b, first using

TABLE 4.12. ANOVA for accumulating analysis

Source	df	S	MS	F-ratio	ρ%
A	3	0.785 p	0.262		
B	3	45.708	15.236	36.3**	23.2
C	3	0.785 p	0.262		
D	3	4.638	1.546	3.7*	1.8
E	3	1.703p	0.568		
F	3	5.516 p	1.839	4.4*	2.2
G	3	0.785 p	0.262		
H	3	0.785 p	0.262		
I	3	5.516	1.839	4.4*	2.2
F × I	3	0.785 p	0.262		
J	3	0.476 p	0.159		
K	3	4.638	1.546	3.7*	1.8
L	3	1.354 p	0.451		
e_1	6	5.128 p	1.709		
($e_{1,\text{pooled}}$)	(30)	(12.586)	(0.4195)		68.8
e_2	144	113.388	0.787		
Total	189	192			100

the frequencies for Category (I),

$$\bar{B}_2 = \frac{3}{32} = 0.0938 \approx -9.85 \text{ dB}$$

$$\bar{D}_1 = \frac{6}{32} = 0.1825 \approx -6.37 \text{ dB}$$

$$\bar{F}_2 = \frac{4}{32} = 0.125 \approx -8.45 \text{ dB}$$

$$\bar{I}_2 = \frac{7}{32} = 0.2188 \approx -5.53 \text{ dB}$$

$$\bar{K}_1 = \frac{5}{32} = 0.1563 \approx -7.32 \text{ dB}.$$

Also for the overall mean

$$\bar{T} = \frac{11}{64} = 0.1719 \approx -6.83 \text{ dB}.$$

Therefore for category (I)

$$\begin{aligned}(\text{decibel value of } \hat{\mu}_{(\mathrm{I})}) = &(\text{decibels of } \bar{B}_2) + (\text{decibels of } \bar{D}_1)\\ &+ (\text{decibels of } \bar{F}_2) + (\text{decibels of } \bar{I}_2)\\ &+ (\text{decibels of } \bar{K}_1) - 4(\text{decibels of } \bar{T})\\ = &-10.2\ \text{dB}.\end{aligned}$$

So

$$\hat{\mu}_{(\mathrm{I})} = 0.087 \text{ or } 8.7\%.$$

Similarly , using the frequencies for Category (II) from Table 4.11b

$$\bar{B}_2 = \frac{14}{32} = 0.4375 \approx -1.09\ \text{dB}$$

$$\bar{D}_1 = \frac{21}{32} = 0.6563 \approx 2.81\ \text{dB}$$

$$\bar{F}_2 = \frac{21}{32} = 0.6563 \approx 2.81\ \text{dB}$$

$$\bar{I}_2 = \frac{21}{32} = 0.6563 \approx 2.81\ \text{dB}$$

$$\bar{K}_1 = \frac{21}{32} = 0.6563 \approx 2.81\ \text{dB}$$

and

$$\bar{T} = \frac{46}{64} = 0.7188 \approx 4.075\ \text{dB},$$

and so

$$\text{decibel value of } \hat{\mu}_{(\mathrm{II})} = -6.15\ \text{dB},$$

that is

$$\hat{\mu}_{(\mathrm{II})} = 0.195 \quad \text{or} \quad 19.5\%.$$

From the above values of $\hat{\mu}_{(\mathrm{I})}$ and $\hat{\mu}_{(\mathrm{II})}$ we can estimate $\hat{\mu}_{\mathrm{I}}$ and $\hat{\mu}_{\mathrm{II}}$. Indeed

$$\hat{\mu}_{\mathrm{I}} = \hat{\mu}_{(\mathrm{I})} = 8.7\% \quad \text{and} \quad \hat{\mu}_{\mathrm{II}} = \hat{\mu}_{(\mathrm{II})} - \hat{\mu}_{(\mathrm{I})} = 10.8\%.$$

Following similar steps through the logit (omega) transformation, we find

$$\hat{\mu}_{(\mathrm{III})} = 15.6\%.$$

Unfortunately, the difference $(\hat{\mu}_{(\mathrm{III})} - \hat{\mu}_{(\mathrm{II})})$ yields a negative value. This is one of the (few) drawbacks of the logit transformation. In such a case we can accept that $\hat{\mu}_{(\mathrm{III})} = \hat{\mu}_{(\mathrm{II})}$. This can be confirmed by estimating $\hat{\mu}_{(\mathrm{II})}$

and $\hat{\mu}_{(\mathrm{III})}$ without the logit transformation (i.e. using only the proportions for each source level), in which case the same proportion is obtained for both $\hat{\mu}_{(\mathrm{II})}$ and $\hat{\mu}_{(\mathrm{III})}$.

So

$$\hat{\mu}_{(\mathrm{III})} = 19.5\% \qquad (\text{i.e. } \hat{\mu}_{\mathrm{III}} = 0\%),$$

and since (IV) is the last cumulative category we have

$$\hat{\mu}_{(\mathrm{IV})} = 100\%,$$

and

$$\hat{\mu}_{\mathrm{IV}} = \hat{\mu}_{(\mathrm{IV})} - \hat{\mu}_{(\mathrm{III})} = 0.805 \text{ or } (80.5\%).$$

Therefore, in the long run, under the optimal conditions, on average 80.5% of the tests will yield results belonging to category IV.

For the calculation of confidence limits one could use

$$\hat{\mu}_c \pm \sqrt{F(1, \mathrm{df_e}; \alpha) \times MS_\mathrm{e} \times \hat{\mu}_c \times (1 - \hat{\mu}_c) \times \frac{1}{n_\mathrm{e}}} \tag{4.30}$$

as far as cumulative category (c) is concerned. Indeed, since the weights W_c were taken into account by multiplying the error variance of category (c) by the reciprocal of $p_c(1 - p_c)$ as the weight, one can simply estimate with

$$MS_\mathrm{e} \times \hat{\mu}_c(1 - \hat{\mu}_c)$$

to obtain the error variance for $\hat{\mu}_c$.

For example, since from Table 4.12

$$MS_\mathrm{e} = MS_{e_1} = 0.4195$$
$$F(1, 30; 5\%) = 4.17$$

and (see Section 4.4.2)

$$n_\mathrm{e} = \frac{\text{total number of degrees of freedom}}{\text{sum of degrees of freedom for CF, B, D, F, I, and K}}$$
$$= \frac{192}{3 \times 6} = 10.67,$$

the confidence limits for $\hat{\mu}_{(\mathrm{II})}$ are given from

$$0.195 \pm \sqrt{4.17 \times 0.4195 \times 0.195(1 - 0.195) \times \frac{1}{10.67}}$$

or

$$0.195 \pm 0.16.$$

If the values for the limits are negative or exceed 1 (100%), the calculation should be performed in the omega scale following the steps described in Section 4.6.3 [i.e. making use of eqn (4.27)].

If the estimated $\hat{\mu}_c$ is itself 0 or 1, instead of (4.30), Taguchi suggests the following

$$\hat{\mu}_c \pm \sqrt{F(1, \mathrm{df_e}; \alpha) MS_\mathrm{e} \times \frac{1}{2n_\mathrm{e}}\left(1 - \frac{1}{2n_\mathrm{e}}\right) \times \frac{1}{n_\mathrm{e}}}$$

for the calculation of confidence limits.

4.7.2 *Critique*

An obvious disadvantage of the accumulating analysis is the fact that the frequencies of the cumulative categories are not independent, also that the observations are not normally distributed.

Consequently, a certain approximation is involved in the significance levels suggested by the ANOVA of Table 4.12. As a result, spurious effects can be detected and small real effects can be detected for the wrong reason. This was demonstrated by Hamada and Wu (1986) through the analysis of simulation results for various situations, and also through the reanalysis of the data from two real experiments.

The main problem with accumulating analysis is the result of neglecting the inherent nature of ordered categorical data which makes them harder to analyse than continuous data. During the analysis (in Section 4.7.1) it was shown that always

$$TSS = (\text{number of classes being analysed}) \times (\text{total number of measured values})$$

Therefore, always

$$TSS = \text{constant}.$$

Then, for example in the two–factor main effects setting,

$$S_\mathrm{e} = \text{constant} - S_\mathrm{A} - S_\mathrm{B},$$

and consequently, when testing for significance of A, the distribution of the F_A ratio statistic for A depends on factor B.

Hamada and Wu (1986) have shown that for fractional factorial designs, the statistic used in testing for significance of a particular factor, depends on *all* the other factors. Consequently the statistics used in the accumulating analysis are no longer comparing distributions, but mixtures of distributions. If the 'testing' detects a difference between these mixtures of distributions, this does not mean that a particular factor has an effect, because whether the mixtures of distributions are different depends on all the other factors.

The above suggest important details which must be considered before the 'accumulation analysis' technique is used:

(i) In the design phase, a balanced design should always be used.
(ii) The ordered categories should be refined as much as possible.
(iii) Application of the technique to only continuous data should be avoided. Indeed by grouping the continuous data, valuable information for detecting effects is lost; furthermore the transformation into ordered categories can change the location of the mixtures of distributions with the consequence that either real effects can be missed or spurious effects detected.
(iv) When a fractionated design is used, the confounding patterns must be considered, and additional runs must be made so that the effects in question can be unconfounded (see also Hamada and Wu 1986).

4.8 'Experimental regression' analysis

Multiple regression analysis is a powerful technique for fitting the best linear model possible on to a series of data values (see Chapter 2 and Appendix B). The limitations of this technique arise when the available data cannot adequately be represented by a linear model (even after a transformation), or when the unknown population parameters whose estimation is required, are more numerous than the observed values—a very common occurrence in situations when experimenting is difficult.

For these cases, Taguchi recommends a 'multivariable successive approximation' method, which makes use of orthogonal arrays, and is free of the limitations usually accompanying a regression analysis. The method uses a relatively small number of observed values to construct a satisfactory model, not necessarily linear, which could even involve more paramaters than the actual number of data values. The basic steps of the technique can be briefly outlined as follows. Initially a tentative theoretical response model is decided upon based on past knowledge and experience or on simple plots of the available data. In successive stages, few levels (usually three) are chosen for each parameter, and the model's response for various combinations of the levels (of all the parameters) is evaluated. The rows of an appropriate orthogonal array will indicate the particular level-combinations to be chosen.

At every stage, an 'optimal' set of parameter levels is determined, and this is the one which produces the minimum sum of squares of the deviations of the theoretical response values from the actual response values. By choosing new levels near these 'optimal' for each parameter, the procedure is repeated until a convergence is achieved to a model which provides a reasonable fit to the observed data values.

Insufficient convergence will indicate either that the range of the unknowns was incorrect, or it was taken too wide, or perhaps that the form of the initial (tentative) model needs changing.

The above procedure can be demonstrated by the following example:

Example: The error function, erf(x), can be approximated by several different models (Abranowitz and Segun 1968). One of them has the form

$$Y = 1 - [1 + Ax + Bx^2 + Cx^3 + Dx^4]^{-4} \tag{4.31}$$

Let us consider the problem when Y is known for some values of x and we seek the unknown constants A, B, C, and D. The usual method of fitting a multiple regression model through the least squares method is not appropriate here since Y is not linear in the unknowns A, B, C, and D.

Table 4.13 gives eight (known) values for the error function Y under various values of x. With such a restricted set of real data values, some tentative knowledge of the ranges in which A, B, C, and D belong, is required. Let us assume that the approximate ranges for the unknowns are as shown in Table 4.14.

TABLE 4.13

x	0.00004	0.0005	0.002	0.006	0.02	0.6	1.3	1.9
$Y = \text{erf}(x)$	0.000045	0.000564	0.002757	0.00677	0.022565	0.603865	0.934008	0.9927

TABLE 4.14

Unknown	Range
A	0.20–0.30
B	0.22–0.24
C	0.0–0.001
D	0.05–0.10

Then, considering A, B, C, and D as factors, we create three equidistant levels for each factor as follows:

A(1) = 0.20,	A(2) = 0.25,	A(3) = 0.30
B(1) = 0.22,	B(2) = 0.23,	B(3) = 0.24
C(1) = 0.0,	C(2) = 0.0005,	C(3) = 0.001
D(1) = 0.05,	D(2) = 0.075,	D(3) = 0.10

We let A, B, C, and D correspond to columns 2, 3, 4, and 5 of the orthogonal array $OA_{18}(2 \times 3^7)$. Then, trial numbers 1 through 18 in the array will give 18 different equations as shown in Table 4.15.

TABLE 4.15

Trial	Combination	Equation $\hat{Y}$
1	A(1)B(1)C(1)D(1)	$1+[1+(0.2)x+(0.22)x^2+(0)x^3+(0.05)x^4]^{-4}$
2	A(1)B(2)C(2)D(2)	$1+[1+(0.2)x+(0.23)x^2+(0.0005)x^3+(0.075)x^4]^{-4}$
⋮	⋮	⋮
18	A(3)B(3)C(2)D(1)	$1+[1+(0.3)x+(0.24)x^2+(0.0005)x^3+(0.05)x^4]^{-4}$

If the equation in Trial 1 fits well to the actual data, the values of $\hat{Y}$ thus calculated for every x of the upper row of Table 4.13, will be close to the actual values of Y in the lower row of Table 4.13; consequently the sum of squares SS_e of the differences $(Y - \hat{Y})$ for Trial 1 will be small in comparison with the SS_e of the other trials. The SS_e for all trials are shown in Table 4.15a.

TABLE 4.15a. SS_e's for 1st approximation

Trial	A	B	C	D	SS_e
1	1	1	1	1	0.00727266
2	1	2	2	2	0.00473433
3	1	3	3	3	0.00317562
4	2	1	1	2	0.00039200
5	2	2	2	3	0.00037885
6	2	3	3	1	0.00076459
7	3	1	2	1	0.00016563
8	3	2	3	2	0.00023693
9	3	3	1	3	0.00069076
10	1	1	3	3	0.00455490
11	1	2	1	1	0.00617916
12	1	3	2	2	0.00394844
13	2	1	2	3	0.00057389
14	2	2	3	1	0.00111353
15	2	3	1	2	0.00036821
16	3	1	3	2	0.00011763
17	3	2	1	3	0.00047256
18	3	3	2	1	0.00027346

TABLE 4.15b. Level totals for 1st approximation

Parameter	Level 1	Level 2	Level 3
A	0.02986511	0.00409108	0.00195688
B	0.01357672	0.01311526	0.00922109
C	0.01587536	0.01007461	0.00996310
D	0.01576904	0.01029745	0.00984658

We can now determine which of the levels for each parameter is the best at this stage, by comparing between totals in each level. For example,

$$A_1 = \text{Total of } SS_e\text{'s for level A(1)}$$
$$= 0.00727266 + 0.00473433 + \cdots + 0.00394844 = 0.02986511.$$

We can similarly find A_2 and A_3; all the results are shown in Table 4.15b. It is clear that at this stage the best parameter setting (producing the minimum sum of squares of errors) is

A(3)B(3)C(3)D(3)

i.e.

A	B	C	D
0.3	0.24	0.001	0.1 .

We can now retake new parameter levels close to their current optimum, and repeat the approximation procedure. The current optimum could be considered the new middle level (level 2), say $[2]_N$, with the other new levels defined by

$$[2]_N \pm \frac{H_0}{2}, \tag{4.32}$$

where H_0 is the length between the old levels. For example, for the second successive approximation the following factor levels could be used.

A(1) = 0.275, A(2) = 0.3, A(3) = 0.325
B(1) = 0.235, B(2) = 0.24, B(3) = 0.245
C(1) = 0.00075, C(2) = 0.001, C(3) = 0.00125
D(1) = 0.0875, D(2) = 0.1, D(3) = 0.1125

Note that the new 'third' levels consist of values beyond the original parameter ranges of Table 4.14. Taguchi actually suggests that any new levels are kept within the original ranges. However, this presupposes that the original 'tentative' ranges are the *true* ranges. The authors believe that the creation of 'new' levels using (4.32) is a much less restrictive approach.

A simple computer program can handle the creation of new levels and as many successive approximations as required. In our example, the levels for the tenth approximation turned out to be

A(1) = 0.27627	A(2) = 0.27637	A(3) = 0.27647
B(1) = 0.23330	B(2) = 0.23332	B(3) = 0.23334
C(1) = 0.0010147	C(2) = 0.0010156	C(3) = 0.0010166
D(1) = 0.078955	D(2) = 0.07900	D(3) = 0.07905

Note that the most of the new levels are once again within the original ranges of Table 4.14.

The SS_e's for each of the 18 trials of the OA_{18} are shown in Table 4.16a, whereas the sum totals for each factor level are shown in Table 4.16b. Note that the lack of model fit depicted through the level totals of the SS_e's of Table 4.16b is much smaller than the lack of fit depicted by Table 4.15b. Taguchi suggests that the successive approximations should continue until there is no significant difference between the level averages (for the SS_e's) in the parameters.

TABLE 4.16a. SS_e's for 10th approximation

Trial	A	B	C	D	SS_e
1	1	1	1	1	0.00000153501
2	1	2	2	2	0.00000157015
3	1	3	3	3	0.00000160699
4	2	1	1	2	0.00000155048
5	2	2	2	3	0.00000159007
6	2	3	3	1	0.00000152687
7	3	1	2	1	0.00000151129
8	3	2	3	2	0.00000155302
9	3	3	1	3	0.00000159484
10	1	1	3	3	0.00000159982
11	1	2	1	1	0.00000153788
12	1	3	2	2	0.00000157353
13	2	1	2	3	0.00000158514
14	2	2	3	1	0.00000152235
15	2	3	1	2	0.00000155991
16	3	1	3	2	0.00000154694
17	3	2	1	3	0.00000158828
18	3	3	2	1	0.00000152302

TABLE 4.16b. Level totals for 10th approximation

Parameter	Level 1	Level 2	Level 3
A	0.00000942	0.00000933	0.00000952
B	0.00000933	0.00000936	0.00000939
C	0.00000937	0.00000935	0.00000936
D	0.00000916	0.00000935	0.00000957

The question as to when a difference of level averages, say $|\bar{A}_2 - \bar{A}_1|$, could be considered significant or not, should be judged in comparison with the error-variance V_e inherent within the actual data values of Table 4.13 which can be estimated as follows.

First calculate the total (corrected) sum of squares of the available data values as in the usual way, i.e.

$$TSS = \sum_{j=1}^{n} Y_j^2 - \frac{(\sum_{j=1}^{n} Y_j)^2}{n},$$

where n is the total number of values; in our case

$$TSS = 2.223205 - \frac{(2.562855)^2}{8} = 1.402177.$$

Then, an estimate of the total variation V_T can be obtained by

$$V_T = \frac{TSS}{n-1} = 0.200311.$$

Finally, Taguchi proposes that

$$V_e = 0.00005 \times V_T = 0.00001.$$

Another method he suggests, is to plot the data in a graph, draw in a smooth curve, find the sum of squares of the differences from the observed values, and then divide by n to obtain V_e. Having estimated V_e, the successive approximation should continue until the difference in level averages becomes less than $2 \times V_e$.

The authors believe that the procedure should continue even beyond this stage, assuming, of course, an unrestricted computer capability. Indeed, the successive approximation and resetting of new levels could continue until any further iteration fails to alter the parameter values (to a prequoted number of decimal places).

For our example, no further iteration (beyond the 10th) succeeded in altering the parameter values to 4 decimal places; we can therefore accept the following estimates:

$$\hat{A} = 0.2764, \qquad \hat{B} = 0.2333, \qquad \hat{C} = 0.0010, \qquad \hat{D} = 0.0789.$$

Remark: It is known that the best parameter values for erf(x) as given by (4.31) are, in fact,

$$A = 0.278393, \qquad B = 0.230389, \qquad C = 0.000972, \qquad D = 0.078108.$$

These are not too different from the estimated ones obtained using Taguchi's 'experimental regression' technique which was based only on 8 data values. Of course, the successful convergence might be mainly due to the fact that the original 'tentative' ranges for the parameters (Table 4.14) were very reasonable.

Nevertheless, in the absence of a better alternative, the use of orthogonal arrays through 'multivariable successive approximation' seems to be a tenable method for fitting complicated models using only a limited number of actual observations.

5

RESPONSE SURFACE METHODS AND DESIGNS

5.1 Introduction

Both the modelling and experimental designs of Chapters 2 and 3 can be extended to more complex situations. If $\mathbf{x} = (x_1, \ldots, x_p)$ are the independent variables, or factors, we can consider an arbitrary experimental design region R (sometimes called a 'design space').

Writing the model as

$$Y = Y(\mathbf{x}) = f(\mathbf{x}, \boldsymbol{\theta}) + \varepsilon$$

where $\boldsymbol{\theta} = (\theta_1, \ldots, \theta_p)$ we attempt to fit the model to the data to give

$$\hat{Y} = \hat{Y}(\mathbf{x}) = f(\mathbf{x}, \hat{\boldsymbol{\theta}}).$$

In the response surface methodology Y is called the response and $\hat{Y}$ the fitted surface. Whereas in Chapter 2 we were more interested in $\boldsymbol{\theta}$ or the effects of individual factors in the model, response surface methods concentrate on the *shape* of Y (or $\hat{Y}$). A useful model is the general quadratic model

$$Y = \theta_0 + \sum_{i=1}^{p} \theta_i x_i + \sum_{i=1}^{p} \theta_{ii} x_i^2 + \sum_{i<j}^{p} \sum^{p} \theta_{ij} x_i x_j + \varepsilon.$$

In two dimensions ($p = 2$) this becomes

$$Y = \theta_0 + \theta_1 x_1 + \theta_2 x_2 + \theta_{11} x_1^2 + \theta_{22} x_2^2 + \theta_{12} x_1 x_2 + \varepsilon.$$

This is a quadratic model with a first-order interaction corresponding to the appropriate 3^2 factorial design as mentioned in Chapter 3.

Although the experimental design region will often be a p-dimensional rectangle

$$a_i \le x_i \le b_i \qquad (i = 1, \ldots, p)$$

it is important to understand that this need not always be so. Certain regions may be excluded for physical reasons, structural reasons, reasons of safety and so on. In certain special cases the x_i's may be constrained by the very nature of the problem. One such example is the mixture experiment when all the x_i's are proportions:

$$0 \le x_i \le 1 \quad \text{and} \quad \sum_{i=1}^{p} x_i = 1.$$

When observations are taken on, say, the surface of a plate or wafer, the geometry itself may prescribe the experimental region. For example there may be holes in the plate. In other cases non-linear constraints on the x_i's can give strangely shaped regions. We see this in the design centring method of Chapter 7.

In these general settings quite a lot will be required of a design other than the estimation and testing of parameters and effects. Here are some general aims.

(i) Fitting $\hat{Y}$ to Y accurately.

(ii) Finding the maximum (or minimum) of $Y(\hat{Y})$.

(iii) Finding the direction of maximum increase in $Y(\hat{Y})$.

(iv) Finding the $\mathbf{x}$ which keeps $Y(\hat{Y})$ on target c.

(v) Find the general shape of Y.

(vi) Elimination of variables x_i which do not affect Y.

In addition to such general aims response surface methods reflect, more than the elementary factorial methods which they incorporate, the desire to proceed sequentially. Scientists and engineers tend to employ some version of the 'hypothetico-deductive' method which dates back to Francis Bacon:

> the true method of experience . . . first lights the candle, and then by means of the candle shows the way; commencing as it does with experience duly ordered and digested, not bungling or erratic, and from it deducing axioms, and from established axioms again new experiments . . . .

Thus we may indeed change our hypotheses (models) as we proceed through a series of stages. As an example, discussed further in Chapter 6, we may conduct a fairly large experiment to discover some optimum setting of the x_i and then do a small confirmatory experiment around the optimum setting to verify or improve our initial estimates.

5.2 Model fitting

When the model is linear in the parameters and the errors satisfy the required statistical conditions (see Chapter 2) we may again use least squares analysis. For example, consider the two-dimensional quadratic model of the last section. Suppose we take observations $Y_1, \ldots, Y_n$ at $(x_{11}, x_{21}), \ldots, (x_{1n}, x_{2n})$ respectively. Then the least squares analysis would seek to minimize

$$\sum_{i=1}^{n} \{Y_i - f[(x_{1i}, x_{2i}), \theta]\}^2$$

$$= \sum_{i=1}^{n} [Y_i - (\theta_0 + \theta_1 x_{1i} + \theta_2 x_{2i} + \theta_{11} x_{1i}^2 + \theta_{22} x_{2i}^2 + \theta_{12} x_{1i} x_{2i})]^2.$$

The solution can be expressed in the matrix notation of Chapter 2

$$\hat{\boldsymbol{\theta}} = (\mathbf{X}^T\mathbf{X})^{-1}\mathbf{X}^T\mathbf{Y}$$

where

$$\hat{\boldsymbol{\theta}} = (\hat{\theta}_0, \hat{\theta}_1, \hat{\theta}_2, \hat{\theta}_{11}, \hat{\theta}_{22}, \hat{\theta}_{12})^T$$

$$\mathbf{Y} = (Y_1, \ldots, Y_n)^T$$

and

$$\mathbf{X} = \begin{bmatrix} 1 & x_{11} & x_{21} & x_{11}^2 & x_{21}^2 & x_{11}x_{21} \\ 1 & x_{12} & x_{22} & x_{12}^2 & x_{22}^2 & x_{12}x_{22} \\ \vdots & \vdots & \vdots & \vdots & \vdots & \vdots \\ 1 & x_{1n} & x_{2n} & x_{1n}^2 & x_{2n}^2 & x_{1n}x_{2n} \end{bmatrix}.$$

The matrix $\mathbf{X}^T\mathbf{X}$ can be written conveniently in terms of the 'moments' of the x values:

$$m_{10} = \bar{x}_1 = \frac{1}{n}\sum x_{1i}$$

$$m_{01} = \bar{x}_2 = \frac{1}{n}\sum x_{2i}$$

$$m_{20} = \frac{1}{n}\sum x_{1i}^2$$

$$m_{02} = \frac{1}{n}\sum x_{2i}^2$$

$$m_{11} = \frac{1}{n}\sum x_{1i}x_{2i}$$

etc. Thus

$$\mathbf{X}^T\mathbf{X} = n\begin{bmatrix} 1 & m_{10} & m_{01} & m_{20} & m_{02} & m_{11} \\ m_{10} & m_{20} & m_{11} & m_{30} & m_{12} & m_{21} \\ m_{01} & m_{11} & m_{02} & m_{21} & m_{03} & m_{12} \\ m_{20} & m_{30} & m_{21} & m_{40} & m_{22} & m_{31} \\ m_{02} & m_{12} & m_{03} & m_{22} & m_{04} & m_{13} \\ m_{11} & m_{21} & m_{12} & m_{31} & m_{13} & m_{22} \end{bmatrix}$$

$$= n\mathbf{M}.$$

We refer to $\mathbf{M}$ as the 'moment matrix' and $\mathbf{X}^T\mathbf{X}$ as the 'information matrix'. The theory tells us that the covariance matrix of the parameter estimates is given by

$$\text{cov}(\hat{\boldsymbol{\theta}}) = \sigma^2(\mathbf{X}^T\mathbf{X})^{-1} = \sigma^2\frac{1}{n}\mathbf{M}^{-1}$$

where σ^2 is the error variance.

In this and similar situations the structure of the experiment affects the structure of $\mathbf{M}$ and hence of $\mathbf{M}^{-1}$. We have seen this already with the orthogonality principle in Chapter 3. Consider, for example, the 3^2 full-factorial design centred (after shifting and rescaling the x-values) at $(0, 0)$. We have the nine values $(\pm 1, \pm 1)$, $(\pm 1, 0)$ $(0, \pm 1)$ and $(0, 0)$ itself. For this design

$$\mathbf{M} = \frac{1}{9}\begin{bmatrix} 9 & 0 & 0 & 6 & 6 & 0 \\ 0 & 6 & 0 & 0 & 0 & 0 \\ 0 & 0 & 6 & 0 & 0 & 0 \\ 6 & 0 & 0 & 6 & 4 & 0 \\ 6 & 0 & 0 & 4 & 6 & 0 \\ 0 & 0 & 0 & 0 & 0 & 4 \end{bmatrix}$$

all moments with odd powers being zero.

If we rearrange the re-runs of the original model as

$$1,\ x_1^2,\ x_2^2,\ x_1,\ x_2,\ x_1x_2$$

we can see this more clearly, the new moment matrix being

$$\mathbf{M} = \frac{1}{9}\begin{bmatrix} 9 & 6 & 6 & 0 & 0 & 0 \\ 6 & 6 & 4 & 0 & 0 & 0 \\ 6 & 4 & 6 & 0 & 0 & 0 \\ 0 & 0 & 0 & 6 & 0 & 0 \\ 0 & 0 & 0 & 0 & 6 & 0 \\ 0 & 0 & 0 & 0 & 0 & 4 \end{bmatrix}$$

The blocks of zeros now reveal the orthogonality structure. This helps us invert the original matrix by hand also to give

$$\mathbf{M}^{-1} = \begin{bmatrix} 5 & 0 & 0 & -3 & -3 & 0 \\ 0 & \frac{3}{2} & 0 & 0 & 0 & 0 \\ 0 & 0 & \frac{3}{2} & 0 & 0 & 0 \\ -3 & 0 & 0 & \frac{9}{2} & 0 & 0 \\ -3 & 0 & 0 & 0 & \frac{9}{2} & 0 \\ 0 & 0 & 0 & 0 & 0 & \frac{9}{4} \end{bmatrix}$$

Of special importance in response surface methods is the variance of the estimated response $\mathrm{var}(\hat{Y})$ which tells us how well we are predicting or interpolating $E(Y)$ over the whole region R. We may even be interested in extrapolating for points outside R although it is generally recognized that this is risky.

From Appendix B we see that

$$\mathrm{var}(\hat{Y}(\mathbf{x})) = \sigma^2 f(\mathbf{x})^T(\mathbf{X}^T\mathbf{X})^{-1}f(\mathbf{x})$$

where $f(\mathbf{x})^T$ is the vector of functions in the model. For the quadratic model

$$f(\mathbf{x}) = (1, x_1, x_2, x_1^2, x_2^2, x_1x_2).$$

For the 3^2 design $\begin{pmatrix} 0 & 0 \\ \pm 1, & \pm 1 \end{pmatrix}$ we can compute

$$\mathrm{var}(\hat{Y}) = \frac{\sigma^2}{9}\left[5 - \frac{9}{2}(x_1^2 + x_2^2) + \frac{9}{4}x_1^2x_2^2 + \frac{9}{2}(x_1^4 + x_2^4)\right].$$

The contours of this function are given in Fig. 5.1a. The lower the value of $\mathrm{var}(\hat{Y})$ the better our interpolation and we can see that we do rather better along the corners.

The analysis of variance based on the models proceeds as in Chapter 2 and we can calculate the mean sums of squares and the residual sums of

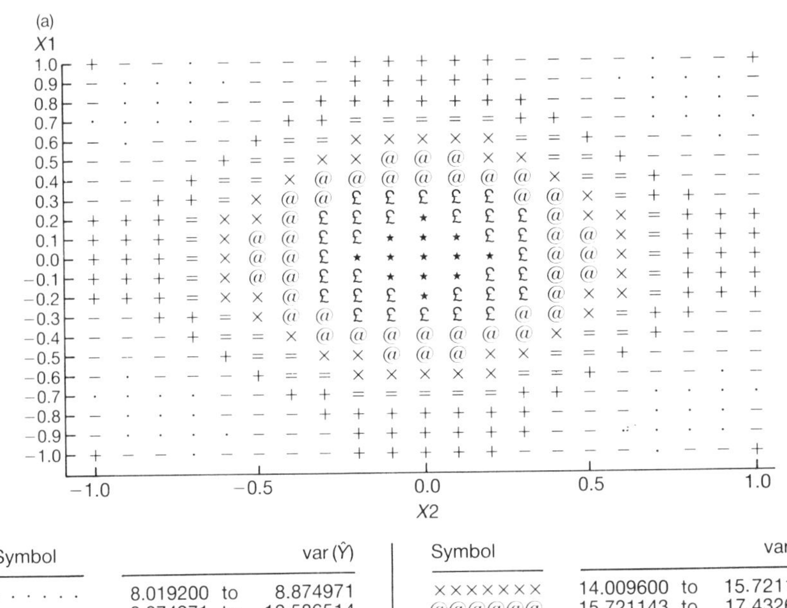

(a)
X1
X2
Symbol
var (Ŷ)
. . . . . . 8.019200 to 8.874971
– – – – – – 8.874971 to 10.586514
+ + + + + + 10.586514 to 12.298057
= = = = = = 12.298057 to 14.009600
× × × × × × × 14.009600 to 15.721143
@ @ @ @ @ @ 15.721143 to 17.432686
£ £ £ £ £ £ 17.432686 to 19.144229
* * * * * * 19.144229 to 20.00000

Fig. 5.1a. Contour plot of var($\hat{Y}$).

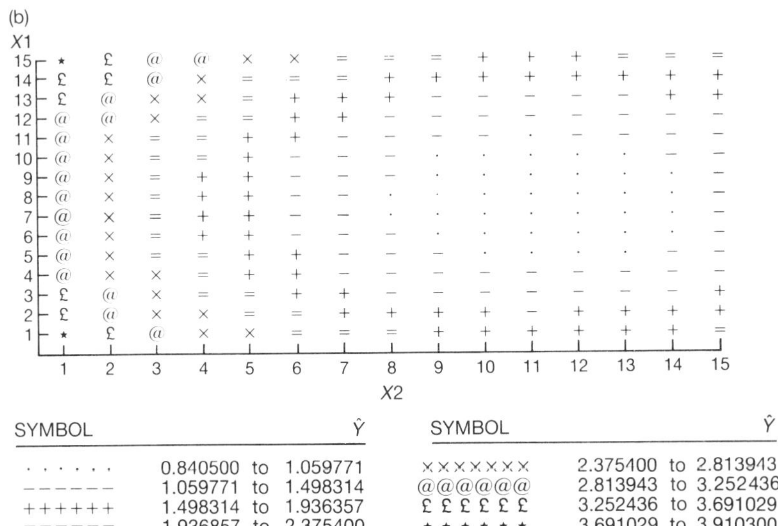

Fig. 5.1b. Contour plot of $\hat{Y}$.

squares in the usual way. For example, we estimate σ^2 by

$$s^2 = \frac{1}{n-m} \times (\text{residual } SS)$$

$$= \frac{1}{n-m} \sum_{i=1}^{n} (Y_i - \hat{Y}_i)^2$$

where $\hat{Y}_i$ is the fitted response at the ith observation and n is the number of observations and m the number of parameters. The typical procedure would be to eliminate non-significant factors or parameters (using analysis of variance) and present the resulting model. It is advisable to start with a model of sufficient complexity to incorporate the 'true' model as a submodel. This risks forcing a large number of observations on the experimenter, so that every parameter in the large model is estimable, that is can be properly estimated. For example, for the quadratic model in p dimensions there are

$$m = 1 + p + p + \frac{p(p-1)}{2} = \tfrac{1}{2}(p+1)(p+2)$$

parameters. In practice, there has to be some trade-off between the cost of observations and the strength of belief in existence of terms in the model.

Example 5.1: The data in Table 5.1 gives the resistance Y in ohms of a certain device according to the location (x_1, x_2) on the wafer from which the device chip is cut. The location was chosen at random.
The analysis showed a good fit for a quadratic model with no interaction:

$$\hat{Y} = 4.46 - 0.306x_1 - 0.434x_2 + 0.0198x_1^2 + 0.0193x_2^2.$$

The t values for the separate terms were

Term	Estimate	t
Constant	4.4580	8.20
x_1	−0.30643	−3.25
x_2	−0.43424	−4.43
x_1^2	0.19796	3.57
x_2^2	0.19292	3.32

The analysis of variance table for the full regression is as follows, with a

TABLE 5.1

Observation	x_1	x_2	Y
1	9	4	1.25
2	6	10	0.502
3	4	11	1.41
4	15	7	1.68
5	9	5	1.57
6	8	15	0.659
7	1	6	2.02
8	14	4	3.32
9	14	14	1.61
10	12	9	1.52
11	13	14	1.83
12	5	8	1.51
13	7	1	2.92
14	13	8	1.45
15	4	6	1.87
16	1	6	2.00
17	9	9	0.873
18	4	11	1.46
19	13	2	3.17
20	4	12	1.39

significant F-value of 12.8

Source	df	S	MS
Regression	4	7.9708	1.9927
Residual	15	2.3406	0.1560
Total (corrected)	19	10.3114	

The contours of the fit are roughly circular, reflecting the geometry of the wafer. There is a minimum of $\hat{Y}$ at (7.73, 11.24) somewhat off-centre which is approximately (7, 7). (See Fig. 5.1b.)

5.3 Special response surface designs

A number of special designs, not standard functions of 2^k or 3^k factorial designs, have been suggested in early work on response surface and are in common use today. Some of these have appealing properties in terms of model fitting and the shape of the variance function var $[\hat{Y}(\mathbf{x})]$. They consist of composites of factorial fractions with additional symmetrically placed points. The commonest such additions are

(i) Centre points.
(ii) Star points.

These additions result in the so-called central composite designs.

Referring to the 3^2 experiment of the last section we may instead take a 2^n factorial $(\pm 1, \pm 1)$ and add one or more observations at the centre point $(0, 0)$ and star points at $(\pm c, 0)$ and $(0, \pm c)$ for some constant c typically greater than one. This is shown in Fig. 5.2a.

As a special case if $c = \sqrt{2}$ and we take, say, one centre point we have the design in Fig. 5.2b. For this design we can compute the variance function as

$$\operatorname{var}[\hat{Y}(x)] = \sigma^2\left[1 - \frac{7}{8}(x_1^2 + x_2^2) + \frac{11}{32}(x_1^2 + x_2^2)^2\right].$$

The contours of this function are circles, which means that interpolation is equally efficient in every direction. A design with this property is called *rotatable.*

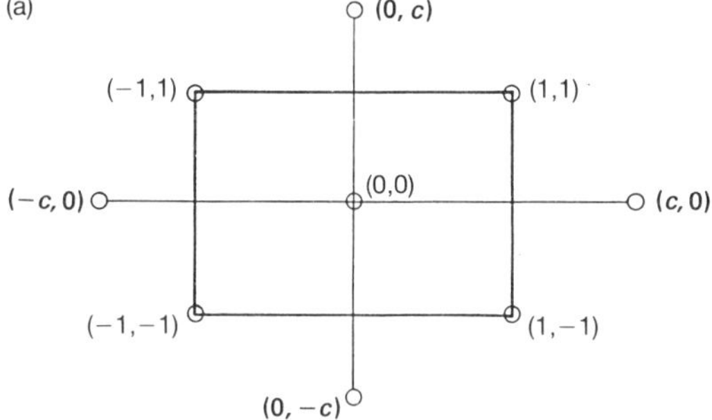

FIG. 5.2a. Star-points design.

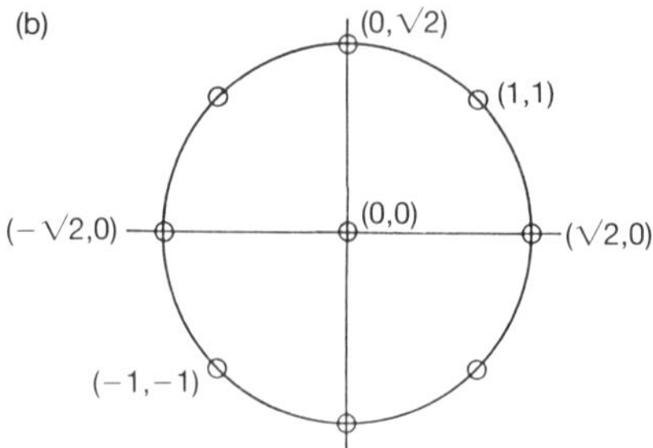

FIG. 5.2b. Rotatable design.

Here is an example in three dimensions:

Factorial	$(\pm 1, \pm 1, \pm 1)$	8 points
Star	$(\pm c, 0, 0)$	6 observations
	$(0, \pm c, 0)$	
	$(0, 0, \pm c)$	
Centre	$(0, 0, 0)$	one or more points

with one centre point this makes $n = 15$ observations which is considerably less than the 27 required by the 3^3 factorial design with

$$\begin{bmatrix} 0 & 0 & 0 \\ \pm 1, & \pm 1, & \pm 1 \end{bmatrix}.$$

For the quadratic model of Section 5.2 with $p = 3$, $m = 10$ the value of c required to make the design rotatable with one point can be shown to be $8^{\frac{1}{4}} = 1.682$.

It is instructive to compare different designs in the three-dimensional case by looking at their diagrams in Fig. 5.3. In Example 5.4 we discuss some suitable points for a mixture experiment with three factors.

5.4 Optimum experimental designs

Until about thirty years ago the question 'how good is the experimental design?' would probably have been answered by referring to balance, orthogonality, estimability and similar structural features of the design. These are very important and indeed much of this book is devoted to designs with these nice properties. However, it is possible to set up portmanteau criteria for deciding when a design is good or bad. These are usually based on how well the parameters are estimated or how well Y itself is estimated.

5.4.1 Comparing two experimental designs

Let us return briefly to the weighing design problem of Chapter 2 (see section 2.2.1). Recall the three weighings:

Weighing 1 object A

Weighing 2 object B

Weighing 3 objects A *and* B.

The estimates were

$$\hat{\theta}_A = \tfrac{1}{3}(2Y_1 - Y_2 + Y_3)$$

$$\hat{\theta}_B = \tfrac{1}{3}(-Y_1 + 2Y_2 + Y_3)$$

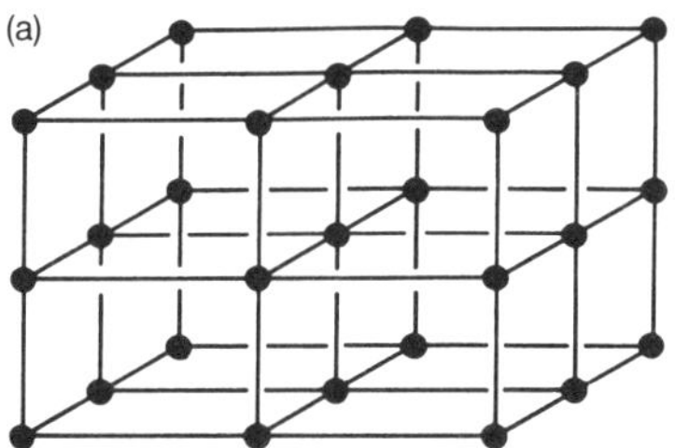

FIG. 5.3a. 3^3 factorial design.

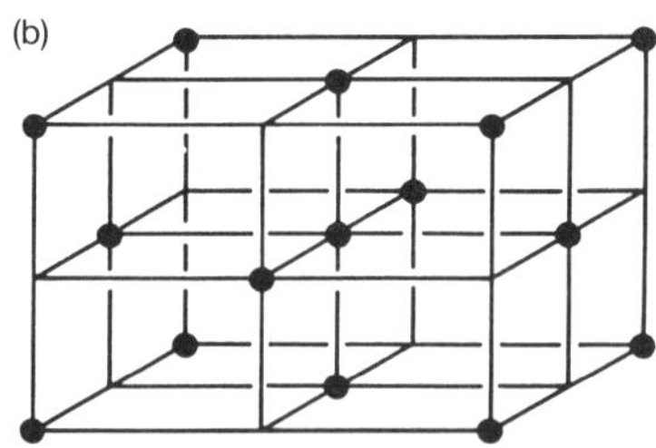

FIG. 5.3b. Central composite design ($c = 1$).

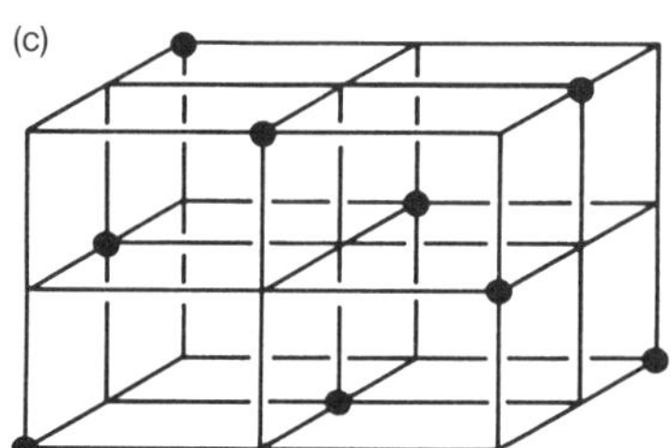

FIG. 5.3c. 3×3 Latin square.

and

$$\operatorname{cov}(\hat{\boldsymbol{\theta}}) = \begin{bmatrix} \operatorname{var}(\hat{\theta}_A) & \operatorname{cov}(\hat{\theta}_A, \hat{\theta}_B) \\ \operatorname{cov}(\hat{\theta}_A, \hat{\theta}_B) & \operatorname{var}(\hat{\theta}_B) \end{bmatrix}$$

$$= \sigma^2 \begin{bmatrix} \frac{2}{3} & -\frac{1}{3} \\ -\frac{1}{3} & \frac{2}{3} \end{bmatrix}.$$

Let us compare this with the alternative designs

Weighing 1 A

Weighing 2 A (again)

Weighing 3 B.

This gives

$$\hat{\theta}_A = \tfrac{1}{2}(Y_1 + Y_2)$$
$$\hat{\theta}_B = Y_3$$

and

$$\mathrm{cov}(\hat{\boldsymbol{\theta}}) = \sigma^2 \begin{bmatrix} \frac{1}{2} & 0 \\ 0 & 1 \end{bmatrix}.$$

We can set up some specific criteria to judge how good these designs are. Here are two: (i) compare $\mathrm{trace}[\mathrm{cov}(\hat{\boldsymbol{\theta}})] = \mathrm{var}(\hat{\theta}_A) + \mathrm{var}(\hat{\theta}_B)$; (ii) compare $\det[\mathrm{cov}(\hat{\boldsymbol{\theta}})] = \mathrm{var}(\hat{\theta}_A) \times \mathrm{var}(\hat{\theta}_B) - [\mathrm{cov}(\hat{\theta}_A, \hat{\theta}_B)]^2$. The smaller the values the better. Then

	trace	$\sqrt{\det}$
Design 1	$\dfrac{4\sigma^2}{3}$	$\dfrac{\sigma}{\sqrt{3}}$
Design 2	$\dfrac{3\sigma^2}{2}$	$\dfrac{\sigma}{\sqrt{2}}$

Thus, by both criteria, Design 1 is better. We may want to compare two parameters. In this simple example this would mean estimating $\phi = \theta_A - \theta_B$, the difference in weights. The estimate of ϕ is $\hat{\phi} = \hat{\theta}_A - \hat{\theta}_B$ which is $Y_1 - Y_2$ for Design 1 and $\tfrac{1}{2}(Y_1 + Y_2) - Y_3$ for Design 2 with variances respectively $2\sigma^2$ and $3\sigma^2/2$. Thus Design 2 is better for the specific purpose of estimating the contrast $\theta_A - \theta_B$. Indeed for this the third weighing in Design 1 is useless!

5.4.2 *Optimality criteria*

We can elevate the ideas contained in this example to a full theory of the optimal design of experiments. Here then is a list of common criteria.

D-optimality	minimize $\det[\mathrm{cov}(\hat{\boldsymbol{\theta}})]$
A-optimality	minimize $\mathrm{trace}[\mathrm{cov}(\hat{\boldsymbol{\theta}})]$
G-optimality	minimize $\max_R \mathrm{var}[\hat{Y}(\mathbf{x})]$
Average optimality	minimize $\mathrm{average}_R\, \mathrm{var}[\hat{Y}(\mathbf{x})]$

Returning to our two-dimensional quadratic model we can compare the 3^2 design which removes the centre point to replicate at one corner. See Fig. 5.4.

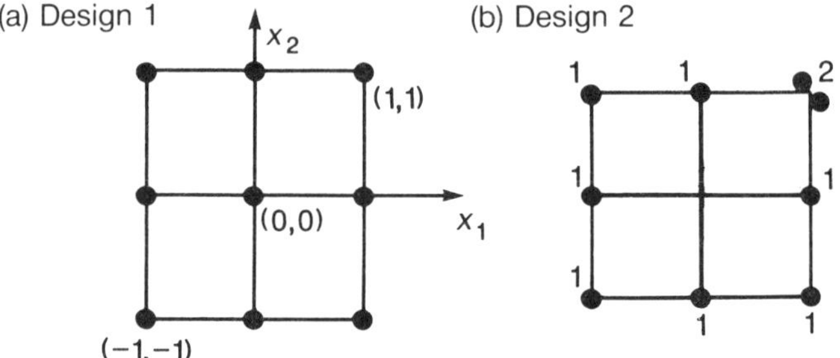

FIG. 5.4a. Good 3^2 design. FIG. 5.4b. Bad 3^2 design.

Comparing the criteria we have

	D	A	G	Ave.
Design 1	102.5	19.25	7.250	6.000
Design 2	125.8	28.50	10.940	6.761

It is obvious that Design 2 is inferior in comparison with Design 1.

One advantage of using an optimality criterion is that we can deal with situations in which the design region R has a strange shape and there is no obvious factorial fraction which fits into the design region. To avoid the difficulty that a change in sample size affects the solution dramatically, it has been convenient to think sometimes in terms of the moment matrix $\mathbf{M}$ of the last section, rather than $\mathbf{X}^T\mathbf{X}$. This is equivalent to thinking in terms of the proportion of observations at certain points in R; see, for example, Fig. 5.5.

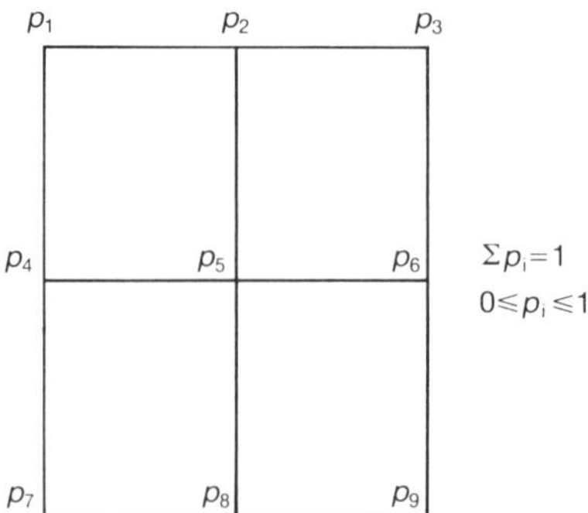

FIG. 5.5. Proportions on the design region.

Here is a simple example: For straight-line regression in the region $[-1, 1]$ we have

$$\mathbf{M} = \frac{1}{n}\begin{bmatrix} n & \Sigma x_i \\ \Sigma x_i & \Sigma x_i^2 \end{bmatrix} = \begin{bmatrix} 1 & m_1 \\ m_1 & m_2 \end{bmatrix}$$

where $m_1 = \frac{1}{n}\Sigma x_i$, $m_2 = \frac{1}{n}\Sigma x_i^2$. Considering a proportion p_- at -1 and p_+ at $+1$ we can write

$$\mathbf{M} = \begin{bmatrix} 1 & p_+ - p_- \\ p_+ - p_- & 1 \end{bmatrix}$$

$$\mathbf{M}^{-1} = \frac{1}{1-(p_+ - p_-)^2}\begin{bmatrix} 1 & -p_+ + p_- \\ -p_+ + p_- & 1 \end{bmatrix}$$

Then

$$\det(\mathbf{M}) = 1 - (p_+ - p_-)^2,$$

$$\det(\mathbf{M}^{-1}) = [1 - (p_+ - p_-)^2]^{-1}.$$

The minimum is achieved at $p_- = p_+ = \frac{1}{2}$, i.e. half the observations at 1 and half at -1. Similarly

$$\frac{n}{\sigma^2}\operatorname{var}[\hat{Y}(\mathbf{x})] = (1, x)\mathbf{M}^{-1}\begin{bmatrix} 1 \\ x \end{bmatrix} = \frac{1}{1-(p_+ - p_-)^2}[1 - 2(p_+ - p_-)x + x^2].$$

This achieves a maximum in $[-1, 1]$ at $x = 1$ if $p_- \geq p_+$ and at $x = -1$ if $p_+ \geq p_-$. The maximum value in either case is

$$\frac{2(1 + |p_+ - p_-|)}{1 - (p_+ - p_-)^2} = 2(1 - |p_+ - p_-|)^{-1},$$

which is again minimized when $p_- = p_+ = \frac{1}{2}$ to give G-optimality. A general result (called the General Equivalence Theorem) tells us that, provided we deal with moment matrices, D-optimality and G-optimality are always equivalent. This is fortunate because it connects a criterion based on parameter estimation (D) with one based on the fitted response (G). It is the foundation for a very rich theory. (See the bibliography at the end of the chapter.) The General Equivalence Theorem says that not only D and G optimality are equivalent but the value of $\max_{R} \operatorname{var}[\hat{Y}(\mathbf{x})]$ is precisely $m\sigma^2/n$ where m is the number of parameters. We just saw that this was the case for straight line regression.

Here are some non–standard examples to show how the theory works in each case. We exhibit the D (and G)–optimum designs.

Example 5.2: Consider the model

$$Y = \theta_0 + \theta_1 x_1 + \theta_2 x_2 + \varepsilon$$

over the quadrilateral with corners P[2, 2], Q[1, −1], R[−1, 1], and S[−1, −1]. The D and G optimum designs puts proportions of observations $\frac{10}{32}$, $\frac{9}{32}$, $\frac{9}{32}$, and $\frac{4}{32}$ at P, Q, R, and S respectively. If we compute the variance function for the hypothetical experiment with these proportions of observations at the points we obtain

$$\mathbf{M} = \frac{1}{32}\begin{bmatrix} 32 & 16 & 16 \\ 16 & 62 & 26 \\ 16 & 26 & 62 \end{bmatrix}$$

$$\mathbf{M}^{-1} = \frac{1}{9}\begin{bmatrix} 11 & -2 & -2 \\ -2 & 6 & -2 \\ -2 & -2 & 6 \end{bmatrix}.$$

From this we have

$$\operatorname{var}[\hat{Y}(\mathbf{x})] = \frac{\sigma^2}{9n}(11 - 4x_1 - 4x_2 + 6x_1^2 + 6x_2^2 - 4x_1x_2).$$

This reaches a maximum of $3\sigma^2/n$ at each of P, Q, R, and S.

Example 5.3: Consider the same model as before but with the quadrilateral region with corners $P(\frac{1}{3}, \frac{1}{3})$, Q(1, −1), R(−1, 1) and S(0, 0). The region is essentially too squashed to allow any observations at P. The optimum design places proportion $\frac{1}{3}$ at Q, R, and S.

Example 5.4: This is an example of a mixture experiment. Suppose there are three ingredients x_1, x_2, and x_3 whose proportions add up to one: $x_1 + x_2 + x_3 = 1$, x_1, x_2, $x_3 \geq 0$. Assume a quadratic model of the form

$$Y = \theta_1 x_1 + \theta_2 x_2 + \theta_3 x_3 + \theta_4 x_1 x_2 + \theta_5 x_1 x_3 + \theta_6 x_2 x_3 + \varepsilon.$$

Notice that no constant term is allowed since we can write θ_0 as $\theta_0(x_1 + x_2 + x_3)$. Similarly, for $x_1^2 = x_1(1 - x_2 - x_3)$ so that quadratic terms can be written in terms of linear components and interactions. Let us set ourselves a difficult problem in which we are not allowed to use the 'corners' (1, 0, 0), (0, 1, 0) and (0, 0, 1). So consider the hexagonal region with corners $P(\frac{1}{3}, \frac{2}{3}, 0)$, $Q(\frac{2}{3}, \frac{1}{3}, 0)$, $R(\frac{1}{3}, 0, \frac{2}{3})$, $S(\frac{2}{3}, 0, \frac{1}{3})$, $T(0, \frac{1}{3}, \frac{2}{3})$, and $U(0, \frac{2}{3}, \frac{1}{3})$.

The optimum design places (approximately) proportions 0.022 at each of $(\frac{2}{3}, \frac{1}{6}, \frac{1}{6})$, $(\frac{1}{6}, \frac{2}{3}, \frac{1}{6})$, and $(\frac{1}{6}, \frac{1}{6}, \frac{2}{3})$, 0.130 at P, Q, R, S, T, and U, and 0.155 at $(\frac{1}{3}, \frac{1}{3}, \frac{1}{3})$.

5.4.3 *Iterative improvement of a design*

A useful feature of the optimum design idea is that it can tell us how to improve a bad design if we are allowed some additional observations. As an example, consider the design with four points in two dimensions $(\pm 1, 0)$, $(0, \pm 1)$ which might arise from using one variable at a time around the centre $(0, 0)$. For the linear model

$$Y = \theta_0 + \theta_1 x_1 + \theta_2 x_2 + \varepsilon$$

the value of the D-criterion is

$$\det\left\{\begin{bmatrix} 4 & 0 & 0 \\ 0 & 2 & 0 \\ 0 & 0 & 2 \end{bmatrix}^{-1}\right\} = \frac{1}{16}.$$

Compare with the best design which is the 2^2 factorial: $(\pm 1, \pm 1)$ (assuming that R is the square itself) for which

$$\det\left\{\begin{bmatrix} 4 & 0 & 0 \\ 0 & 4 & 0 \\ 0 & 0 & 4 \end{bmatrix}^{-1}\right\} = \frac{1}{64}.$$

The first design is not advisable. If, however, the first experiment had already been performed one might ask: given *one* additional point where should we place it? Let the general point be (x_1, x_2).

Then

$$\mathbf{X} = \begin{bmatrix} 1 & 1 & 0 \\ 1 & 0 & 1 \\ 1 & -1 & 0 \\ 1 & 0 & -1 \\ 1 & x_1 & x_2 \end{bmatrix} \text{ and } \mathbf{X}^T\mathbf{X} = \begin{bmatrix} 5 & x_1 & x_2 \\ x_1 & 2 + x_1^2 & x_1 x_2 \\ x_2 & x_1 x_2 & 2 + x_2^2 \end{bmatrix}$$

and so $\det(\mathbf{X}^T\mathbf{X}) = 20 + 8x_1^2 + 8x_2^2$. This reaches a maximum at any of the points $(x_1, x_2) = (\pm 1, \pm 1)$ and so a minimum for the determinant of $\text{cov}(\hat{\boldsymbol{\theta}}) = \sigma^2(\mathbf{X}^T\mathbf{X})^{-1}$. We can proceed to add more points. The following useful result lies behind the equivalence of D and G optimality (in the moment matrix case) and is also useful for improving bad designs:

The best place to add a single point to improve on $\det(\mathbf{X}^T\mathbf{X})$ *is the place where* $\text{var}[\hat{Y}(\mathbf{x})]$ *is a maximum (using the old design).*

Thus a useful rule of thumb is to add additional observations in a careful way at the points with the *worst* estimate of $E[Y(\mathbf{x})]$.

This augmentation method is the basis for a range of optimization algorithms which have been fully implemented. The best of these algorithms, called *exchange algorithms* actually adds 'good' points and removes 'bad' points to home in an optimum design, before any experimentation takes place!

5.5 Other sequential methods

5.5.1 Evolutionary operation (EVOP)

A method developed in the 1960s by G. E. P. Box and co-workers goes under a broad heading of evolutionary operation (EVOP). It is a simple kind of stochastic optimization which provides a step-by-step solution to a search for an optimum in a large region R. It is particularly useful for on-line process control and where off-line experimentation is expensive. The simple two-stage process

$$\text{experiment} \rightarrow \text{confirmatory experiment}$$

is really a special case.

At each stage a rather simple design is used sufficient to estimate the direction of increase (or decrease) in $Y(\mathbf{x})$. For example, if we fit a linear model to a process which depends on the X-variables: X_1: temperature and X_2: pressure and the fit is

$$\hat{Y} = \hat{\theta}_0 + \hat{\theta}_1 X_1 + \hat{\theta}_2 X_2$$

then the gradient of $\hat{Y}$ is given by

$$\left[\frac{\partial \hat{Y}}{\partial X_1}, \frac{\partial \hat{Y}}{\partial X_2}\right] = (\hat{\theta}_1, \hat{\theta}_2).$$

The direction of steepest ascent is precisely that of vector $(\hat{\theta}_1, \hat{\theta}_2)$.

Proceeding a certain step-length in that direction (but staying in R) one would lay down a second experiment to re-estimate the gradient and continue until there was no further improvement (within a given cost). For two variables, simple five-point designs of the form $(\pm 1, \pm 1)$ and $(0, 0)$ suitably centred and rescaled, have proved very useful with the 'central-composite' designs of Section 5.3 for higher dimensions. On the basis of such designs, a simple experiment is run within the operable range of the process *as it is currently running*. Of course, it is assumed that the variables to be controlled can be 'set' within a short distance of the current settings *without* disturbing full production. When the data at all design points have been gathered, one 'cycle' is said to have been completed. Subsequent cycles, if needed, are completed until a significant result is detected, for example, until a significant increase in yield has been realized, as a result of the effect of a particular controllable factor or interaction. At this point one 'phase' is said to have been completed and the decision is usually taken to change the current operating conditions to whatever conditions have been determined to be the optimal. This procedure can go on continually during production, with the objective being to always move in the direction of an optimum response.

The method arose out of a need to study the behaviour of a production process—a somewhat different emphasis to that in this book. However, although it is designed to be used for process control, we suggest that it might also be used for tolerance design (see Chapter 7) and other preproduction methods and we therefore give a brief example here.

Example 5.5: In a certain forging process two temperatures are identified as having a critical effect on the percentage of parts with no faults (faults include blistering, crushing, porosity, flakiness, and so on). The two temperatures are metal temperature (x_1) and base temperature (x_2). The first-phase experiments gave the results below.

Phase 1	x_1 °C	x_2 °C	Y%
	740	250	56
	750	270	56
	740	270	71
	750	250	70
	745	260	68

The fitted model from Phase 1 was

$$Y = 68.2 + 0.950(x_1 - 745) + 0.525(x_2 - 260).$$

The centre for the second phase was taken at (765, 275) and the results were

Phase 2	x_1 °C	x_2 °C	Y%
	760	270	89
	770	280	97
	760	280	95
	770	270	96
	765	275	94

5.5.2 Stochastic approximation

For target attainment $Y = c$ a range of mathematical methods have appeared in the literature and we shall mention only one here, the Robbins–Munro sequential method of stochastic approximation. The foundation for this is Newton's method for finding the solution of an equation

$$Y = f(x) = 0$$

where $f(x)$ is any suitably smooth function. In the deterministic case

Newton's method for solving $f(x) = 0$ uses the iteration

$$x_{n+1} = x_n - \frac{f(x_n)}{f'(x_n)} \tag{5.1}$$

$\{f'(x) = \mathrm{d}Y/\mathrm{d}x\}$. If

$$Y_n = f(x_n) + \varepsilon_n$$

so that we observe Y with error at each stage, (5.1) will not work unless we dampen it. Under certain conditions the following procedure always converges (in probability):

$$x_{n+1} = x_n + \alpha_n Y_n$$

where $\Sigma\alpha_n = \infty$ and $\Sigma\alpha_n^2 < \infty$. The most obvious choice of an α_n procedure is $\alpha_n = \text{constant} \times 1/n$ since $\Sigma \frac{1}{n} \to \infty$, $\Sigma\, 1/n^2 < \infty$.

The method is useful when very little is known about $f(x)$ and many observations are allowed, because convergence is slow.

Example 5.6: The following is a hypothetical example. The function is

$$Y = \tfrac{1}{150}(3 - x)(5 + x)(10 - x).$$

The computer was set to find the root $x = 3$ starting at $x_1 = 0.5$ using the error distribution $N(0, 0.7^2)$ for the ε_n and $\alpha_n = 2/n$. The first 20 pairs $(x, f(x_n))$ are given showing slow but definite convergence of x_n but erratic behaviour of $f(x_n)$.

x_n	$f(x_n)$
0.5	0.686
1.872	0.613
2.486	−0.285
2.296	0.904
2.747	0.346
2.886	−0.551
2.702	0.223
2.766	−0.612
2.613	−0.535
2.494	0.312
2.518	0.872
2.663	0.130
2.683	1.145
2.846	0.399
2.900	0.462
2.958	0.700
3.040	1.014
3.153	−0.252
3.126	0.387
3.095	0.671

5.6 Spatial experiments

The experimental designs described earlier in this chapter relate to situations in which we are willing to assume some kind of model for the process or product being tested. This is not always the case, even if we take a rather large model to incorporate any reasonable model as a submodel. Taking larger and larger models can in any case lead to larger sample sizes and even numerical stability problems. Another disadvantage with factorials and the optimum designs discussed above is that they tend to place points at the extremes of the design region R. This contrasts with the simulation techniques in Chapter 7, where the random location of x or the pseudo-random methods of Section 7.5 tend to spread observations throughout the region.

There are methods which spread observations throughout the region and use models which are more adaptable than factorial polynomial models. One of these is the method of spatial experimental design which we summarize briefly here.

The basic model for spatial designs uses an additional term $\delta(x)$. Thus

$$Y(\mathbf{x}) = f(\mathbf{x}, \boldsymbol{\theta}) + \delta(\mathbf{x}) + \varepsilon,$$

where $\delta(\mathbf{x})$ is the autocorrelation process. This means that for two locations, $\mathbf{x}_1$ and $\mathbf{x}_2$,

$$\text{cov}[\delta(\mathbf{x}_1), \delta(\mathbf{x}_2)] \neq 0.$$

For example, $\delta(\mathbf{x})$ may be a so-called isotropic process in which the above covariance depends only on a distance between $\mathbf{x}_1$ and $\mathbf{x}_2$. Prediction for these models, namely obtaining $\hat{Y}$, requires a somewhat more sophisticated analysis. A typical $\hat{Y}$ would be the so-called best linear unbiased predictor, familiar from the theory of time series forecasting:

$$\hat{Y} \text{ achieves } E(\hat{Y}) = E(Y),$$

and minimizes

$$\text{MSE}(\hat{Y}) = E(\hat{Y} - Y)^2.$$

A typical criterion would be to choose the locations, that is, a set of $\mathbf{x}$

$$\mathbf{x}_1, \ldots, \mathbf{x}_n$$

so that using

$$Y(\mathbf{x}_1), \ldots, Y(\mathbf{x}_n),$$

$$J = \int_R \text{MSE}(\hat{Y}) \, d\mathbf{x}$$

is a minimum. Designs based on such a theory tend to spread out throughout the region R rather than sticking to the extreme points. Such methods have been used in the computer experiments mentioned in Chapter 7 as an alternative to Monte Carlo methods.

5.7 Bibliography

The pioneering work on response surface design and methodology was carried out by G. E. P. Box and co-workers, notably N. R. Draper, W. G. Hunter, and J. S. Hunter. The book of Box, Hunter, and Hunter (1978) (referred to in Chapter 3) contains much material. A very comprehensive account appears in Box and Draper (1987). This latter book contains much of the less mathematical work from the early papers and has a full list of references. Evolutionary operation is described in the original text of Box and Draper (1969). The journal *Technometrics* contains a wealth of papers and examples going back many years on response surfaces and several excellent survey articles. An article describing the state of the art after the initial surge of work in the area is Mead and Pike (1975).

The theory of optimum experimental design is dominated by the work of J. Kiefer, now all available in his collected papers (1984). This volume contains a short survey by the second author of the present book and comments by other authors. There are similar review articles by Atkinson (1982) and Cook and Nachtsheim (1980). The generation of optimum designs using algorithms was a development from the original work of Kiefer and Wolfowitz and is due to Federov, Wynn, Pazman, and others. An early text is Federov (1972) and the book by Silvey (1980) contains a concise account. A recent theoretical account of algorithms is the book of Pazman (1980). Much of the early computer development was carried out by Mitchell (1974) whose DETMAX algorithm is now a standard approach. A theoretical account is contained in Wu and Wynn (1978).

The original paper of Robins and Munro (1951) contains the algorithm on stochastic approximation. A more recent book is Kushner and Clark (1978).

The use of spatial experimental design methods in computer experiments is reviewed at length in Mitchell *et al*. (1989)

6

OFF-LINE QUALITY CONTROL PRINCIPLES

In this chapter we concentrate on the latest phase in the evolution of quality systems, 'quality by design', where off-line quality control techniques belong. Taguchi's contributions to this phase are outlined, and an evaluation and critique of his sometimes unorthodox methods is made. Ways to statistically 'improve' a Taguchi analysis are suggested, the most important of which is a technique for choosing a data transformation which can ensure the satisfaction of statistical assumptions and the selection of the appropriate performance measures.

Some ways of modelling variability are also outlined. We conclude the chapter by demonstrating how proper use of some of Taguchi's proposed techniques, in conjunction with appropriate statistical methodology, can lead to successful characterization and optimization of complex multi-response processes with minimum cost.

6.1 A general view

Quality control methods such as control charts, reliability studies, cause and effect diagrams, process capability studies and Statistical Process Control (SPC) are known as 'on-line quality control methods', since they concentrate on the manufacturing (on-line) stage in order to reduce manufacturing imperfections in the product and keep the manufacturing process in control. Among these SPC is a powerful cost saving and quality enhancing approach to reducing variability within the production phase. However, SPC cannot compensate for poor quality in design. If there is large variability due to uncontrollable factors during manufacture, prohibitively costly process control schemes may be required to improve the process capability, and they cannot guarantee a product robust to deterioration and variability due to uncontrollable environmental factors. This means that additional expense might be incurred due to service costs under warranty and, most importantly, due to the loss of market share because of customer dissatisfaction. However, if the quality concept is moved further upstream to the design process and product development stage, these costly eventualities can be avoided. The need for costly process control, mass inspection, and service costs is minimized if one optimizes product and process design to ensure product robustness. Building in quality at the design stage represents the latest phase in the evolution of quality systems which are understood to have traversed four, generally overlapping, phases (see also Fuchs 1986):

1. **Inspection**, which the manufacturer had to rely on as the last step before shipping the product. Inspection still has a valid and useful place in any quality control system for separating the 'bad' from the 'good', but its role is increasingly viewed as a tool for 'confirmation' of good quality rather than for 'rectification' of a bad process.
2. **Quality control**, where measurements and inspection were included in the manufacturing cycle and where statistical methods were applied for the first time to determine whether a process is in control, or if an entire batch can be considered good on the basis of the findings of the analysis of only a small sample. The techniques involved are mainly due to W. A. Shewhart whose 1925 paper in the *Journal of the American Statistical Association* introduced the 'control chart' to the world. (See also Shewhart 1931.)
3. **Quality improvement**, where process control and inspection data are statistically analysed to determine the source of the problems in a constant effort towards manufacturing process improvement. (See also Deming 1982).
4. **Quality by design**, which embodies all the above and more by designing quality into the product and the process prior to the manufacturing stage. It incorporates the idea that a product can be made robust to the variations in the user's environment, and the process which produces the product can be made robust to the normal variations in materials, components, and manufacturing before normal production begins. The responsibility for quality fundamentally rests with the 'design team' and not with the manufacturing operatives.

It is this idea of 'robust design' at minimum cost through statistical methods on which Taguchi concentrates.

6.2 Taguchi's approach to 'quality by design'

Taguchi's fundamental concept rests on the importance of *economically* achieving high quality, low variability and consistency of functional performance. The main objective of his methodology is 'quality-costs minimization', where quality costs mean anything from experimental, reject, inspection, rework and service-under-warranty costs, to losing market share due to inferior products. Taguchi's methodology is heavily based on statistical methods, especially statistically designed experiments, and concentrate on minimizing the deviation from the target caused by uncontrollable factors, which he calls 'noise factors'.

Noise could be external or internal. External sources of noise, Taguchi's 'outer noise', are the variations in environmental conditions such as temperature, humidity, vibrations, dust or human variations in operating a product. Internal sources of noise, Taguchi's inner noise, are the deviations of the actual characteristics of a product from the corresponding nominal settings; these could result from manufacturing process imperfections, causing variability between products, or from deterioration, over time, of internal elements and components. There is a special type of uncontrollable factor which Taguchi called an 'indicative factor'. This is a factor whose levels are determined by the conditions of use, so it is not possible for the researcher to select them. Its interaction with a controllable factor can be considered a controllable source, and so its main role could be in the adjustment of the levels of the controllable factors.

In order to minimize the effects of noise sources, Taguchi suggests that certain 'counter measures by design' have to be taken, the main forms of which are the following.

1. System design (or 'primary' design).
2. Parameter design (or 'secondary' design).
3. Allowance or Tolerance design (or 'tertiary' design).

6.2.1 System design

This is where one applies special scientific and engineering knowledge to produce a basic functional prototype design, having surveyed the relevant technology, researched the customers' needs and understood the manufacturing environment. System design is a specialist's territory, relying heavily on knowledge from technology fields, past experience and intuition.

6.2.2 Parameter design

This is what Taguchi considers to be the most important 'counter-measure', because it can provide the means of both reducing costs and improving quality. By making effective use of experimental design and statistical techniques, one can identify the settings of easy to control product (or process) parameters that reduce the sensitivity of engineering designs to the sources of noise. If, at this stage, the product can be designed so that its output characteristics are resistant to all sources of noise, then it will function satisfactorily despite variability in its component parts and environmental conditions, and its cost will be low. Reducing the effect of the noise sources rather than controlling them, is a very cost-effective way of improving engineering designs and developing stable and reliable products or manufacturing processes.

Taguchi suggests that during the parameter design stage, only easy to change parameters and low cost levels of these parameters are studied. He considers it as foolish to experiment by varying expensive parameters, when robustness can be achieved by using low-cost ones.

6.2.2.1 Separation of factors

The effect of both internal and external noise factors can change with the settings of these easy to change parameters which we shall call 'controllable factors' (also known as 'design factors'). Through parameter design we can identify those settings of the controllable factors that reduce performance variation, caused by the noise factors, while keeping the response of interest on target.

The controllable factors are separated into two main groups:

(1) those which affect the variability in a response, called 'variability-control' factors (also known simply as 'control' factors); and
(2) those which affect only the (mean) level of a response, called 'target-control' factors (also known as 'signal' factors).

Of course, one could consider a third group of factors being those controllable factors which *do not* affect either the mean response or the variability in the response. Information about those 'neutral' factors is still important, since cost savings can result by setting them at their most economic levels.

It is the concentration on variability which distinguishes Taguchi's approach from earlier procedures. The aim is to reduce variability by changing the variability-control factors, while maintaining the required average performance through appropriate adjustments to the target-control factors. The following example given by Taguchi (1986) is a simple illustration of the separation of variables. Consider an electrical power circuit where the characteristic of interest is the output voltage with a target value of Y_0. Assume that the voltage is largely determined by the gain of a transistor in the circuit whose nominal value (x) can be controlled. Suppose that the effect of the transistor gain on the output voltage is non-linear, as shown in Fig. 6.1. A transistor with a gain of x_0 would produce the required output voltage of Y_0. The effect of a variation about the nominal value x_0 on the resulting variation about Y_0 is

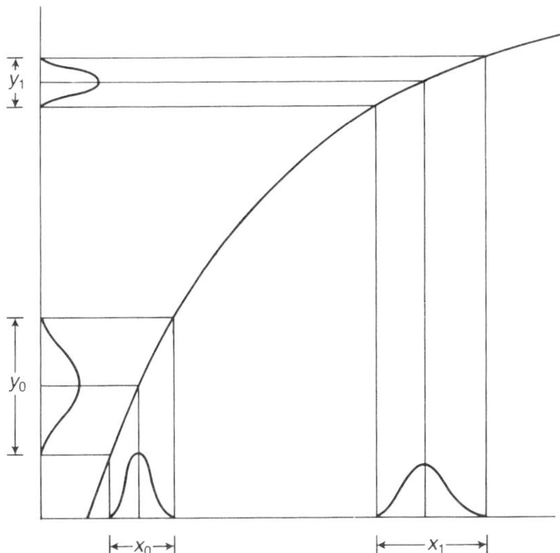

FIG. 6.1. Effect of transistor gain on output voltage.

indicated by bands straddling the nominal values. However, if the circuit designer chooses a nominal gain of x_1, then, owing to the non-linearity of response, it can be seen that the variation about the corresponding voltage, Y_1, is much reduced. Now suppose that there is a resistor in the circuit which has a linear effect on the voltage at all levels of transistor gain. Then the resistance of this component can be chosen so that the difference between the voltage Y_1 and the desired voltage Y_0 is eliminated. The response is then on target and the variability in response is minimized. Thus, transistor gain is a variability-control factor and resistance is a target-control factor. In more complicated cases the identification of such factors requires the use of experimental design methodology and response surface techniques as discussed in Chapters 3 and 5 respectively.

The essence of Taguchi's parameter design is to exploit the non-linear effects of the product (or process) controllable parameters in order to reduce the product's (or process) performance variation. Of course, the question arises as to what extent does controlled experimentation allow determination of the variation behaviour in a wider environment. Taguchi's answer is surprisingly simple. Try to *produce* variability in the 'laboratory' by mimicking the variability of the production process or the user's environment. In other words, create a *simulated* environment to study the variability of a product design. This requires an appropriate 'Parameter design experimental set-up'.

6.2.2.2 Experimental set-up

A parameter design experiment should consist of two parts which can be represented by a Controllable Factor Array (CFA) (Taguchi's 'Inner array') and a Noise Factor Array (NFA) (Taguchi's 'outer array'). The columns of a CFA will represent the controllable factors and each row (also called a 'trial run') a specific product design, i.e. a combination of settings of the controllable factors. The columns of a NFA will represent noise factors with rows representing different combinations of noise factor levels. A combination of CFA and NFA as indicated in Fig. 6.2 will constitute a complete parameter design experiment. For constructing CFAs and NFAs, Taguchi recommends the use of 'orthogonal arrays' (see Chapter 3).

The NFA is a selective rather than a random subset of the noise space. The test levels of the noise factors should be appropriately chosen so that the noise space is adequately covered. If the distribution of a noise factor N_i is known, Taguchi recommends the selection of the following levels for this factor: $(m_i - s_i)$ and $(m_i + s_i)$ if N_i is assumed to have a linear effect; $(m_i - s_i\sqrt{3/2})$, m_i, $(m_i + s_i\sqrt{3/2})$ if N_i is assumed to have a curvilinear effect, where m_i and s_i are the estimates respectively of the mean and standard deviation of the distribution of N_i.

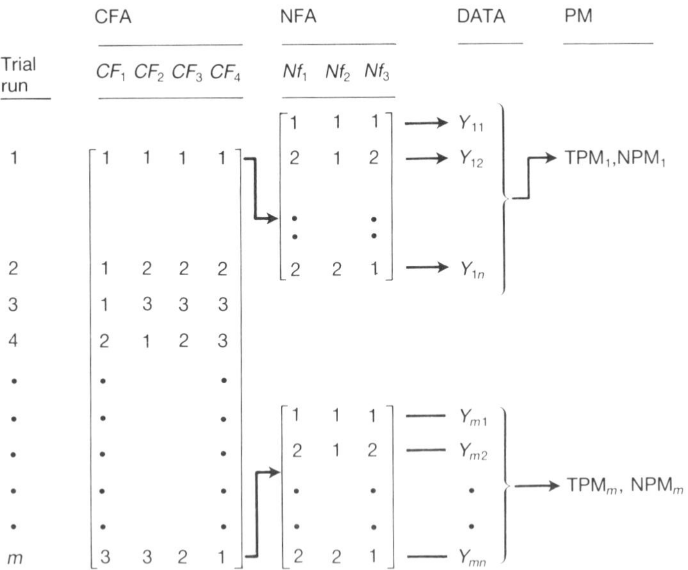

FIG. 6.2. Experimental set-up.

These choices of the test levels are apparently based on the assumption that noise factors have approximately symmetrical distributions. If the distribution of the noise factors is not known, or worse even, if the noise factors cannot be 'simulated' for experimentation purposes, Taguchi recommends random selection of repeat observations within each trial run of the CFA.

A parameter design experiment can be conducted either through physical experiments or through computer simulation trials when the response can be numerically evaluated on the basis of a known response model relating product performance characteristics with controllable and noise factors.

When physical experiments are used, it is sometimes impossible or very expensive to carry out an experiment according to the set up of Fig. 6.2. This is where use of screening or highly fractional experimental designs studying the factors at two or at most three levels is what Taguchi strongly recommends. When numerical evaluation of a response function is possible, assuming some computing capability, more informed analyses can be performed in the presence of both internal and external noise factors.

6.2.2.3 Performance measures

Let us return to the experimental set-up of Fig. 6.2.

For each of the m rows of the CFA, the n rows of the NFA will provide at least n observations on the response of interest. These observations will then be used to compute certain statistical measures called 'Performance Measures' (PM) whose analysis will provide the means to estimate and then to minimize the effect of the noise factors on the product's performance. Indeed, an analysis of the Noise Performance Measure (NPM) which is a measure of the process variability, will identify the controllable factors that affect the variability in the response—the 'variability-control factors'—and also the optimal combined setting of those factors to minimize this variability. Also, an analysis of the Target Performance Measure (TPM), which is a measure of the process mean, will reveal which of the controllable factors, which *are not* variability control factors, have a large effect on the mean response. Such factors, the 'target-control factors', can be used to bring the mean response on to the target value, while simultaneously keeping the variability at its minimum.

Many measures for the NPM have been suggested. When there is a target value to be achieved for the response, Taguchi recommends the use of the Signal-to-Noise Ratio (SNR), which estimates μ/σ (the inverse of the coefficient of variation) with μ being the process mean and σ the process standard deviation. For practical purposes we compute

$$\text{SNR} = 10 \log_{10}(\bar{Y}^2/s^2),$$

where

$$\bar{Y} = \sum_{j=1}^{n} Y_j/n, \qquad s^2 = \sum_{j=1}^{n} (Y_j - \bar{Y})^2/(n-1),$$

and $Y_1, \ldots, Y_n$ approximate a random sample of sample size n from the distribution of the response Y for a specific trial run. When 'the smaller the better' is a characteristic of the response (as, for example, when leakages, power losses, defective items, etc., have a target value of zero), Taguchi recommends the use of

$$\eta = -10 \log_{10}\left(\frac{1}{n}\sum_{j=1}^{n} Y_j^{(2)}\right),$$

whereas when the response is 'the larger the better', as a NPM Taguchi recommends

$$\theta = -10 \log_{10}\left(\frac{1}{n}\sum_{j=1}^{n} Y_j^{(-2)}\right).$$

(See Section 6.5.2 for a justification for using the logarithmic scale. Note that the minus sign in η and θ is used by convention so that the NPM is always maximized.)

When the performance characteristic is measured on a binary scale (good or bad, success or failure), Taguchi suggests

$$Z = 10 \log_{10}\{p/(1-p)\},$$

where p is the proportion of 'good'.

For the analysis of the NPM and TPMs, the usual ANOVA techniques (see Chapter 2) can be used.

6.2.2.4 The Taguchi procedure

There are certain steps that Taguchi suggests should be taken in carrying out experimental studies. These steps which are outlined below should be an integral part of the parameter design stage.

1. Define the problem. Provide a clear statement of the problem to be solved. It is important to establish just what the experiment is intended to achieve.
2. Determine the objective. Identify the output characteristics (responses) to be studied and eventually optimized (preferably measurable and with good additivity), and determine the method of measurement.

3. Set up a 'brainstorming' session. This is a very important stage for performing an experimental study. Statisticians will have to get together with the engineers who are closely related to the production process or the product under consideration, in order to determine the controllable and uncontrollable (noise) factors, to define the experimental range and the appropriate factor levels. Taguchi believes that it is generally preferable to consider as many factors (rather than many interactions) as economically feasible for the initial screening.
4. Design the experiment. Select appropriate designs (see Chapter 3) and construct the CFA and NFA, assigning the controllable factors and their interactions to the columns of the CFA, and assign the noise factors to the NFA columns.
5. Set up the experiment and collect the data.
6. Analyse the data. Evaluate the performance measures (TPM and NPM) for each trial run of the CFA, and analyse them using the usual Statistical techniques of Chapter 2 or the non-standard techniques (favoured by Taguchi) of Chapter 4.
7. Interpret the results. Identify the Variability-Control Factors (VCF) and Target-Control Factors (TCF) and select their optimal levels; the optimal levels being, for the VCFs the ones which maximize the NPM (minimize variability in the response), and for the TCFs the ones which bring the mean response nearest to the target value. Predict the process performance under the optimal conditions using the 'estimation techniques' described in Section 4.4.
8. Always, always, always run a confirmatory experiment to verify predicted results. It is necessary to confirm by some follow-up experiments that the new parameter settings improve the performance measures over their value at the initial settings. A successful confirmation experiment will alleviate concerns about the possibilities of a wrong choice for the effects to be studied and the experimental design, wrong assumptions of no-interactions, or improper assumptions underlying the response model.

If the predicted results (from step 7) are not confirmed, or the results are otherwise unsatisfactory, additional experiments may be required and a reiteration of steps 3 to 8 might be necessary.

6.2.3 Allowance or tolerance design

If the influences of inner or outer sources of noise cannot be successfully reduced by use of parameter design, Taguchi suggests the 'allowance' design where the same steps as before can be followed but where additional factors are considered that were previously excluded because of cost or the difficulty of the necessary experimentation.

If that also fails, the tolerances of the product's components are considered and tolerance (re-)design is advocated. This would mean retaining the optimum nominal levels for the factors (as identified by parameter design), but reducing the tolerances of certain crucial factors (components) in an optimal and cost-effective way so that the overall variability in the response is reduced to acceptable levels. A trade-off can be involved by relaxing the tolerances of certain non-crucial components. This approach is a 'reassignment of tolerances through scientific methods' rather than by convention, which has been the common practice in industry up to now.

In order to redesign the tolerances, Taguchi's approach is to make use of the 'contribution ratios' (see Section 4.3). An approach using response surface techniques (see Chapter 5) can also be followed. We will revert to both approaches for tolerance design in Chapter 7.

6.3 Evaluation and critique

Taguchi has introduced some new quality engineering ideas that should have a profound effect on all aspects of manufacture from design through production to product testing and acceptance. The advantages that these ideas may bring make it important that they are fully understood and further developed. The essentials of his ideas may be summarized as follows.

1. Formulation of a measure of the variability of a product or process (NPM). Use of ANOVA techniques applied to this NPM to analyse the nature of performance variation and to split it into identifiable components. This extends conventional ANOVA, where only the *mean* response is investigated.
2. Use of statistical methods to produce a product design and production process that are chosen to minimize performance variation (maximize quality). This includes both 'parameter design' (selecting optimal values for the significant design factors) and 'tolerance design', ensuring that performance is insensitive to changes in 'internal', 'production' or 'external' noise factors.
3. The importance of 'closeness to target' as a performance criterion, rather than 'within specification'.
4. The realization that substantial cost savings can result from the ability to produce designs whose performance is tolerant to variations

in component parameters or in manufacturing and operating environments.

5. Switching the main emphasis of quality control from on-line (during production) to off-line (before production) with the aim of replacing detailed continuous monitoring by a few well-designed preparatory experiments.

The technique is a straightforward well-integrated system for implementing statistical experimental designs: it helps to encourage more experimentation and closer association between statisticians and engineers, and it increases statistical awareness in industry.

But just as there is a variety of results in experiments there is also a variety in the interpretation of the method when it comes to application. In many of the published applications, the following points became apparent.

(i) An unnecessarily limited list of experimental designs is offered: for example, the $OA_{18}(2 \times 3^7)$ orthogonal array (one of the designs to be found in Appendix D) seems to be one consistently used in many of the published applications; the factors of interest are even manipulated to fit this design rather than choosing the right design to fit the important factors at their necessary levels. However, a great variety of designs does exist (see Appendices C and D). Combinations of these designs can be constructed to permit factors with differing numbers of levels to appear together. Non-orthogonal arrays can also be used (see Chapters 3 and 5).

(ii) Interactions are not adequately dealt with. In fact, interactions among design and noise factors are never considered. Admittedly, Taguchi plays down the importance of interactions—but he recognizes their presence and he recommends methods for reducing them (see Section 4.2). Otherwise, important factors might be wrongly disregarded. We should not forget that a significant interaction effect means that all the factors involved (in the interaction) are significant, despite the fact that their main effects in the ANOVA layout might not show a significant F-ratio. The only safeguard against possible interactions seems to be a small follow-up confirmatory experiment. But such an experiment can only confirm the optimal new settings of the factors considered and it cannot 'rediscover' important factors that were disregarded in the first place.

(iii) The choice of the performance measures is unconvincing. From all the NPMs that Taguchi suggests, which one should be used? In published applications the SNR, η or θ (see Section 6.2.2.3) are used with no satisfactory justification. But, the signal-to-noise ratio SNR could produce a mean bias if the standard deviation and the mean are not linearly connected (see Box 1986, Logothetis 1988); also the usefulness of measures η and θ is extremely doubtful for any type of data (see Box and Ramirez 1986, Logothetis 1987*a*, Box 1988). A performance measure invariant to any type of data or situation could be of great value.

(iv) The importance of data transformation does not seem to be appreciated or exploited enough. Through a simple transformation one could, in many instances, overcome the problems depicted in (i), (ii), and (iii) above (see Section 6.4).

(v) 'Noise factors' are not fully investigated prior to the application of the Taguchi method, although they are the basis and the most important aspect of the technique. Variability could be either systematic or random. Sources of systematic variability cannot really be considered as the 'noise factors' that Taguchi speaks of; they are more like the 'indicative factors' (factors 'whose levels are not selectable but we need to know the SNRs for each level . . .') that Taguchi refers to in Chapter 9 of Taguchi and Wu (1980). (See also Section 6.2.) Systematic variability has to be investigated to discover whether it is caused by a factor not already considered or by an interaction with a design factor. If, for example, the rate of oxide growth on a silicon wafer is significantly faster at the bottom of the wafer than at any other wafer site, this would be indicative of a possible inhomogeneity in furnace conditions, requiring further investigation.

Whenever possible, steps should be taken to isolate and/or eliminate any systematic variations before applying the Taguchi technique. Results could be strikingly different between cases of including and excluding systematic variation. A published demonstration of this, can be found in Pignatiello and Ramberg (1985), and in Logothetis (1987*a*,*b*).

One should also be aware of losing information contained in the data about relationships among the noise variables when one overall NPM is used for all noise factors. For example, at some parameter settings a noise variable such as temperature could have a strong effect, the other noise factors being negligible; while at other parameter settings, a different noise factor, say humidity, could dominate. An overall NPM would not reveal this information. A separate analysis for each noise factor could be worthwhile and more informative.

Taguchi has promoted the use of statistical design of experiments for product design and process improvement. But he has only given a starting point and further research is needed involving amendments, extensions and refinements. A doctrinaire recitation and blind application of the Taguchi methodology cannot lead anywhere. Two versions of the Taguchi approach could lead to two completely different conclusions as the examples of Section 6.4 will show.

Some amendments are perhaps required on Taguchi's non-standard methods of analysis of the available data. For example, an obvious disadvantage of Taguchi's accumulating analysis (see Section 4.7) is the fact that the frequencies of the cumulative categories are not independent. Also, Taguchi's 'life-data' in his 'minute' analysis (see Section 4.6.4) are not statistically independent. A statistical analysis, therefore, under such circumstances can be seriously invalidated. Nair (1986), Hamada and Wu (1986), and Box, Bisgaard, and Fung (1988) provide some noteworthy constructive criticisms and offer some alternatives.

As with any innovation, there have been and will always be controversial issues regarding Taguchi's approach. Controversy within the statistical community has concentrated upon the strict statistical validity of Taguchi's alternative, non-standard, techniques. Nevertheless, this should not be the real issue in considering Taguchi's contributions. Rather we should be concentrating upon the unique opportunity he provides to achieve optimization of products and processes, prior to production, by the use of simple statistical methodologies. He has created a long-awaited and much welcomed momentum for the use of statistical experimental design in industry, and made statistics accessible to the non-expert user. This is perhaps his major contribution to the industrial world. We simply need to modify some of Taguchi's methodologies in terms consistent with basic statistical principles in an attempt to answer some 'open questions' and bridge some gaps that still exist between the Western approaches to analysis and those of Taguchi.

In the next section we will attempt to answer one such an open question concerning the type of the performance measures one should analyse for the identification of the variability-control and target-control factors.

6.4 The role of data transformation

It has been recognised for some time that data-transformation methods have an important role in statistical analysis. They are often regarded as a prerequisite for an efficient application of techniques such as analysis of variance (ANOVA) and multiple regression analysis, to ensure that crucial statistical assumptions, such as normality of error distributions, constancy of error variance, and independence of observations (see Chapter 2) are satisfied.

Since the ANOVA and regression techniques are an integral part of the 'Taguchi analysis' for off-line quality control, it seems important to investigate whether data-transformation can contribute towards a more statistically reliable application of the Taguchi methodology.

In order to apply the Taguchi methodology we need to define proper Performance Measures (PM), i.e. measures reflecting the mean response and the variability in response. The analysis of these measures will show which of the design factors affect only the mean response—the target-control factors, and which affect the variability in the response—the variability-control factors (see Section 6.2). Ideally the performance measures should be chosen so that the classification produces as little overlap as possible between variability-control and target-control factors. Often it is possible to find a transformation that can achieve this aim. Such transformations also tend to eliminate excessive skewness caused by dependence between mean and variance in the original scale, thus making the assumption of normality more tenable. Also, in practice, it is often convenient to assume that there are no interactions between the factors that affect performance. This is especially important when dealing with many factors but a small experimental capacity. A Taguchi analysis is likely to be more effective and simpler when carried out in a transformed scale where factorial effects are linear or additive, and justify assumptions of 'no interactions'.

One of the best techniques for choosing a transformation which could simultaneously achieve

(A1) normality of distributions,

(A2) constancy of error variance, i.e. independence between cell mean and cell variance,

(A3) simplicity (linearity) of the model structure,

is due to Box and Cox (1964). However, although a transformation chosen to achieve independence between cell mean and cell variance (A2) often has the effect of improving the closeness to normality, and also often improves the scale as far as (A3) is concerned, it is unreasonable to expect that, for all types of data, there will be a transformation so that (A1), (A2), (A3), are simultaneously satisfied. In many instances, the final choice of transformation will depend on which of the assumptions (A1), (A2), or (A3) is considered most important. In a Taguchi analysis, if the ideal situation (where (A1), (A2), and (A3) are all satisfied) cannot be achieved, the primary purpose of the chosen transformation should be taken to be the satisfaction of (A2), provided it is possible to recognize any incorrect assumptions of 'model linearity'.

With this in mind, we will assess the effectiveness of the Box and Cox transformations for the Taguchi methodology (Section 6.4.1) and suggest a simple method of safeguarding against violation of the (A2) assumption (Section 6.4.2). The method is assessed by considering both published and recent Taguchi applications (Sections 6.4.3, 6.4.4).

Finally, we summarize what data-transformations can offer in the search for more statistically valid conclusions.

6.4.1 *The Box–Cox transformations*

Box and Cox (1964), in an important paper, considered two classes of transformations: A single parameter family indexed by λ and defined by

$$Y^{(\lambda)} = \begin{cases} \dfrac{Y^{\lambda} - 1}{\lambda} & (\lambda \neq 0) \\ \log(Y) & (\lambda = 0) \end{cases} \tag{6.1}$$

which holds for $Y > 0$; and a two-parameter family indexed by λ_1 and λ_2:

$$Y^{(\lambda)} = \begin{cases} \dfrac{(Y + \lambda_2)^{\lambda_1} - 1}{\lambda_1} & (\lambda_1 \neq 0) \\ \log(Y + \lambda_2) & (\lambda_1 = 0) \end{cases} \tag{6.2}$$

which hold for $Y > -\lambda_2$. Estimation of the parameter λ was discussed from a sampling theory and Bayesian point of view. The fundamental assumption made was that for some λ, the transformed observations defined by (6.1) [or by (6.2) in the shifted location case] can be treated as independently normally distributed with constant variance σ_{λ}^2 and with expectations defined by a model of simple structure (linear). The cases for different λ's were made comparable by working with the normalized transformation

$$z^{(\lambda)} = Y^{(\lambda)}/J^{1/n}$$

where $J = J(\lambda; Y)$ is the Jacobian of the transformation defined by

$$J(\lambda; Y) = \prod_{i=1}^{n} \left| \frac{\mathrm{d}Y_i^{(\lambda)}}{\mathrm{d}Y_i} \right| .$$

The resulting normalized values were then expressed as [corresponding to (6.1) and (6.2) respectively]:

$$z^{(\lambda)} = \begin{cases} \dfrac{Y^{\lambda} - 1}{\lambda(\dot{Y})^{\lambda - 1}} & \lambda \neq 0 \\ \dot{Y}\log(y) & \lambda = 0, \end{cases} \tag{6.1'}$$

where $\dot{Y}$ is the geometric mean of the observations, or in the shifted location case:

$$z^{(\lambda)} = \begin{cases} \dfrac{(Y + \lambda_2)^{\lambda_1} - 1}{\lambda_1\{\mathrm{gm}(Y + \lambda_2)\}^{(\lambda_1 - 1)}} & \lambda_1 \neq 0 \\ \mathrm{gm}(Y + \lambda_2)\log(Y + \lambda_2) & \lambda_1 = 0, \end{cases} \tag{6.2'}$$

where $\mathrm{gm}(Y + \lambda_2)$ is the sample geometric mean of the $(Y + \lambda_2)$'s.

Invoking standard least-squares theory, the maximized log likelihood function with respect to λ, can be found to be proportional to $\{S(\lambda; z)\}^{-n}$, where $S(\lambda; z)$ is the residual sum of squares of $z^{(\lambda)}$, assuming it can be retrieved; in particular except for a constant

$$L_{\max}(\lambda) = -\frac{1}{2} n \log\left\{\frac{S(\lambda; z)}{n}\right\}$$

and so the maximum likelihood estimate for λ is obtained by minimizing $S(\lambda; z)$ with respect to λ, assuming that $S(\lambda; z)$ exists. In practice, in the majority of the cases, values for λ of $\frac{1}{2}$ (square-root transformation), 0 (log-transformation), -1 (inverse), 2 (square), or 1 (no transformation) are the common ones to be used although any real value for λ is possible.

If α is a level of significance, a $100 \times (1 - \alpha)\%$ confidence interval can be found for λ by calculating a critical sum of squares SS_c from

$$SS_c = S_{\min}(\lambda; z)\left\{1 + \frac{t_\nu^2(\alpha/2)}{\nu}\right\}$$

where $S_{\min}(\lambda; z)$ is the minimum residual sum of squares with respect to λ, with the associated ν degrees of freedom and $t_\nu = t(\nu; \alpha)$ is the corresponding value from the t-tables (see also Box, Hunter, and Hunter 1978).

The contribution of each of the requirements (A1), (A2), (A3) to the estimation of λ can be expressed in terms of maximized likelihoods as (see Box and Cox 1964, Section 5):

$$L_{\max}(\lambda/A, H, N) = L_{\max}(\lambda/N) + \log L_1(\lambda; z) - \frac{1}{2} n \log\left\{1 + \frac{\nu_1}{\nu_2} F(\lambda, z)\right\}, \tag{6.3}$$

where A, H, and N denote respectively the constraints to the simpler linear model (without interactions or high degree terms), to a homoscedastic (constant variance) model and to a model with normal distributions,

- $L_{\max}(\lambda/N)$ is the criterion for testing for normality,
- $L_1(\lambda; z)$ is the criterion for testing constancy of variance given normality, and
- $F(\lambda; z)$ is the criterion for testing simplicity (absence of interactions or high-order terms) given normality and constancy of variance, defined by

$$F(\lambda; z) = \frac{S_1(\lambda; z)/\nu_1}{S_2(\lambda; z)/\nu_2},$$

where $S_2(\lambda; z)$ is the residual sum of squares assuming a complex model (with ν_2 degrees of freedom) and $S_1(\lambda; z)$ is the extra sum of squares corresponding to the interaction and high-order terms (with ν_1 degrees of freedom) (clearly the residual sum of squares corresponding to the simplest linear model can be calculated as

$$S_1(\lambda; z) + S_2(\lambda; z) \text{ with } \nu_1 + \nu_2 \text{ degrees of freedom.}$$

Thus $F(\lambda; z)$ is a descriptive measure of non-additivity and, as Box and Cox (1964) comment, can be considered as such, independently of H and N.

From (6.3), it is clear that a large degree of non-linearity will greatly contribute to the choice of λ. We then run the risk that over-simplification, i.e. a 'model linearity oriented' transformation, could induce a mean bias in the error variance, i.e. violation of assumption (A2); and if, as in the Taguchi case, we are more interested in the satisfaction of (A2), we would wish to safeguard against this happening. In the next section a simple safeguard will be described. We will then demonstrate that a properly chosen data-transformation can indeed lead to a simplified, clear-cut, and efficient Taguchi analysis by achieving the following requirements:

(R1) formulation of a proper measure for NPM, i.e. one that reflects the variability (not biased by the mean) in the product or process;

(R2) satisfaction of statistical assumptions such as independence between cell mean and cell variance, normality of error distributions, etc.;

(R3) simplicity (linearity or additivity) of the response model thus making prior assumptions of 'no interactions' safer.

A robust procedure for choosing the appropriate data transformation, without presupposing the existence of a retrievable residual sum of squares, is described next.

6.4.2 Safeguarding against a mean-bias, the 'β-technique'

In order to be able to take clear-cut decisions for improving a product or a process, we need clear-cut results from a Taguchi analysis of the available data. One of the drawbacks is often the ambiguity in classification of variability-control and target-control factors which prevents a distinction between the factors affecting the variation and those affecting only the mean response. A large overlap in classification is often the result of violation of assumption (A2), i.e. when the variance of the measurements within an experimental trial (cell) tends to change with the mean level of these measurements. In order to make the variance independent of the mean we need a suitable change of scale, a suitable data-transformation. If the functional relationship between the cell variance and the cell mean is known, this determines the type of transformation to use.

Indeed, suppose that $\sigma_Y^{(i)}$ and $\mu_Y^{(i)}$ are respectively the ith trial standard deviation and mean in the original scale of the ith trial observations Y, and that they are connected by a functional relationship:

$$\sigma_Y^{(i)} = f(\mu_Y^{(i)}),$$

which is assumed to be the same for all trials.

Let $T(Y)$ be the function required to transform the data in such a way that the new cell variance of the transformed data is independent of the mean. Using the Taylor series expansion near $[\mu, T(\mu)]$ we have

$$T(Y) = T(\mu) + (Y-\mu)T'(\mu) + \tfrac{1}{2}(Y-\mu)^2 T''(\mu) + \cdots.$$

We can now derive an approximation to the variance of the transformed variable as

$$\text{var}[T(Y)] = [T'(\mu)]^2 \text{var}(Y-\mu) + \text{higher-order terms},$$

or approximately

$$\sigma_T^2 = [T'(\mu)]^2 \sigma_Y^2.$$

So, approximately

$$\sigma_T = (T'(\mu))\sigma_Y, \tag{6.4}$$

i.e.

$$T(\mu) = \int \frac{\sigma_T}{\sigma_Y} \, \mathrm{d}\mu$$

and since T is supposed to make the variance σ_T^2 constant, c^2 say, and since $\sigma_Y = f(\mu)$, we have

$$T(\mu) = \int \frac{c}{f(\mu)} \, \mathrm{d}\mu. \tag{6.5}$$

Of course, the roughness of the approximation used here is only too apparent. For example, if X is normally distributed, then the variance of $T(X) = X^2$ as given by (6.4) is $4\sigma^2\mu^2$, while actually it is $4\sigma^2\mu^2 + 2\sigma^4$. Nevertheless, (6.5) can still direct us as to what approximately T should be in special cases. For example, if

$$\sigma_Y = f(\mu) = \alpha \cdot \mu^\beta, \tag{6.6.a}$$

then from (6.5)

$$T(\mu) = \begin{cases} C_1 \cdot \mu^{1-\beta} & \text{if } \beta \neq 1 \\ C_1 \cdot \log(\mu) & \text{if } \beta = 1 \end{cases} \qquad (C_1\text{: a constant}),$$

i.e. we need a transformation to the power of $(1-\beta)$ if $\beta \neq 1$, and the log-transformation if $\beta = 1$. In other words, if $\lambda = 1-\beta$ we need a transformation to the power of λ if $\lambda \neq 0$, and the log-transformation if $\lambda = 0$.

This is equivalent to the 1-parameter Box–Cox family of transformations (1) (see Section 6.4.1) with $\lambda = 1-\beta$. Indeed, if we consider the Box–Cox family of power-transformations

$$T(Y) = Y^{(\lambda)},$$

from (6.4) we have

$$\sigma_\lambda \propto \lambda \mu^{\lambda-1} \sigma_Y \text{ (}\propto\text{ means ‘proportional to’)}$$

and if $\sigma_Y = \mu^\beta$ then

$$\sigma_\lambda \propto \mu^{\lambda-1+\beta}.$$

So, for σ_λ not to depend on μ $(d\sigma_\lambda/d\mu = 0)$, λ must be chosen so that $\lambda = 1-\beta$.

One way to establish whether (6.6a) is a reasonable approximation is to estimate β through the least squares estimator for b in the relationship between sample mean $\bar{Y}_i$ and sample standard deviation s_i:

$$\log(s_i) = \log(a) + b \log(\bar{Y}_i) + \varepsilon_i \qquad i = 1, \ldots, n \qquad (6.7a)$$

(n is the number of experimental trials) with its significance demonstrated through the t-ratio for b (see Section 2.1.4).

When the data contains non-positive values (in which case $\log(\bar{Y}_i)$ or $\log(Y_i)$ cannot be defined), one could consider the relationship

$$\sigma = f(\mu) \approx \alpha(\mu + \beta_1)^\beta \qquad (6.6b)$$

so that $Y > -\beta_1$; the parameters β, β_1 can be estimated respectively from b and b_1 which produce the best coefficient of determination in the regression analysis

$$\log(s_i) = a + b \log(\bar{Y}_i + b_1). \qquad (6.7b)$$

In the sequel, we shall show that the estimate of β could have considerable use in minimizing the overlap in the classification of variability-control and target-control factors.

Notice that in many cases the ‘β-transformation’ (suggested through estimating β from (6.7)—the ‘β-technique’), is identical to the λ-transformation (suggested using the Box–Cox technique), which means that the simple procedure (6.7) is good enough for estimating a transformation simultaneously satisfying (R1), (R2), and (R3). In such cases the analysis should be done in the transformed scale. In cases where there is a disagreement, (provided β is statistically significant) we should favour the β-transformation (i.e. considering the satisfaction of (R1) and (R2) as more important) keeping in mind that this disagreement is an indication of the presence of strong interactions and/or higher-order terms. One could argue in favour of always using the simple procedure (6.7) without going through the more complicated Box–Cox procedure, which could be less robust in the presence of outliers, assumes the existence of a retrievable residual sum of squares and could induce a mean bias. But the Box–Cox technique has many advantages:

(i) It is indicative of the presence of interactions, a test for which has been overlooked in many Taguchi applications in the past.

(ii) It makes full use of the information provided by the available data.

(iii) It assures that the simplicity (linearity or additivity) and normality of the model are valid assumptions.

(iv) It can provide reliable and shorter confidence intervals for λ.

(v) It can deal with cases with no replication in each cell [the possibility of applying the Taguchi methodology in non-replicated experiments is investigated by Box and Meyer (1986b)].

It is the authors' opinion that the Box–Cox technique should be used where possible, with the β-technique acting as a safeguard against a mean bias. Of course, if the functional relationship between σ_Y and μ is known to be other than the power one (6.6), then (6.5) could be used to provide information of the type of transformation T needed. Therefore the following procedure is suggested:

Procedure

(i) If the β-transformation is identical to the λ-transformation ($T_\beta = T_{BC} = T$), then the data should be transformed appropriately and the following performance measures should be analysed

$$\text{TPM} = \mu_T$$

$$\text{NPM} = -10 \log_{10}(\sigma_T^2)$$

where μ_T and σ_T^2 are respectively the trial mean and variance in the transformed scale $T(Y)$.

(ii) If the β-transformation is not the same as the λ-transformation, then analyse

$$\text{TPM} = \mu$$

$$\text{NPM} = 10 \log_{10}\left\{\frac{f(\mu)}{\sigma}\right\}^2,$$

where $f(\mu)$ is defined by (6.6) (and estimated through (6.7)) and where μ and σ are the trial-mean and standard deviation in the original scale. [Generally, this also covers the case when f is known, but (6.5) is difficult to solve.] Taguchi's SNR is the special case when $f(\mu) = \alpha\mu$, i.e. when $\beta \approx 1$, $\lambda \approx 0$ and $T(Y) = \log(Y)$.

6.4.3 *Case-studies*

In this section we will assess the effectiveness of a data transformation on data from previously published applications of the Taguchi methodology. If the data values are available, both the Box–Cox technique and the β-technique will be applied to determine the transformation. If only the trial means and variances are available, the β-technique will be applied and procedure (ii) will be followed.

Case-study 1

One of the first applications of the Taguchi methodology in USA had the objective of obtaining a microprocessor window size of 3.5 μm with minimum variability in order to improve the photolithographic process in integrated circuit fabrication (see Phadke *et al.* 1983). Part of the data analysed was the window line width prior to the plasma etch stage of the process ('pre-etch line width').

For this stage, eight factors were identified as important in controlling window sizes: mask dimension (A), viscosity (B), spin-speed (C), bake temperature (D), bake time (E), aperture (F), exposure time (G), and developing time (H).

Two levels were chosen for factors A, B, and D and three levels for all other factors. B and D were combined to form a three-level factor BD and the 7 factors were assigned to the $OA_{18}(2 \times 3^7)$ orthogonal array as shown in Table 6.1.

The wafer factor and the wafer site factor were considered to be the noise factors; so the objective was to make the process insensitive to variation between wafers and among wafer sites (see Logothetis 1987*b* for a more detailed consideration of these noise factors). For each trial run, five measurements were taken from each of two wafers and the resulting data is given in Table 6.2. The sample mean $\bar{Y}$ and sample standard deviation s for each experimental trial was calculated and the signal-to-noise ratio SNR $= \log(\bar{Y}/s)$ was analysed; factors A and F were then identified as variability-control factors and the analysis of $\bar{Y}$ identified B, C, and G as target-control factors.

Thus, Phadke *et al.* used the signal-to-noise ratio (SNR), μ/σ, as their NPM measure. But the SNR is appropriate only if the original data

TABLE 6.1. OA_{18} Orthogonal array

	Factors							
Trial run	1 A	2 BD	3 C	4 E	5 F	6 G	7 H	8
1	1	1	1	1	1	1	1	1
2	1	1	2	2	2	2	2	2
3	1	1	3	3	3	3	3	3
4	1	2	1	1	2	2	3	3
5	1	2	2	2	3	3	1	1
6	1	2	3	3	1	1	2	2
7	1	3	1	2	1	3	2	3
8	1	3	2	3	2	1	3	1
9	1	3	3	1	3	2	1	2
10	2	1	1	3	3	2	2	1
11	2	1	2	1	1	3	3	2
12	2	1	3	2	2	1	1	3
13	2	2	1	2	3	1	3	2
14	2	2	2	3	1	2	1	3
15	2	2	3	1	2	3	2	1
16	2	3	1	3	2	3	1	2
17	2	3	2	1	3	1	2	3
18	2	3	3	2	1	2	3	1

requires a log-transformation (see Section 6.4.2 above), because only then is the NPM independent of μ (corresponding to a relationship of the form $\sigma = \alpha\mu^{\beta}$ with $\beta = 1$). Of course, we can very easily examine whether there is indeed such a relationship between the cell-mean and cell-standard deviation, by estimating β through the simple linear regression

$$\log(\sigma_i) = \log(\alpha) + \beta \log(\mu_i) \qquad i = 1, \ldots, 18,$$

using

$$\log(s_i) = \log(a) + b \log(\bar{Y}_i) + \varepsilon_i \qquad i = 1, \ldots, 18.$$

Invoking simple regression theory (see Section 2.1.4) we find that for the data of Table 6.2:

$$\hat{\beta} = b = 0.0113$$

with a t-ratio

$$t(b) = 0.017.$$

Therefore we can accept that $\beta = 0$.

TABLE 6.2. Experimental data

	Line-width control feature photoresist—Nanoline tool (micrometres)					
Experiment No.	Top	Centre	Bottom	Left	Right	Comments
1	2.43	2.52	2.63	2.52	2.5	
1	2.36	2.5	2.62	2.43	2.49	
2	2.76	2.66	2.74	2.6	2.53	
2	2.66	2.73	2.95	2.57	2.64	
3	2.82	2.71	2.78	2.55	2.36	
3	2.76	2.67	2.90	2.62	2.43	
4	2.02	2.06	2.21	1.98	2.13	
4	1.85	1.66	2.07	1.81	1.83	
5	—	—	—	—	—	Wafer broke
5	1.87	1.78	2.07	1.8	1.83	
6	2.51	2.56	2.55	2.45	2.53	
6	2.68	2.6	2.85	2.55	2.56	
7	1.99	1.99	2.11	1.99	2.0	
7	1.96	2.2	2.04	2.01	2.03	
8	3.15	3.44	3.67	3.09	3.06	
8	3.27	3.29	3.49	3.02	3.19	
9	3.0	2.91	3.07	2.66	2.74	
9	2.73	2.79	3.0	2.69	2.7	
10	2.69	2.5	2.51	2.46	2.4	
10	2.75	2.73	2.75	2.78	3.03	
11	3.2	3.19	3.32	3.2	3.15	
11	3.07	3.14	3.14	3.13	3.12	
12	3.21	3.32	3.33	3.23	3.10	
12	3.48	3.44	3.49	3.25	3.38	
13	2.6	2.56	2.62	2.55	2.56	
13	2.53	2.49	2.79	2.5	2.56	
14	2.18	2.2	2.45	2.22	2.32	
14	2.33	2.2	2.41	2.37	2.38	
15	2.45	2.50	2.51	2.43	2.43	
15	—	—	—	—	—	No wafer
16	2.67	2.53	2.72	2.7	2.6	
16	2.76	2.67	2.73	2.69	2.6	
17	3.31	3.3	3.44	3.12	3.14	
17	3.12	2.97	3.18	3.03	2.95	
18	3.46	3.49	3.5	3.45	3.57	
18	—	—	—	—	—	No wafer

TABLE 6.3

λ	RSS	λ	RSS
4.00	8.8887	0.50	5.1429
3.00	6.6841	0.25	5.2679
2.00	5.4635	0.00	5.4502
1.50	5.1608	−0.50	6.0053
$\hat{\lambda} = 1.00$	$5.0522 = RSS_{min}$	−1.00	6.8569
0.75	5.0719	−2.00	9.8651

So the β-technique indicates that no transformation is needed. We can confirm this by applying the Box–Cox technique: indeed, if we transform the original data Y into $z^{(\lambda)}$ [see (6.1)′ in Section 6.4.1] and then for some values of λ in the range $(-4, +4)$ we tabulate the residual sum of squares (RSS) for the ANOVA of the corresponding transformed data assuming an additive model (i.e. a model involving only the main effects of the factors A, BD, C, E, F, G, and H), we obtain the results of Table 6.3.

The plot of λ vs. RSS can be seen in Fig. 6.3. We note that the value of λ corresponding to the minimum of RSS, which is also equivalent to the maximum likelihood value for λ (see Section 6.4.1) is the value $\lambda = 1$, which means that the original data needs no transformation; this in turn

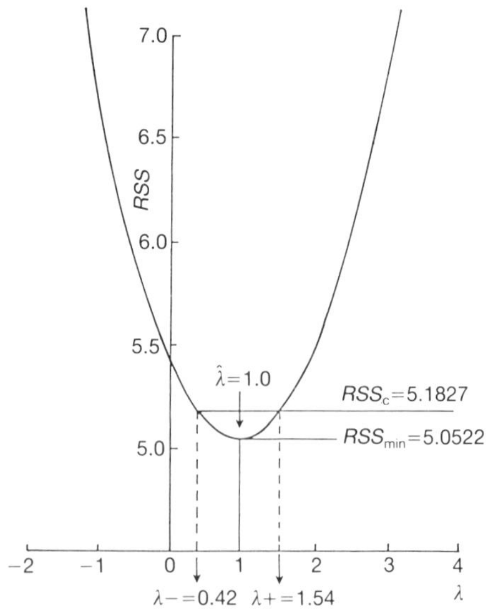

FIG. 6.3. Lambda plot (case-study 1).

suggests that the SNR, used by Phadke *et al.* (1983), which is effective when log-transformation is needed ($\lambda = 0$), is not appropriate as a NPM measure in this case. The value $\lambda = 0$ is not even included in the 95% confidence interval for λ which is found by calculating a critical sum of squares RSS_c by (see Section 6.4.1):

$$RSS_c = RSS_{min}\left[1 + \frac{t_v^2(0.025)}{v}\right]$$

where v is the number of degrees of freedom associated with the RSS. In fact,

$$RSS_c = 5.0522\left[1 + \frac{(1.975)^2}{151}\right] = 5.1827$$

and the 95% confidence limits can now be read from Fig. 6.3 as being $\lambda_- = 0.42$ and $\lambda_+ = 1.54$, which certainly do not include the value $\lambda = 0$. Therefore, instead of the SNR we should use simply

$$\text{NPM} = -10\log_{10}(s^2)$$

whose analysis for the data of Table 6.3 gives the results of Table 6.4.

According to the results of Table 6.4, the factors A, F and G affect the variability in the pre-etch line-width and so they can be considered as variability-control factors. This differs from the reported results in Phadke *et al.* (1983) in the role of the factor G (exposure time) which was eventually chosen as the only target-control factor, perhaps wrongly, since G does affect the variability to some degree. This demonstrates the importance of choosing the appropriate NPM for the available data and recognition of the appropriate transformation that the original data needs can help in this choice.

TABLE 6.4. Pooled ANOVA for NPM

Source	df	S	MS	F-ratio
A	1	49.84	49.84	6.5 significant at 5% level
E	2	37.12	18.56	2.4
F	2	84.04	42.02	5.5 significant at 5% level
G	2	49.11	24.56	3.2 significant at 10% level
Error	10	76.35	7.64	
Total	17	296.46		

Case-study 2

The process studied by Pignatiello and Ramberg (1985) involved the heat treatment of leaf springs for trucks. There were four design factors of interest (with 2 levels each); furnace temperature (B), heating time (C), transfer time (D) and 'hold-down' time (E) (labelled A, B, C, and D in Section 2.3.2).

The engineers felt that the quench oil temperature would be difficult to control during production and therefore treated it as a noise factor (O) with two levels, a low level (130–150 °F) and a high level (150–170 °F). This factor can be considered as a 'fixed' noise factor (its levels have not been chosen at random) which when analysed, showed a highly significant effect (the usefulness of a noise-factors analysis has been noted in Logothetis 1987*b*). The strength of its effect convinced the engineers that an attempt to control 'O' at least between 'high' and 'low' was worth while in comparison with alternative strategies.

So, eventually, the difficulty of controlling the oil quench temperature *within* its high or low level was considered as the noise. The quality characteristic of interest was the free height of the leaf spring in the unloaded condition, with a target value of 8 inches. According to a $\frac{1}{2}$ replicate of a 2^5 factorial design, three observations were taken for each trial run with the results shown in Table 6.5 (cf. Table 2.12a).

TABLE 6.5

O	D	C	B	E	Data		
−	−	−	−	−	7.78	7.78	7.81
−	−	−	+	+	8.15	8.18	7.88
−	−	+	−	+	7.50	7.56	7.50
−	−	+	+	−	7.59	7.56	7.75
−	+	−	−	+	7.94	8.00	7.88
−	+	−	+	−	7.69	8.09	8.06
−	+	+	−	−	7.56	7.62	7.44
−	+	+	+	+	7.56	7.81	7.69
+	−	−	−	−	7.50	7.25	7.12
+	−	−	+	+	7.88	7.88	7.44
+	−	+	−	+	7.50	7.56	7.50
+	−	+	+	−	7.63	7.75	7.56
+	+	−	−	+	7.32	7.44	7.44
+	+	−	+	−	7.56	7.69	7.62
+	+	+	−	−	7.18	7.18	7.25
+	+	+	+	+	7.81	7.50	7.59

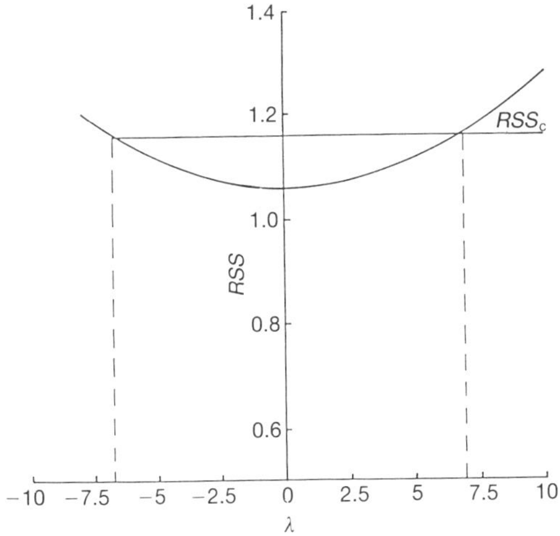

FIG. 6.4. Lambda plot (case-study 2).

A Box–Cox analysis of the above gave the λ-plot of Fig. 6.4. Note that the graph is extremely flat in the neighbourhood of the minimum for *RSS* which is at $\hat{\lambda}_{BC} = 0$. Indeed, a 95% confidence interval for λ_{BC} can be found to include anything from $\lambda_- = -6.8$ to $\lambda_+ = 6.9$.

The β-technique gives a non-significant value of

$$\hat{\beta} = b = 6.08 \text{ (i.e., } \hat{\lambda}_\beta = 1 - b = -5)$$

with a t-ratio of 0.94, which suggests an even wider confidence interval.

These results suggest the following:

(a) Any transformation to the power of λ from -7 to $+7$ approximately, will produce similar results for this data; 'no transformation' ($\lambda = 1$) is as good as the log-transformation ($\lambda_{BC} = 0$) and also as good as a transformation to the power of -5 ($\approx \lambda_\beta$).

(b) the 'disagreement' between the Box–Cox λ ($\hat{\lambda}_{BC} = 0$) and the β-technique's $\lambda(\hat{\lambda}_\beta \approx -5)$ is an indication of the possible presence of interactions.

Pignatiello and Ramberg (1985) analysed the SNR (which corresponds to $\hat{\lambda} = 0$). Their analysis showed the effects of the factor B and of the interactions D × O, B × C × O, and C × D × O to be significant. Identical results can be obtained by analysing $-10 \log(s^2)$ (corresponding to $\hat{\lambda}_{BC} = 1$) or $10 \log(\bar{x}^6/s)$ (corresponding to $\hat{\lambda}_\beta = -5$), thus confirming (a) and (b) above.

Pignatiello and Ramberg (1985) ended their paper with an 'open question', as to whether one should always use $-10\log(s^2)$ as a NPM measure, suggesting there is still a gap within the Taguchi analysis as far as the NPM choice is concerned. An analysis of transformations bridges the gap.

Case-study 3

Pao *et al.* (1985) attempted to optimize (minimize) the mean response time of a VAX 11-780 machine under UNIX System Release 5.0 operation. There were eight design factors of interest, namely: disk-drive configuration (A), distribution of key system and user files among disks (B), memory size (C), systems buffer size (D), number and selection of 'sticky bits' set (E), number of KMCs (devices for computer-to-terminal communications) used (F), size of 'INODE' table (G), and sizes of certain other system tables (H).

Factors (A) and (F) were each studied at two levels, and the others at 3 levels. The orthogonal array of Table 6.1 was used and the factors (A)–(H) were assigned respectively to columns 6, 2, 3, 4, 5, 1, 7, and 8. For column 6, level 3 was made identical to level 2 to accommodate the 2-level factor A. The response time of a set of commands was measured 120 times for each trial run. A single performance measure was analysed:

$$\eta_i = -10\log_{10}(\text{mean squared response time}) \qquad i = 1, \ldots, 18.$$

This performance measure is suggested (see Section 6.2.2.3) for the cases when the ideal situation is 'the lower the better' and can be expressed as

$$\eta_i = -10\log\frac{1}{N}\sum_{j=1}^{N} Y_{ij}^2 \qquad i = 1, \ldots, 18 \quad (N = 120).$$

But note that

$$\frac{1}{N}\sum_{j=1}^{N} Y_{ij}^2 = (\bar{Y}_i)^2 + \frac{N-1}{N}s_i^2 \qquad i = 1, \ldots, 18. \tag{6.8}$$

So, this criterion confounds location effects arising from changes in ith cell mean ($\bar{Y}_i$) with dispersion effects arising from changes in the ith cell variance (s_i^2) and could result in an ambiguity in classification between control and signal factors.

The paper makes available only the values $\bar{Y}_i$ and η_i, $i = 1, \ldots, 18$ for each of the 18 trials runs, and using (6.8) we can obtain the results of Table 6.6. The $\log(s)$ vs. $\log(\bar{Y})$ plot of Fig. 6.5 shows an obvious linear relationship. This is confirmed by the β-technique (6.7) which gives the following

$$\log(s) = -1.45 + 1.61 \times \log(\bar{Y}),$$

TABLE 6.6

Trial run	Mean ($\bar{Y}$)	Standard deviation (s)
1	4.65	2.7718
2	5.28	3.9500
3	3.06	1.3587
4	4.53	3.1800
5	3.26	1.3431
6	4.55	3.2741
7	3.37	1.9249
8	5.62	3.9414
9	4.87	2.3747
10	4.13	2.3408
11	4.08	2.7107
12	4.45	2.5485
13	3.81	2.2314
14	5.87	3.6009
15	3.42	1.7175
16	3.66	1.8284
17	3.92	1.9400
18	4.42	1.9983

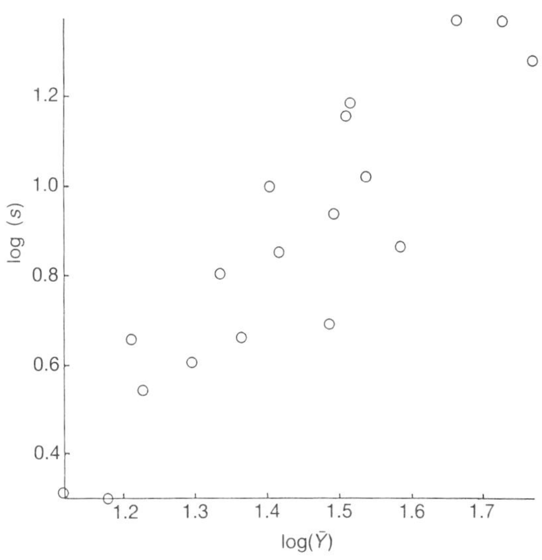

FIG. 6.5. Plot of log(s) vs. log($\bar{Y}$) (case-study 3).

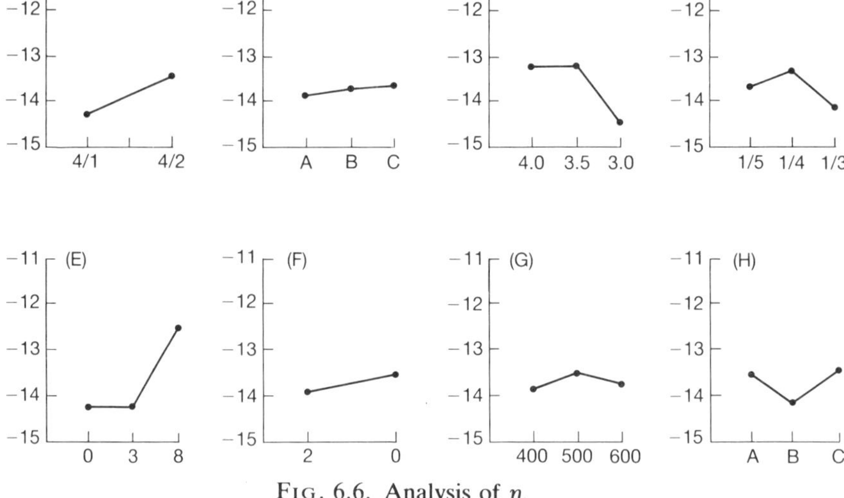

FIG. 6.6. Analysis of η.

with a t-value for the regression coefficient of

$$t(b) = 8.7.$$

This shows a highly significant relationship between the cell mean and variance of the form

$$\sigma \approx C\mu^{1.5} \qquad (\text{i.e. } \beta \approx 1.5)$$

This shows that for a clear-cut Taguchi analysis we either need to transform the original data by raising it to the power $\lambda \approx 1 - \beta = -0.5$ (inverse square root), i.e. Y into $1/\sqrt{Y}$, or analyse

$$\text{NPM} = 10 \log(\bar{Y}^{1.5}/s) \tag{6.9}$$

for the determination of the variability-control factors.

Pao *et al.* (1985) analysed η and from the plots of the factorial effects (the ANOVA of η does not show any significant effect) they concluded that factors A, C, D, and E are significant. These plots are given in Fig. 6.6. But if we carry out a similar analysis for the mean response $(-\bar{Y})$, we find almost identical results (see Fig. 6.7); this implies that η is greatly confounded with $\bar{Y}$, and so there is no way of telling whether the A, C, D, or E effects are essentially associated with dispersion and not with location effects.

A measure of dispersion unbiased by the mean is (6.9) whose analysis gives the results of Fig. 6.8. We note a strong dispersion effect for factor B (with level 3 as the optimal), C (with level 2 as the optimal), E (with

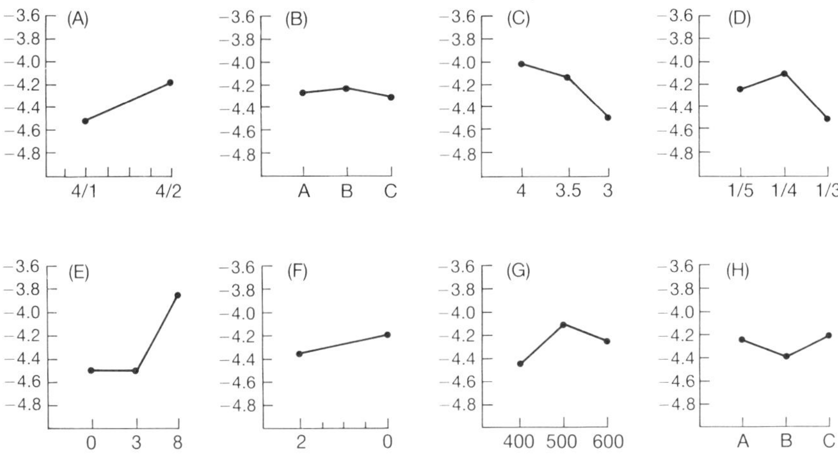

FIG. 6.7. Analysis of the mean response ($-\bar{Y}$).

level 3 as the optimal), G (with level 1 as the optimal), and H (with level 1 as the optimal); these factors could be considered as variability-control factors and set at their optimal levels, whereas A and D could be considered as the target-control factors. (Of course, a complete analysis should have included the analysis of the mean response in the scale of the suggested inverse square root transformation.)

The optimum factor levels suggested by the above analysis compared

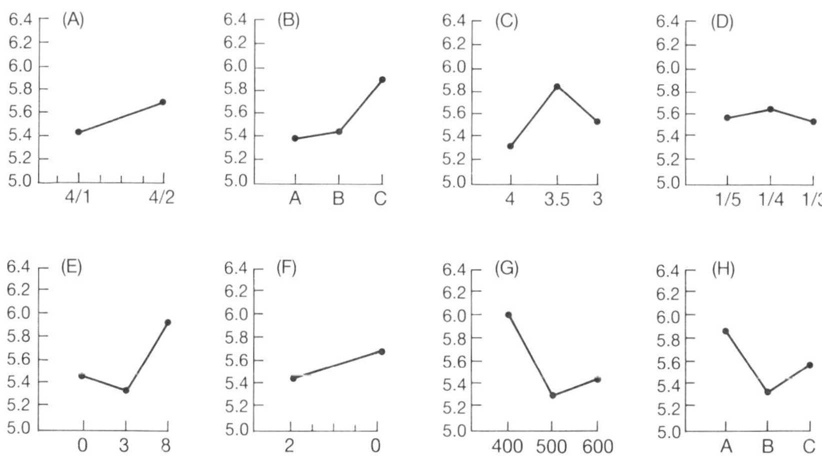

FIG. 6.8. Analysis of NPM $= 10\log_{10}(\bar{Y}^{1.5}/s)$.

with those suggested in Pao *et al.* (1985) are as follows:

Factors	Optimum levels	Pao's suggested levels
A	2	2
B	3	1
C	2	1
D	2	2
E	3	3
F	2	2
G	1	3
H	1	3

The difference in the suggested levels (in 4 out of 8 factors) is due to the analysis of the NPM as defined in (6.9), rather than of η. The analysis of a proper performance measure, separately for the noise and for the mean response, can provide a more comprehensive analysis of the dispersion and location effects, which can lead to more valid results.

Case-study 4

Barker (1986) considered an experiment to improve the strength of a small plastic part (the butterfly) in the *carburettor* of a lawn mower engine. Six design factors were identified. These were: Feed rate; first screw r.p.m.; second screw r.p.m.; gate size; first temperature and second temperature, which were assigned—in their 3 levels—respectively to the columns 1, 2, 5, 9, 10, and 12 of the $OA_{27}(3^{13})$ array (see Appendix D). The effect of the variation due to the tolerances of the factors was studied by considering the above 6 factors also as noise factors and assigning them to the OA_{18} array; for each trial run this noise factor array provided 19 observations on the response of interest which is the breaking strength of the 'butterfly'. As NPM, Barker (1986) used

$$\theta = -10 \log_{10} \frac{1}{n} \sum_{i=1}^{n} \frac{1}{Y_i^2}$$

which is 'suggested' for the cases when the ideal situation is 'the larger the better' (see Section 6.2.2.3).

Only the cell mean and cell standard deviations are available and these are given along with the design factor array in Table 6.7.

The β-technique gives a value of

$$\hat{\beta} = b = -0.49 \approx -0.5$$

TABLE 6.7

Run	1 Feed Rate	2 First r.p.m.	5 Second r.p.m.	9 Gate Size	10 First Temp.	12 Second Temp.	$\bar{Y}$	s
1	1	1	1	1	1	1	87.4	31.5
2	1	1	2	2	2	2	115.6	24.6
3	1	1	3	3	3	3	106.2	36.3
4	1	2	1	2	2	3	101.5	34.6
5	1	2	2	3	3	1	117.6	35.4
6	1	2	3	1	1	2	115.2	24.3
7	1	3	1	3	3	2	131.1	30.7
8	1	3	2	1	1	3	93.9	35.4
9	1	3	3	2	2	1	134.3	35.5
10	2	1	1	2	3	2	111.6	21.9
11	2	1	2	3	1	3	108.6	30.5
12	2	1	3	1	2	1	111.9	29.3
13	2	2	1	3	1	1	105.7	28.0
14	2	2	2	1	2	2	118.3	20.1
15	2	2	3	2	3	3	133.1	34.0
16	2	3	1	1	2	3	104.1	33.2
17	2	3	2	2	3	1	144.5	35.4
18	2	3	3	3	1	2	133.5	21.4
19	3	1	1	3	2	3	82.5	41.6
20	3	1	2	1	3	1	85.8	42.0
21	3	1	3	2	1	2	120.4	33.5
22	3	2	1	1	3	2	99.3	36.7
23	3	2	2	2	1	3	99.1	41.5
24	3	2	3	3	2	1	115.3	41.1
25	3	3	1	2	1	1	96.2	40.9
26	3	3	2	3	2	2	121.6	36.7
27	3	3	3	1	3	3	120.8	46.4

with a t-value of -1.66 which is significant to at least 10% level. This suggests a transformation of the original data to the power of

$$\lambda = 1 - \beta = 1.5$$

or a noise performance measure of the form

$$\text{NPM} = 10 \log_{10} \frac{\bar{Y}^{-0.5}}{s}.$$

The analysis, for identifying the variability-control factors, gives the following results

ANOVA for NPM

Source	*SS*	*F*-ratio	Probability	Optimal level
Feed rate	7.89	28.4	0.0001*	2
First r.p.m.	2.15	7.7	0.0055*	1
Second r.p.m.	0.66	2.4	0.1296	
Gate	0.59	2.1	0.1576	
First temp.	1.99	7.2	0.0072*	1
Second temp.	6.23	22.4	0.0001*	2

* Highly significant.

These differ from the results reported by Barker (1986) who, through the analysis of θ, found the following variability-control factors (with their optimal levels):

Feed rate (2), First r.p.m. (3), Second r.p.m. (3), and Second temp. (2).

Case-study 5

Up to now we have examined various published Taguchi case studies, which, because they were using unsuitable performance measures (SNR, η, θ), had some or all of the requirements R1, R2, and R3 (see Section 6.4.1) unsatisfied.

We finish this section by considering the analysis of an experimental study reported by Quinlan (1985) to reduce post-extrusion shrinkage of a speedometer casing. We finish with this application because of the striking difference between the results reported by Quinlan (1985) and the results one obtains after a transformation analysis.

The shrinkage values for four samples taken from 3000-ft lengths of product manufactured at each of sixteen combinations of fifteen 2-level factors are shown in Table 6.8. The design is a highly fractional design which does not leave any degrees of freedom for the residual sum of squares when one analyses a performance measure. Quinlan (1985) analysed the statistic η; by pooling together the sum of squares of the smallest magnitude, 8 factors (E, G, K, A, C, F, D, H) were found to have a large effect on η.

Again, there is no way of telling whether these factors are variability-control or target-control factors because of the confounding between the dispersion and location effects in the definition of η. A Box–Cox analysis on the data of Table 6.8 produces the results of Fig. 6.9, from which it is clear that we can accept $\hat{\lambda} \simeq 1$ (i.e. no transformation). This is not

TABLE 6.8

Run	H	D	−L	B	−J	F	N	A	−I	−E	M	−C	K	G	−O	Y_1	Y_2	Y_3	Y_4
1	−1	−1	1	−1	1	−1	−1	−1	1	1	−1	1	−1	−1	1	0.49	0.54	0.46	0.45
2	1	−1	−1	−1	−1	−1	1	−1	−1	1	1	1	1	−1	−1	0.55	0.60	0.57	0.58
3	−1	1	−1	−1	1	1	1	−1	1	−1	1	1	−1	1	−1	0.07	0.09	0.11	0.08
4	1	1	1	−1	−1	1	−1	−1	−1	−1	−1	1	1	1	1	0.16	0.16	0.19	0.19
5	−1	−1	1	1	−1	1	1	−1	1	1	−1	−1	1	1	−1	0.13	0.22	0.20	0.23
6	1	−1	−1	1	1	1	−1	−1	−1	1	1	−1	−1	1	1	0.16	0.17	0.13	0.12
7	−1	1	−1	1	−1	−1	−1	−1	1	−1	1	−1	1	−1	1	0.24	0.22	0.19	0.25
8	1	1	1	1	1	−1	1	−1	−1	−1	−1	−1	−1	−1	−1	0.13	0.19	0.19	0.19
9	−1	−1	1	−1	1	−1	−1	1	−1	−1	1	−1	1	1	−1	0.08	0.10	0.14	0.18
10	1	−1	−1	−1	−1	−1	1	1	1	−1	−1	−1	−1	1	1	0.07	0.04	0.19	0.18
11	−1	1	−1	−1	1	1	1	1	−1	1	−1	−1	1	−1	1	0.48	0.49	0.44	0.41
12	1	1	1	−1	−1	1	−1	1	1	1	1	−1	−1	−1	−1	0.54	0.53	0.53	0.54
13	−1	−1	1	1	−1	1	1	1	−1	−1	1	1	−1	−1	1	0.13	0.17	0.21	0.17
14	1	−1	−1	1	1	1	−1	1	1	−1	−1	1	1	−1	−1	0.28	0.26	0.26	0.30
15	−1	1	−1	1	−1	−1	−1	1	−1	1	−1	1	−1	1	−1	0.34	0.32	0.30	0.41
16	1	1	1	1	1	−1	1	1	1	1	1	1	1	1	1	0.58	0.62	0.59	0.54

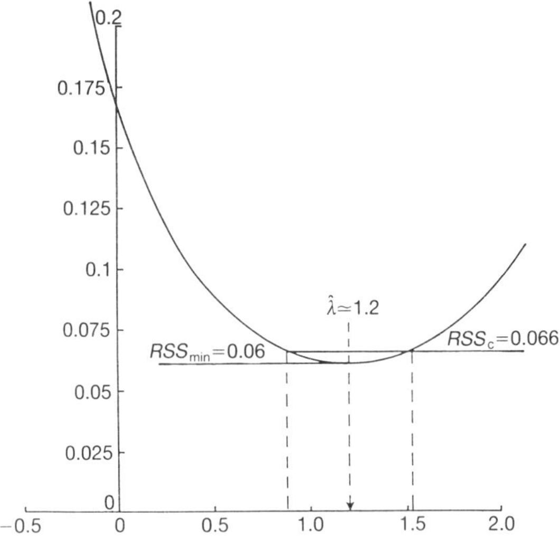

FIG. 6.9. Lambda plot (case-study 5).

contradicted by a β-analysis which produces a non-significant value of $b = -0.2$ so that we can accept $\beta \simeq 0$ (i.e. no transformation). Thus analysis of $\bar{Y}$ and the

$$\mathrm{NPM} = -10 \log(s^2)$$

is indicated. (An analysis with the exact value $\hat{\lambda} = 1.2$ produces identical results.) A stepwise regression procedure at 5% level of significance showed that no factor significantly affects NPM and only factors E and G significantly affect $\bar{Y}$. So it is clear that only factors E and G are significant and these are essentially only associated with location and not with dispersion effects [Box and Ramirez (1986) reached similar conclusions with the use of half-normal plots in the log-transformed scale].

The striking difference between the above conclusions and the ones reported by Quinlan (1985) shows how the proper choice of the NPM can simplify the analysis and can avoid misleading conclusions in a rather complicated experiment.

Remark: In all the case studies examined so far we found significant differences from the published results. It is of course impossible to ascertain whether a process modified on the basis of our findings would have produced a better yield than that already reported. Nevertheless, the potential of a proper data-transformation for achieving (R1), (R2), and (R3), cannot be overlooked in striving for statistically valid conclusions in a Taguchi analysis.

6.4.4 The case $T_\beta \neq T_{BC}$

In the case-studies examined in Section 6.4.3, whenever it was possible to apply both the Box–Cox and the β-technique, no significant disagreement was found to exist in the type of data transformation suggested by either technique. We will now examine a case-study where such a disagreement exists. The experiments for this application were designed and analysed at GEC, and were carried out at Ferranti Electronics (see also Logothetis and Haigh 1987, 1988). The study had as an objective the characterization and optimization of a newly installed multi-response process used to define the interconnect tracks on an integrated circuit.

Controllable variables

The following controllable factors were studied (all set at three levels):

F1: the pressure of the reaction chamber (in milliTorr);

F2: the RF power level (in watts);

F3: the temperature of the lower electrode on which the wafer is supported (in °C);

F4: the flow rate of boron trichloride (in standard cubic centimeters per minute—sccm);

F5: the flow rate of chlorine (in sccm);

F6: the flow rate of silicon tetrachloride (in sccm).

Response variables

Three responses of interest were as follows:

A: The etch rate of the aluminium–silicon alloy layer which forms the interconnect tracks (see Fig. 6.10). A value over 5000 Å/min would be considered satisfactory.

P: The photoresist etch rate (see Fig. 6.10). Its value should be such that the ratio between the A-response and P-response (selectively ratio) is between 3 and 4.

OX: The oxide etch rate (see Fig. 6.10) for the silicon dioxide layer exposed during the over-etch period. It should be as low as possible.

'Uncontrollable' variables

Another objective of the experiment was to determine the effect of the 'over-etch time' (OE) on the responses of interest. It was already

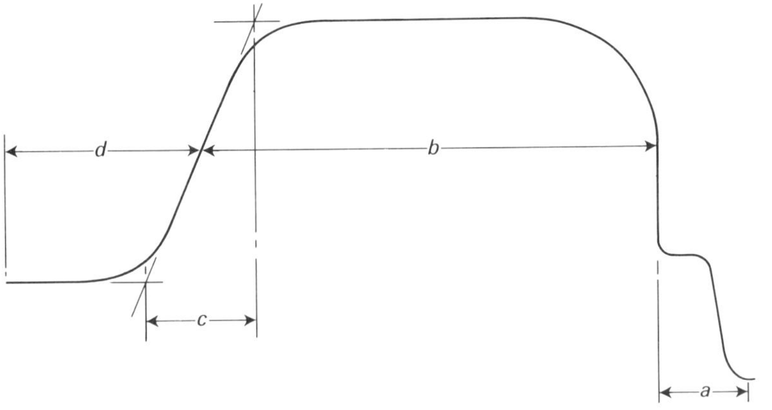

FIG. 6.10. End-point trace diagram. *Key: a,* initiation time; *b,* bulk etch time; *c,* tangent to slope time; *d,* over-etch time.

$$\text{Al/Si etch rate} = \frac{\text{Al step height}}{b} \times 60\ \text{Å/min}$$

$$\text{Resist etch rate} = \frac{\text{Resist step height}}{(a+b+d)} \times 60\ \text{Å/min}$$

$$\text{Oxide etch rate} = \frac{\text{Oxide step height}}{d} \times 60\ \text{Å/min}$$

expected that over-etching directly affects the oxide etch rate (OX), but no other prior engineering information was available. OE is necessary to ensure the clearing of the gaps between the tracks but, ideally, it should be minimized. In order to study the effect of OE, two levels were chosen, one at 30 seconds and the other at 90 seconds. Note that OE, in practice, is not controllable; it is what Taguchi calls 'an indicative factor'; that is, 'a factor whose levels are not selectable but we nevertheless need to know its effect' (see Section 6.2 and Taguchi and Wu (1980) Chapter 9.2).

The final objective of the experiment was to determine which of the design factors affect the variability across the wafer. The Taguchi technique could also show how this variability could be minimized. For this purpose more (than one) measurements have been taken at different sites across each wafer; so the 'wafer-site' was considered as a noise factor.

The experimental design

The six design factors, in their three levels, were assigned onto the columns of the $OA_{18}(3^6)$ array shown in Table 6.9. This orthogonal

TABLE 6.9. $OA_{18}(3^6)$ Orthogonal array

	Factors					
Trial run	F1	F2	F3	F4	F5	F6
1	1	1	1	1	1	1
2	1	2	2	2	2	2
3	1	3	3	3	3	3
4	2	1	1	2	2	3
5	2	2	2	3	3	1
6	2	3	3	1	1	2
7	3	1	2	1	3	2
8	3	2	3	2	1	3
9	3	3	1	3	2	1
10	1	1	3	3	2	2
11	1	2	1	1	3	3
12	1	3	2	2	1	1
13	2	1	2	3	1	3
14	2	2	3	1	2	1
15	2	3	1	2	3	2
16	3	1	3	2	3	1
17	3	2	1	3	1	2
18	3	3	2	1	2	3

design is a highly fractional but efficient design for dealing, at the expense of only 18 trials, with six three-level factors which would otherwise need 729 experiments [see $OA_{18}(2 \times 3^7)$ in Appendix D]. For each of the 18 trials, two wafers were tested, one with 30 seconds and the other with 90 seconds over-etch time, thus also allowing the study of the 'indicative factor', OE. For the purpose of studying (and minimizing) the within-wafer variability with respect to A- and P-responses, five measurements were taken across the wafer at North, Centre, South, East, and West site points. The resulting data are given in Table 6.10.

Analysis of the 'A'-response

For a set of real values for λ in the interval $(-2, 2)$ we transform the data values of the 'A'-response (from Table 6.10a) according to the family of transformations (6.1). Then we tabulate against each value of λ the residual sum of squares (*RSS*) from the ANOVA or the regression analysis assuming the simplest model possible, in this case a linear model involving only the linear effects of the factors F1–F6 and OE.

From Table 6.11 we see that the Box–Cox technique indicates the need

TABLE 6.10a. The A-response data

Trial	OE = 30					OE = 90				
	N	C	S	E	W	N	C	S	E	W
1	4790	4690	4790	4790	4690	5070	5070	4970	5170	4970
2	5580	5580	5120	5470	5470	6050	6050	5700	5810	5810
3	5880	6000	5750	5880	5500	5750	6880	6880	5750	5500
4	6350	5630	6200	6200	6060	6110	6390	5970	6500	6110
5	8920	9120	9120	8920	8920	9350	9550	9350	9350	9350
6	5190	5420	5190	5190	5190	5950	6070	5710	5830	5950
7	12900	12900	12900	13200	12900	12600	12900	12600	12600	12600
8	5450	5450	5210	5210	5210	5450	5580	5450	5450	5450
9	9450	9450	9450	9250	9250	10140	10140	9320	9730	9730
10	5070	4960	4960	4860	4860	5310	5310	5100	5210	5100
11	5540	5540	5770	5420	5310	5680	5680	5570	5570	5570
12	5150	5040	5040	5150	5040	5260	5470	5360	5260	5260
13	4890	5000	4780	4890	4890	5110	5210	5000	5110	5110
14	8520	8520	8330	8150	8150	8780	8780	8600	8780	8780
15	10870	10870	10870	10570	10570	10870	10870	10870	10570	10570
16	12740	12450	12450	12450	12450	11940	11940	11670	11670	11670
17	5840	5840	5710	5710	5710	6230	6510	6230	6230	6230
18	8730	8920	8540	8540	8730	9110	9110	8730	8920	8730

TABLE 6.10b. The P-response data

Trial	OE = 30					OE = 90				
	N	C	S	E	W	N	C	S	E	W
1	1420	1410	1260	1330	1340	1500	1480	1470	1470	1380
2	1830	1650	1550	1550	1650	2120	2110	2100	2040	2050
3	1770	2110	1820	2000	1570	2330	—	—	2240	2250
4	680	790	590	650	540	570	660	550	690	610
5	1540	1630	1600	1490	1510	1580	1630	1690	1690	1710
6	3610	3770	3610	3610	3610	1350	1570	1320	1440	1390
7	460	440	430	640	440	30	280	60	160	120
8	440	370	300	360	370	490	430	420	420	430
9	750	730	720	870	650	1060	1160	790	1030	1040
10	1670	1590	1210	2000	1660	1690	1680	1610	1610	1620
11	1770	1830	2210	1740	1600	2000	—	1970	1790	1860
12	1680	1250	1400	1560	1660	2270	2260	2370	2250	2380
13	380	600	360	530	550	590	630	620	570	580
14	1545	1720	1300	1400	1620	2410	2390	2700	2140	2610
15	2590	2480	2550	2550	2750	2580	2480	2550	2550	2770
16	500	260	390	320	410	330	260	180	—	310
17	360	350	330	390	340	340	330	280	410	420
18	680	730	440	720	660	860	730	—	780	470

TABLE 6.11

λ	*RSS*
2.0	206746233
1.0	106278285
0.5	82916332
0.0	69483168
−0.5	62843692
−0.75	61461473
$\lambda_{BC} = -1.0$	$61164873 = RSS_{min}$
−1.25	61857489
−1.50	63603230

for the inverse transformation ($\lambda_{BC} = -1$). A graphical representation of Table 6.11 is given by Fig. 6.11, where a 95% confidence interval for λ_{BC} can also be seen (from −1.4 to −0.6). However, Fig. 6.12 reveals a direct dependence between the cell variance s_T^2 and the cell mean $\bar{Y}_T$ in the transformed (inverse) scale, and this can be confirmed by the regression analysis of $\log(s_T)$ on $\log(\bar{Y}_T)$ which yields a significant value for the

TABLE 6.12. ANOVA of NPM(−1)

Source	df	Sum of square (*SS*)	*F*-ratio
$F1_L$	1	346.136	20.59**
$F1_Q$	1	5.224	0.31
$F2_L$	1	28.333	1.69
$F2_Q$	1	0.879	0.05
$F3_L$	1	1.3597	0.08
$F3_Q$	1	52.042	3.1*
$F4_L$	1	46.0496	2.74 (*)
$F4_Q$	1	2.192	0.13
$F5_L$	1	100.078	5.95**
$F5_Q$	1	58.578	3.49*
$F6_L$	1	312.77	18.61**
$F6_Q$	1	3.543	0.21
OE	1	0.1211	0.01
Residual	22	369.7662	
Total	35	1327.081	

** Significant to at least 5% level.
* Significant to 10% level/(*) significant after pooling.

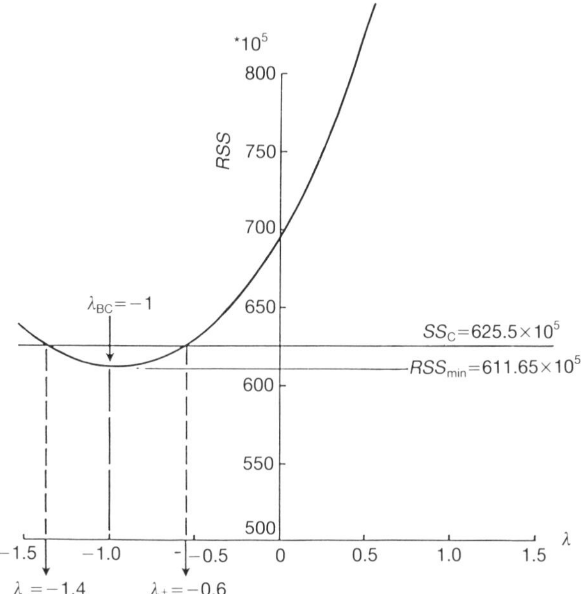

FIG. 6.11. Lambda plot (A-response).

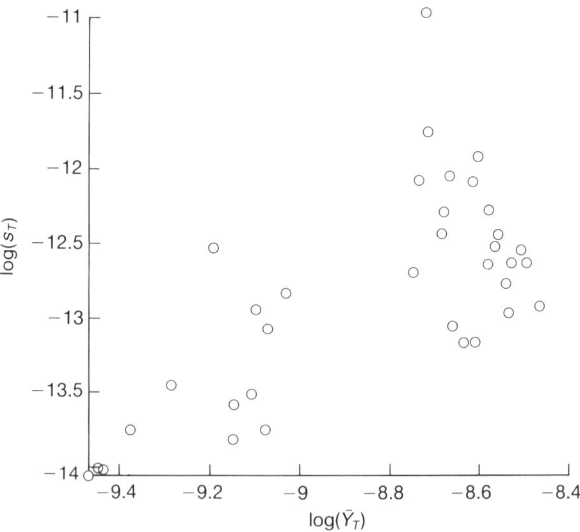

FIG. 6.12. Plot of $\log(S_T)$ vs. $\log(\bar{Y}_T)$.

regression coefficient of

$$\hat{\beta}_T = b_T \simeq 1.52 \quad [t(b_T) = 5.7].$$

This indicates a relationship of the form $\sigma_T \approx \alpha\mu_T^{1.5}$, i.e. violation of (A2). So the application of the β-technique in the new scale shows that an ambiguity might result in the classification between variability-control and target-control factors if we apply the Box–Cox inverse transformation. Indeed, an analysis of $\bar{Y}_T$ shows that all the factors F1–F6 and OE are significant, whereas an analysis of

$$\text{NPM}(-1) = -10\log_{10}(s_T^2)$$

yields the results of Table 6.12. (The subscripts 'L' and 'Q' indicate respectively the linear and quadratic effect.)

Table 6.12 indicates strong linear effects for factors F1, F4, F5, and F6 and strong quadratic effects for factors F3 and F5. Therefore the analysis of the data in the inverse scale has shown that 5 out of 6 controllable factors significantly 'affect' both the mean response *and* the variation in the response, and it is hard to see whether these factors are really associated with location or with dispersion effects.

Now if we apply the β-technique in the original untransformed data we find the following relationship between the untransformed cell (trial) mean $\bar{Y}$ and cell variance s^2:

$$\log(s) = (0.55) + (0.49)\log(\bar{Y}),$$

i.e. $\hat{\beta} = b = 0.5$ with $t(b) = 1.82$. This indicates a relationship of the form

$$\sigma \approx \alpha\mu^{\beta} \quad \text{with} \quad \beta \approx \tfrac{1}{2}$$

in the original scale, and therefore the need for a square-root ($\lambda_\beta = 1 - \beta = \frac{1}{2}$) transformation. If we transform the original data using the square-root transformation, and we analyse the resulting cell variance s_β^2 through

$$\text{NPM}(1/2) = -10\log_{10}(s_\beta^2),$$

we find that only the quadratic effect of F5 and the linear effect of F6 are significant. So, an analysis in the λ_β-scale indicates that only factors F5 and F6 are associated with dispersion effects and can therefore be considered as variability-control factors, whereas factors F1–F4 are all associated with location effects and can be considered as target-control factors. (OE is not significant in this scale.) This greatly simplifies the conclusions that we had after the analysis in the 'inverse' scale, where (A2) was violated and where there was an ambiguity in the classification in 5 out of 6 controllable factors. So, this is an occasion where a 'linear model orientated' Box–Cox transformation has led to a failure in achieving requirements (R2) and (R1). The disagreement between the Box–Cox and the β-technique perhaps indicates the existence of non-linear effects in the scale other than the Box–Cox transformed scale. Indeed, the analysis of the mean response in the λ_β-scale (square root) can show very strong quadratic effects due to the signal factors F2 and F3, and this has to be kept in mind when the optimal levels for these factors are to be decided.

Remarks

(i) Instead of analysing NPM(− 1) and NPM(1/2) one could alternatively have analysed

$$\text{NPM}'(-1) = 20\log_{10}(\bar{Y}^2/s)$$

and

$$\text{NPM}'(\tfrac{1}{2}) = 20\log_{10}(\bar{Y}^{\frac{1}{2}}/s)$$

respectively (in the original scale) with similar results. Note that

$$\text{NPM}'(-1) = \text{NPM}'(1/2) + 20\log_{10}(\bar{Y}^{1.5}),$$

and one can clearly see how a mean bias can be introduced with the use of NPM′(−1), i.e. with the use of the Box–Cox's 'inverse' transformation. For example, one can see graphically how similar the effect of F1 seems to be, on the mean response $\bar{Y}_T$ and on s_T [through NPM′(−1)] in the inverse scale (see Figs. 6.13a and 6.13b). The similarity is due to the fact that the variability performance measure is being biased by the

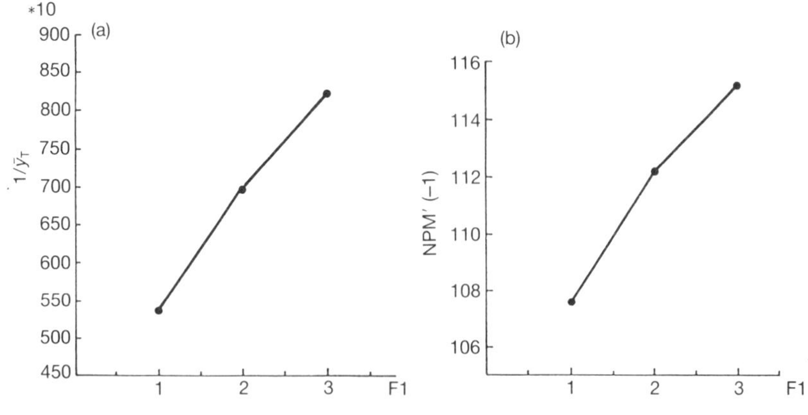

FIG. 6.13.

mean. With the proper transformations to ensure satisfaction of (R2)—in this case the square-root transformation—it turns out that F1 is really associated with location effects and not with dispersion.

(ii) Concerning the uncontrollable indicative factor OE we have established that it does not affect either the mean A-response or the variability in the response. If we analyse the 'fixed' noise factor 'wafer-site' (applying Duncan's multiple comparison test—see Section 2.6.3.5) we can see a significant difference (at the 5% level) between the central site and the rest of the sites in their mean response. This is a characteristic that was to be expected and whose affect was hoped to be minimized by the Taguchi methodology.

Analysis of the P-response

An application of the Box–Cox technique on the data of the P-response (from Table 6.10b) produces the plot of Fig. 6.14. It is clear that the technique indicates the need for a transformation to the power of 0.35 ($\lambda_{BC} = 0.35$). But again, an application of the β-technique on the λ_{BC}-transformed data reveals a relationship between the new cell $\bar{x}_T$ and

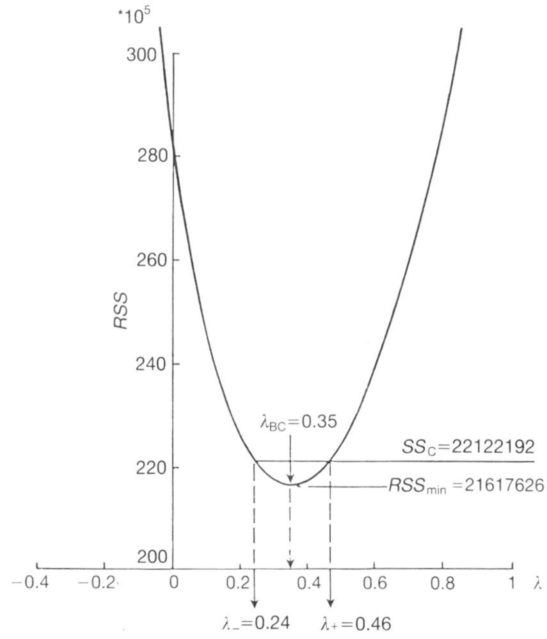

FIG. 6.14. Lambda plot (P-response).

cell-variance S_T^2 of the form

$$\log(S_T) = 2.05 - (1.32)\log(\bar{x}_T) \quad [t(b_T) = -3.7].$$

On the other hand, an application of the β-technique on the original data yields

$$\log(s) = 3.05 + (0.196)\log(\bar{x}) \quad [t(b) = 1.6],$$

i.e. $\sigma \approx \alpha\mu^{\beta}$ where $\beta \approx 0.2$.

This indicates a transformation to the power of $\lambda_\beta = 1 - \beta \approx 0.8$. (A 95% confidence interval for β covers also the case of $\beta = 0$, i.e. no transformation ($\lambda_\beta = 1$). Whether one uses $\lambda_\beta = 0.8$ or 1, the conclusions turn out to be similar.)

An analysis of the mean response both in the λ_β and λ_{BC} scale show significant effects due to factors F1, F2, and F6. To find the factors affecting dispersion we analyse

$$\text{NPM}(0.8) = -10\log_{10}(s_\beta^2)$$

in the λ_β-scale (s_β^2 is the cell variance when the $\lambda_\beta = 0.8$ transformation is used) and

$$\text{NPM}(0.35) = -10\log_{10}(S_T^2)$$

in the λ_{BC}-scale (S_T^2 is the cell variance when the $\lambda_{BC} = 0.35$ transformation is used). The results are as follows:

TABLE 6.13. ANOVA for NPM (0.8)

Source	df	*SS*	*F*-ratio
$F1_L$	1	4.529	0.21
$F1_Q$	1	0.526	0.02
$F2_L$	1	24.231	1.12
$F2_Q$	1	23.228	1.07
$F3_L$	1	1.766	0.08
$F3_Q$	1	0.258	0.01
$F4_L$	1	97.999	4.53**
$F4_Q$	1	10.135	0.47
$F5_L$	1	89.399	4.13**
$F5_Q$	1	51.509	2.38
$F6_L$	1	0.163	0.01
$F6_Q$	1	22.054	1.02
OE	1	84.989	3.93*
Residual	22	476.051	
Total	35	886.837	

TABLE 6.14. ANOVA for NPM (0.35)

Source	df	*SS*	*F*-ratio
$F1_L$	1	254.588	8.9*
$F1_Q$	1	32.328	1.13
$F2_L$	1	12.032	0.42
$F2_Q$	1	35.736	1.25
$F3_L$	1	0.588	0.02
$F3_Q$	1	8.410	0.29
$F4_L$	1	78.371	2.74
$F4_Q$	1	4.568	0.16
$F5_L$	1	61.682	2.16
$F5_Q$	1	28.786	1.01
$F6_L$	1	6.258	0.22
$F6_Q$	1	37.603	1.31
OE	1	76.217	2.66
Residual	22	629.491	
Total	35	1266.66	

According to Table 6.14, the only variability-control factor for the P-response is factor F1. This is rather suspicious, since the effect of F1 on NPS(0.35) is almost identical to the effect of F1 on $\bar{x}_T$ (see Figs 6.15a and 6.15b). On the other hand, according to Table 6.13 factors F4 and F5 are clearly associated with dispersion effects and can be considered as

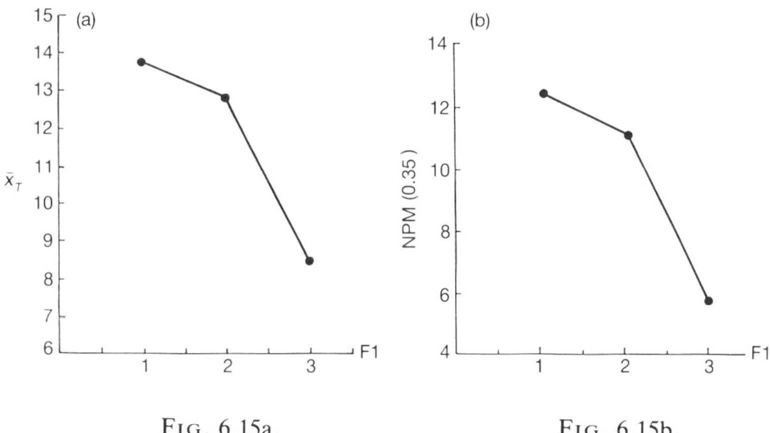

FIG. 6.15a.

FIG. 6.15b.

variability-control factors leaving F1, F2, and F6 as target-control factors. It is clear that the λ_β-transformation which favours (R2) and (R1), has achieved a clear-cut separation between variability-control and target-control factors.

The disagreement with the λ_{BC}-transformation indicates the existence of non-linear effects. A detailed analysis of $\bar{x}_\beta$ (the cell mean in the λ_β-transformed scale) can indeed show the existence of a strong curvilinear effect for F1 and F2. This should be taken into account when one is searching for their optimal levels.

Conclusions of the experiment

According to our results, factors F4–F6, should be considered as variability control factors (i.e. factors affecting the variation in the response), and should be kept at their optimal levels (the levels with the highest noise performance measure). Factors F1 and F2 seem to strongly affect the mean response for both A- and P-response and thus can be considered as target-control factors. Factor F3 can also be used as a target-control factor for the A-response.

The indicative factor OE does not seem to affect the mean response of either A and P, although it seems to have a small effect on the variability in the P-response. But as OE is not controllable, the process optimization has to be achieved by manipulating only factors F1–F6.

On the basis of the previous analyses, a decision was taken to keep the controllable factors F4, F5, and F6 near their optimum levels, and manipulate F1 and F2 in order to bring the responses onto their targets. Some additional experiments have shown that keeping F1 near level 1 and F2 near level 3 (with F3 near level 2) achieves the required targets for the A- and P-responses. Confirmatory experiments under these conditions have shown a 50% decrease in the within-wafer variability, an average value of 5100 Å/min for the A response and a consistent selectivity ratio (A/P) of 3.4 as required. An additional bonus was that the need for high over-etch time was minimized enabling it always to keep below 30 seconds, and this led to very low levels for the oxide etch rate.

6.4.5 Concluding remarks

We advocated data transformation for achieving

(a) minimization of the ambiguity in the determination of the variability-control and target-control factors;

(b) satisfaction of necessary statistical assumptions; and

(c) indication of the possible existence of non-linear effects.

It should lead to a simplified, clear, and statistically valid Taguchi analysis. The Box–Cox technique although a powerful method for choosing a transformation, can still lead to a violation of (b) and to a failure to achieve (a); in such cases the application of the β-technique was advised as a safeguard.

There are many special requirements in multi-response processing; any technique, which can bring some order and ease the task of identifying sources of variation and help to minimize their effect, is invaluable. But equally invaluable is a statistically valid approach for properly applying these techniques. Prior exploration of the available data, analyses of transformations or response model fitting do not require a great knowledge in statistics, but can greatly contribute in the production of clear-cut, more valid, and useful results.

6.5 Estimation of variability

The new emphasis on variability in off-line quality control has led to the reassessment of various methods of estimating the variance σ^2. We have already seen that, if we have m replications at a particular trial (say trial i), then we may estimate σ_i^2 by

$$s_i^2 = \frac{1}{m-1}\sum_{j=1}^{m}(Y_{ij} - \bar{Y}_i)^2 \qquad i = 1, \ldots, n,$$

where $\bar{Y}_i$ is the sample mean of the m observations Y_{ij} $(j = 1, \ldots, m)$ at the ith trial. It is mentioned in Appendix B that s_i^2 is an unbiased estimator of σ_i^2. But we need to go further and use the s_i^2 in some way to truly model the variance as a function $\sigma^2(\mathbf{x})$ of the vector of the controllable variables (factors) $\mathbf{x} = (x_1, \ldots, x_p)$, with the aim of producing a 'response surface' for σ^2; this response surface will summarize the variability in the response Y over $\mathbf{x}$. Let $\mathbf{x}' = (x_1', \ldots, x_k')$.

Recall that (see Chapter 5) we can model the response Y as

$$Y(\mathbf{x}') = \theta_0 + \theta_1 x_1' + \cdots + \theta_k x_k' + \varepsilon(\mathbf{x}')$$

where $\varepsilon(\mathbf{x}')$ is the error with

$$E(\varepsilon(\mathbf{x}')) = 0 \quad \text{and} \quad \operatorname{var}(\varepsilon(\mathbf{x}')) = \sigma^2(\mathbf{x}),$$

which may incorporate variation over noise factors together with experimental error; the expected response

$$\mu(\mathbf{x}') = E(Y(\mathbf{x}')) = \theta_0 + \theta_1 x_1' + \cdots + \theta_k x_k'$$

is the response surface for Y and may be made more complex by inclusion of, say, interaction and/or quadratic terms.

The variable sets $\mathbf{x}$ and $\mathbf{x}'$ may be exclusive subsets of the 'design space' (i.e. of the set of all the controllable factors that affect the process/product's characteristics in some way); at the other extreme $\mathbf{x}$ might be a subset of $\mathbf{x}'$, (or vice versa), in which case we have $\sigma^2(\mathbf{x})$ functionally dependent on $\mu(\mathbf{x}')$. A half-way house is to assume that some common variables exist which affect both σ^2 and μ. We will now provide some justification for using the s_i for modelling purposes of σ^2. In our present formulation we will keep the same variables $\mathbf{x} = (x_1, \ldots, x_p)$ in both 'surfaces' for σ^2 and μ, with the understanding that some may affect only σ^2 and some only μ.

6.5.1 Modelling the trial-variance

Let $\hat{Y}_i$ be the usual least squares estimate of $\mu(\mathbf{x}) = E(Y(\mathbf{x}))$ at the ith trial ($i = 1, \ldots, n$). We may write

$$Y_{ij} = \hat{Y}_i + Y_{ij} - \hat{Y}_i$$

and from this we have

$$\sum_{i=1}^{n}\sum_{j=1}^{m} Y_{ij}^2 = m\sum_{i=1}^{n} \hat{Y}_i^2 + \sum_{i=1}^{n}\sum_{j=1}^{m} (Y_{ij} - \hat{Y}_i)^2. \tag{6.10}$$

The above is the usual resolution of the total (uncorrected) sum of squares into the regression sum of squares and the residual sum of squares *RSS* (see Chapter 2). The last item in (6.10), the *RSS*, can itself be resolved as

$$RSS = m\sum_{i=1}^{n} (\bar{Y}_i - \hat{Y}_i)^2 + \sum_{i=1}^{n}\sum_{j=1}^{m} (Y_{ij} - \bar{Y}_i)^2. \tag{6.11}$$

Since the last term of (6.11) can be written as

$$(m-1)\sum_{i=1}^{n} s_i^2,$$

from (6.10) we have

$$\sum_i\sum_j Y_{ij}^2 = m\sum_i \hat{Y}_i^2 + m\sum_i (\hat{Y}_i - \bar{Y}_i)^2 + (m-1)\sum_i s_i^2. \tag{6.12}$$

Under normality assumptions, the s_i^2's are independent of each other and of the other terms in (6.12), i.e. of

$$m\sum_i \hat{Y}_i^2 \quad \text{and} \quad m\sum_i (\hat{Y}_i - \bar{Y}_i)^2,$$

which provides some justification for using the s_i^2's for modelling purposes. In addition to the need to satisfy normality assumptions, there is an additional problem that should be mentioned: when the σ_i^2's are unequal, (problem of 'heteroscedasticity'), the ordinary least-squares estimates of the parameters in the usual regression model are not the best, and consequently the $\hat{Y}_i$'s are generally suboptimal. However, the loss of efficiency is small if the σ_i^2's are roughly the same.

6.5.2 *Justification for using the log(variance)*

We can show that for a sample $Y_1, \ldots, Y_m$ from a normal distribution with mean μ and variance σ^2, if

$$s^2 = \frac{1}{m-1}\sum_{i=1}^{m}(Y_i - \bar{Y})^2,$$

then

$$V_{m-1} = \frac{(m-1)s^2}{\sigma^2} \tag{6.13}$$

has a chi-square distribution (see Appendix F) with $m-1$ degrees of freedom. This is a fixed, known distribution and does not depend on μ and σ^2. In fact,

$$E(V_{m-1}) = m - 1$$

and

$$\text{var}(V_{m-1}) = 2(m-1).$$

Taking natural logarithms in (6.13) we have

$$\log_e(V_{m-1}) = \log_e(m-1) - \log_e(\sigma^2) + \log_e(s^2).$$

By using a simple transformation argument, the probability density of $\log_e(s^2)$ can be found to be given by

$$h(t) = \left\{\frac{(m-1)}{2\sigma}e^t\right\}^{[(m-1)/2]} \exp\left\{-\frac{(m-1)}{2\sigma^2}e^t\right\}\frac{1}{\Gamma[(m-1)/2]}$$

where $\Gamma(\cdot)$ is the Gamma function (see Appendix F). For large m, one can obtain approximations for the mean (k_1) and variance (k_2) of this distribution as

$$k_1 \approx \log_e(\sigma^2) - \left\{\frac{1}{m} + \frac{1}{3m^2}\right\}$$

$$k_2 \approx \frac{2}{m-1}.$$

Here, the important features are that, the mean of $\log(s^2)$ depends only on $\log(\sigma^2)$ and m, and the variance of $\log(s^2)$ is stable depending only on m. This suggests that we may use $\log_e(s_i^2)$ to model $\log_e(\sigma_i^2)$ and we only need to 'add in' the bias correction term

$$\frac{1}{m}+\frac{1}{3m^2}$$

in order to achieve a stable estimation. Moreover, it can also be shown that $\log(s^2)$ converges to approximate normality as m increases.

This argument goes some way towards justifying the technique of performing a regression analysis of $\log(s_i^2)$ on the variables x_i, in addition to the arguments of Section 6.4. Therefore, a general approach would be to model

$$\text{NPM} = -10\log(S_T^2)$$

using S_T^2, the trial-variance in the (appropriately) transformed scale.

6.5.3 *Use of residuals*

If

$$R_{ij} = Y_{ij} - \hat{Y}_i \qquad i = 1, \ldots, n; j = 1, \ldots, m,$$

we have, of course,

$$RSS = \sum_i \sum_j R_{ij}^2.$$

Now if

$$\bar{R}_i = \frac{1}{m}\sum_{j=1}^{m} R_{ij},$$

then

$$\frac{1}{m-1}\sum_{j=1}^{m}(R_{ij}-\bar{R}_i)^2 = \frac{1}{m-1}\sum_{j=1}^{m}(Y_{ij}-\bar{Y}_i-\hat{Y}_i+\bar{Y}_i)^2 = s_i^2.$$

Thus, the use of s_i^2 is, in fact, a special case of the more general use of residuals to model $\sigma^2(\mathbf{x})$. An alternative then to using regression based on the $\log_e(s_i^2)$, would be one based on the

$$\log_e(R_{ij}^2).$$

The advantage of this approach is that more degrees of freedom are available for the estimation of σ_i^2, $i = 1, \ldots, n$. A disadvantage is that, very small R_{ij} values lead to large negative $\log(R_{ij}^2)$—the so-called problems of 'inliers'—and care has to be taken to adjust very small R_{ij} values upwards to some threshold value.

6.5.4 *Single replicate experiments*

In cases when $m = 1$ (single replicate experimental trials), there are no degrees of freedom to estimate separately $\sigma^2(\mathbf{x})$. In such cases, following a two stage procedure, one could identify, say, a few controllable factors as affecting the mean, and then 'stealing' from the larger number of degrees of freedom produced by refitting a smaller model, in order to estimate σ^2. With p controllable factors, Box and Meyer (1986*b*) suggest that, after the factors which significantly affect the mean response are first identified, say k of them ($k < p$), a response model based on these factors is fitted with the insignificant factors pooled for error. Using the model, predicted values and residuals are found. Then the ratios

$$F_i = \frac{s^2(i_+)}{s^2(i_-)}$$

are calculated where the numerator and denominator represent the variances of residuals at the high and low levels of factor i respectively. The $\log_e(F_i)$'s are examined for extremely high or low values and these extremes will identify factors which affect the variance. Of course, this procedure allows $p - k$ factors to be ignored with the hope that out of the k factors which affect the mean, some of those will also affect the variance. The aliasing between estimation of $\sigma^2(\mathbf{x})$ and $\mu(\mathbf{x})$ when we ignore significant controllable variables, is brought out in the work by Box and Meyer (1986*a,b*). They give an analysis in which prior distributions militate against having more than a few significant controllable factors. The trade-off is critical when $n < p + 1$, in which case variances and means are competing for degrees of freedom. These over-saturated models are of great theoretical interest and a Bayesian methodology seems the only way forward (see Meyer and Box 1987, Meyer 1987).

6.5.5 A response surface approach to Taguchi's procedure

We will illustrate with an example, the identification of variability-control and target-control factors using response surface methodology. The data are taken from Phadke *et al.* (1983), and are the data considered in 'Case-study 1' of Section 6.4.3. (see Table 6.2). This time, we replace missing replicates by repeating the other replicates in each group to obtain a balanced experiment.

The objective was to obtain a microprocessor window size of 3.5 μm with maximum reliability. It is thus a 'response on target' experiment. The first seven factors in the OA_{18} array were used in the pre-etch experiment and these were mask dimension (X_1), viscosity/bake temperature (X_2), spin speed (X_3), bake time (X_4), aperture (X_5), exposure time (X_6), and developing time (X_7). The levels of these variables were set after consultation with design engineers and the response variable was 'line-width', a variable closely related to window size. Table 6.15 gives the sample means $\bar{Y}_i$ and sample standard deviations, s_i, for the eighteen

TABLE 6.15. Microprocessor line-width data

i	1	2	3	4	5	6	7	8	9
$\bar{Y}_i$	2.500	2.684	2.660	1.962	1.870	2.584	2.032	3.267	2.829
s_i	0.0827	0.1196	0.1722	0.1696	0.1102	0.1106	0.0718	0.2101	0.1516
i	10	11	12	13	14	15	16	17	18
$\bar{Y}_i$	2.660	3.166	3.323	2.576	2.306	2.464	2.667	3.156	3.494
s_i	0.1912	0.0674	0.1274	0.0850	0.0987	0.0363	0.0706	0.1569	0.0445

treatment combinations. The means and standard deviations were calculated from the ten replicates at each experimental combination.

Through an analysis of the means in each group (the $\bar{Y}_i$), the linear model

$$\bar{Y}_i = \theta_0 + \theta_i x_{i1} + \cdots + \theta_7 x_{i7} + \varepsilon_i, \qquad i = 1, 2, \ldots, 18,$$

was fitted using ordinary least squares (OLS). In the above model x_{ij} denotes the value of factor X_j in trial i and ε_i is random error. In Table 6.15a we give the breakdown of the regression sum of squares from the analysis of variance. This enables the identification of those factors which contribute significantly to the response.

Only X_1, X_3, X_6, and X_7 contribute significantly to the response. Thus a reduced model was fitted by OLS using the above four factors only. Dropping the subscript i we obtain the fitted response surface

$$\hat{\mu}(\mathbf{x}) = \hat{Y}_x = 2.68 + 0.190X_1 + 0.246X_3 - 0.212X_6 + 0.135X_7.$$

In similar fashion, a response surface was fitted for log(variance) using the values $\ln s_i = \log_e(s_i)$. Only the factors X_1, X_4, X_5, and X_6 contribute significantly to the linear regression. The fitted surface is

$$\log_e(\hat{\sigma}(\mathbf{x})) = \ln(s_x) = -2.24 - 0.176X_1 + 0.156X_4 + 0.276X_5 - 0.219X_6.$$

TABLE 6.15a. Regression sum of squares

Due to	S	df
X_1	0.6475	1
X_2	0.0163	1
X_3	0.7237	1
X_4	0.0004	1
X_5	0.0086	1
X_6	0.5406	1
X_7	0.2187	1
Total	2.1558	7

The above indicates that factors X_1 (mask dimension), X_4 (bake time), X_5 (aperture), and X_6 (exposure time) have an effect on the variance in the response and therefore could be considered as variability-control factors. This result is similar to the one obtained by performing an ANOVA on the

$$\text{NPM} = -10\log(s^2)$$

in Section 6.4.3 (see Table 6.4). The factors X_3 (spin speed), and X_7 (developing time) appear in $\hat{Y}_x$ but not in $\ln(s_x)$ and so they can be used as target-control factors to adjust the response (into target) without affecting the variance.

Note that the reason that we modelled $\ln(s)$, where s^2 is the trial variance of the original data values, is that there was no evidence to suggest that a data transformation was required (see 'Case-study 1', Section 6.4.3). Had there been a need for a transformation T, we might have modelled

$$\ln(s_T)$$

or equivalently

$$\text{NPM} = -10\log(s_T^2).$$

6.6 A demonstration case-study

In this section we show that a proper use of some of Taguchi's proposed techniques in conjunction with the appropriate statistical methods already mentioned in this chapter, can lead to the successful characterization and optimization of multi-response multi-factor processes with minimum experimentation. We will describe in detail an experiment that took place with the purpose of optimizing an advanced plasma etching process used for the etching of n-polysilicon in making silicon integrated circuits. A general procedure to follow, for any intended experimental study in any problem area, will then become apparent.

6.6.1 Objectives, variables involved

n-Polysilicon is etched using a plasma of mixed gases: Cl_2, SF_6, and O_2.

There are two stages:

(a) The first stage: to rapidly break through the native oxide that always exists on the polysilicon surface and must be removed; then etch the bulk of the film with minimal line-width loss, high uniform etch rate, and small photoresist loss. It is not desirable to clear the film at this stage, because the selectivity (ratio of the polysilicon etch-rate to the oxide etch rate) is necessarily low so that the native oxide is easily removable. Such low selectivity may result in an undesirable loss of the thin underlying layer of silicon dioxide.

(b) The second stage is a selective etch stage, employed to clear the remaining film. The requirements at this stage are again minimal line-width loss and high etch rate, but also high selectivity. Since the requirements of the two stages are somewhat different, it was decided to optimize the first separately, and then use the optimum conditions for a set time (30 seconds), prior to each of the second-stage experiments. We will report on the results of the first stage optimization experiments.

Controllable variables

The etcher is a 'triode' machine—it has three electrodes, the top one being earthed, the parallel bottom one driven by 100 kHz rf power, and the side ones driven by 13.56 MHz power. This arrangement allows greater control over the etching performance. The two driving powers represent two design factors. The 100 kHz power can be pulsed and so the 100 kHz duty cycle is a third variable. Five other factors were selected to make up eight controllable factors which are as follows:

A: The temperature of the lower electrode on which the wafer is supported (T).

B: The chamber pressure (Pre).

C: The flow rate of chlorine (Cl_2).

D: The flow rate of sulphur hexafluoride (SF_6).

E: The 13.56 MHz power level (P13).

F: The 100 kHz power level (P100).

G: The duty cycle of the 100 kHz power (D100).

H: The flow rate of oxygen (O_2).

Response variables

Normally the etch depth varies with differing conditions. Because clearing of the film (during the first stage) was not required, two of the three responses of interest had to be weighted by dividing them by the etching depth.

The three responses in order of importance are as follows

LL: the line-width loss of the polysilicon electrode divided by the etch depth. Linewidth loss is calculated by subtracting the final etched dimension from the initial resist dimension (measured using an optical metrology system). This ratio should ideally be less than 0.4 which corresponds to a line-width loss of 0.16 μm when an entire film with an etch thickness of 0.4 μm is to be cleared.

PER: the polysilicon etch rate, which is twice the etch depth resulting from 30 seconds of etching. Any value over 0.4 μm is acceptable.

RL: the resist loss divided by the etch depth (this ratio is equivalent to the inverse of the selectivity to resist). A large loss of resist may adversely affect the performance of the second stage; therefore a low value for this ratio is required, ideally below 0.5.

Achievement of a high uniformity (low variability) for each of the above responses (especially for PER) was also very important.

Uncontrollable variables

High uniformity for the responses of interest could be achieved by minimizing the variability between and within wafers. Clearly two factors that could represent the uncontrollable variables are the following noise factors:

S: the wafer-site factor, representing the variability within wafers.

W: the wafer-factor, representing the variability between wafers.

6.6.2 *Experimental design*

Because the operating conditions had not been previously fully explored, a 'screening' experiment was first designed in order to study the effects of the 8 designed factors A–H while they varied over a wide enough range.

The results of this 'characterization' experiment (which will not be covered in detail), apart from indicating a set of 'tentative' optimum operating conditions, have also shown that the factor D (SF_6) has no effect on any of the three responses of interest: this factor was then kept constant at its nominal level while a new set of 8 additional experimental trials was designed with the purpose of further studying the remaining seven factors at the following levels:

Factors	Levels	
	1	2
F1 T (in °C)	25	15
F2 Pre (in millitorr)	155	120
F3 Cl_2 (in sccm)	10	15
F4 P13 (in watts)	130	100
F5 P100 (in watts)	35	25
F6 D100 (in %)	33	67
F7 O_2 (in sccm)	0	1.19

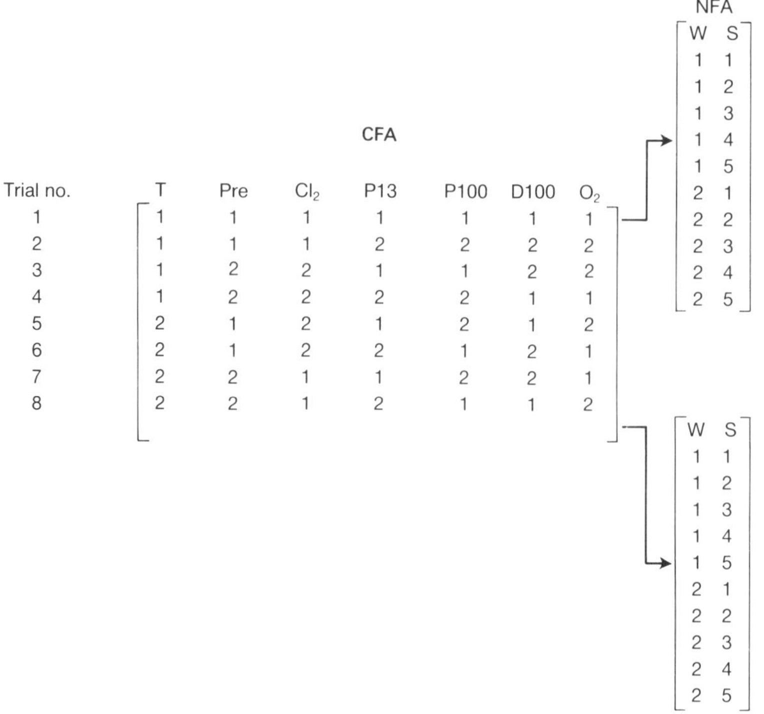

Trial no.	T	Pre	Cl_2	P13	P100	D100	O_2
1	1	1	1	1	1	1	1
2	1	1	1	2	2	2	2
3	1	2	2	1	1	2	2
4	1	2	2	2	2	1	1
5	2	1	2	1	2	1	2
6	2	1	2	2	1	2	1
7	2	2	1	1	2	2	1
8	2	2	1	2	1	1	2

W	S
1	1
1	2
1	3
1	4
1	5
2	1
2	2
2	3
2	4
2	5

W	S
1	1
1	2
1	3
1	4
1	5
2	1
2	2
2	3
2	4
2	5

FIG. 6.16. Experimental set-up.

For each of the 8 experimental trials, two wafers were used, and on each wafer five measurements were taken at North, Centre, South, East, and West sites. The experimental set up is shown in Fig. 6.16 (similar to Fig. 6.2):

The controllable factor array consists of 8 rows corresponding to the 8 experimental trials and seven columns corresponding to the seven 2-level controllable factors. This CFA is a $\frac{1}{16}$th fraction of the full-factorial array which would require 128 experiments. The noise factor array consists of 10 rows corresponding to the 10 data values obtained from the five site measurements on each of the 2 wafers. The two columns of the NFA correspond to the two noise factors: the 2-level wafer factor and the 5-level wafer site factor.

The analysis of the response data for the two most important responses (LL and PER) will be covered in detail. The available data for these responses can be seen in Table 6.16.

6.6.3 Prior exploration of data

The need for an initial exploration of the available data before the implementation of the Taguchi method was investigated in Logothetis (1987*a*). Various exploratory steps were proposed, the main ones of which were the prior analysis of the noise factors and the investigation of data transformations. This was felt to be particularly important with data resulting from poorly designed experiments or in-production/after production monitoring (see Logothetis 1987*b*).

6.6.3.1 Noise factors

Using Duncan's multiple comparison test (see Section 2.6.3.5), an analysis of the 'wafer site' noise factor (which is a 'fixed' factor with 5 levels) has shown evidence (at the 5% significance level) of a 'systematic' variation due to the chosen sites, for both LL and PER responses.

(a) A significant difference was noticed between the West (W) site and the rest of the sites for the LL-response. Further investigations at this stage revealed that the photoresist wall angle is different at the west site, and this always results in a lower linewidth loss value because of the way the measurement is made. Therefore, this systematic difference is not a characteristic of this experiment alone.

(b) A significant difference was also noticed between the 'centre' (C) site and the rest of the sites for the PER-response. This is probably due to the fact that the etching species can be rapidly replenished at the edges of the wafer, whereas at the centre they must diffuse a greater distance before the etching reaction can occur. This establishes a concentration gradient and results in a lower reaction etch rate at the centre. Such non-uniformity is a common occurrence in plasma etching and it is hoped that it will be minimized by the Taguchi technique.

6.6.3.2 Choice of data-transformation and NPM

We will apply the simplest of the two methods of Section 6.4 which is the β-technique (see Section 6.4.2).

Consider first the LL-response data (Table 6.16a). We first calculate the sample mean $\bar{Y}$ and the sample standard deviation s of the 10 observations for each of the 8 trials. For example, for the first trial

$$\bar{Y}_1 = 0.6897, \qquad s_1 = 0.13544$$

etc. (see Table 6.17a). We then perform the simple regression analysis

TABLE 6.16a. LL-response data

	First wafer					Second wafer				
Trial	N	C	S	E	W	N	C	S	E	W
1	0.659	0.800	0.895	0.711	0.538	0.675	0.794	0.650	0.750	0.425
2	0.444	0.394	0.322	0.413	0.406	0.515	0.412	0.375	0.500	0.382
3	0.444	0.554	0.515	0.529	0.444	0.417	0.559	0.588	0.639	0.457
4	0.542	0.727	0.745	0.556	0.429	0.571	0.612	0.640	0.462	0.444
5	0.531	0.553	0.480	0.440	0.486	0.531	0.553	0.480	0.440	0.486
6	0.583	0.697	0.371	0.548	0.247	0.583	0.697	0.371	0.548	0.247
7	0.286	0.632	0.200	0.257	0.152	0.533	0.769	0.414	0.507	0.382
8	0.625	0.511	0.600	0.436	0.367	0.615	0.696	0.640	0.577	0.415

between $\log(s_i)$ and $\log(\bar{Y}_i)$, $i = 1, \ldots, 8$ and we find

$$\log(s) = -2.055 + 0.3756 \log(\bar{Y})$$

i.e. $\hat{\beta} = b \simeq 0.38$, with a t-value for the regression coefficient of

$$t(b) = 0.29$$

which shows no significance. So we accept that $\beta = 0$, i.e. there is no dependence between the cell mean and cell variance. Therefore, there is no need for any transformation, and for the LL-response we can use

$$\text{NPM(LL)} = -10 \log_{10}(s^2).$$

For the PER-response (Table 6.16b) we can similarly find

$$\log(s) = -2.7 + 1.035 \log(\bar{Y})$$

with $t(b) = 1.590$; so, there seems to be a relationship between the cell mean μ and the cell variance σ^2 for the PER-response of the form $\sigma = \alpha\mu^{\beta}$, where $\beta \approx 1$. This indicates the need for the log-transformation on the original PER-data. Then we can use either

$$\text{NPM(PER)} = -10 \log_{10}(S_T^2),$$

where S_T^2 is the cell variance in the log-transformed scale, or

$$\text{NPM}'\text{(PER)} = 10 \log(\bar{Y}/s)^2$$

on the original scale.

Remark: NPM′ is equivalent to Taguchi's signal-to-noise ratio (SNR). Therefore we note that SNR is appropriate only if there is a direct linear relationship between the cell mean and the cell variance ($\beta = 1$), i.e. a

TABLE 6.16b. PER-response data

	First wafer					Second wafer				
Trial	N	C	S	E	W	N	C	S	E	W
1	0.87	0.75	0.81	0.82	0.81	0.85	0.68	0.85	0.77	0.85
2	0.72	0.66	0.62	0.63	0.64	0.66	0.68	0.64	0.68	0.68
3	0.72	0.63	0.66	0.68	0.72	0.72	0.68	0.68	0.72	0.70
4	0.59	0.54	0.51	0.54	0.56	0.56	0.49	0.50	0.52	0.54
5	0.77	0.79	0.79	0.79	0.87	0.86	0.79	0.80	0.84	0.84
6	0.72	0.66	0.70	0.71	0.71	0.86	0.74	0.84	0.80	0.82
7	0.63	0.57	0.60	0.70	0.66	0.60	0.52	0.58	0.67	0.68
8	0.48	0.47	0.50	0.55	0.49	0.52	0.46	0.50	0.52	0.53

need for a log-transformation. If the log-transformation is not made, then we can use the SNR for the determination of the variability-control factors. In our particular case, we transform the PER data to the logarithmic scale and we use the NPM(PER). However, in this case whether we use NPM or NPM′, as defined above, similar conclusions are obtained.

6.6.4 *Determination of variability-control and target-control factors*

One of the main objectives of off-line quality control is the minimization of the variability in the response of interest. This can be achieved by the determination of the variability control factors (the factors which significantly affect this variability) and their optimal levels.

The statistical measures which reflect the variability in the response are the noise performance measures (NPM). The statistical analysis of these measures will determine the variability-control factors. The optimal levels of these factors will be the levels which maximize the value of NPM (minimize the variation). The target-control factors are factors which do not affect the NPM but have a large effect on the mean response. A measure reflecting the mean response is the Target Performance Measure (TPM) which is usually a function of the sample cell mean $\bar{Y}$.

6.6.4.1 Analysis of the LL-response data

In Section 6.6.3 we have shown that the appropriate NPM and TPM measures for the LL-response are:

$$\text{NPM(LL)} = 10\log_{10}(s^2) \tag{6.14a}$$

TABLE 6.17a. TPM and NPM for LL-response data

Trial	$TPM(LL) = \bar{Y}$	$s(LL)$	$NPM(LL) = -10\log_{10}(s^2)$
1	0.6897	0.13544	17.365
2	0.4163	0.05766	24.783
3	0.5146	0.07258	22.784
4	0.5728	0.11069	19.118
5	0.4980	0.04202	27.531
6	0.4892	0.16870	15.458
7	0.4132	0.19800	14.067
8	0.5482	0.10990	19.180

TABLE 6.17b. Level averages for NPM and TPM (LL-response)

	NPM(LL)		TPM(LL)	
Factor	Level 1	Level 2	Level 1	Level 2
F1 (T)	21.01	19.06	0.548	0.487
F2 (Pre)	21.28	18.78	0.523	0.512
F3 (Cl_2)	18.85	21.22	0.517	0.519
F4 (P13)	20.44	19.63	0.529	0.506
F5 (P100)	18.70	21.37	0.560	0.475
F6 (D100)	20.80	19.27	0.577	0.458
F7 (O_2)	16.50	23.57	0.541	0.494

and

$$TPM(LL) = \bar{Y}. \tag{6.14b}$$

Using the data from Table 6.16a, we can easily calculate the above measures and these are shown in Table 6.17a.

We can now obtain the average NPM and TPM for each of the 2 levels of every controllable factor and these are given in Table 6.17b.

A graphical representation of these is given by Figs 6.17a and 6.17b. Figure 6.17a represents the NPM level averages as compared with the overall mean NPM of all 8 experiments (dotted line). We notice that only the level averages of F7(O_2) are significantly different from the overall mean. This can be confirmed by the statistical analysis of NPM which can be made either by the stepwise regression technique (which determines the significant factors in a stepwise manner starting with the most significant first) or by the analysis of variance (ANOVA) technique

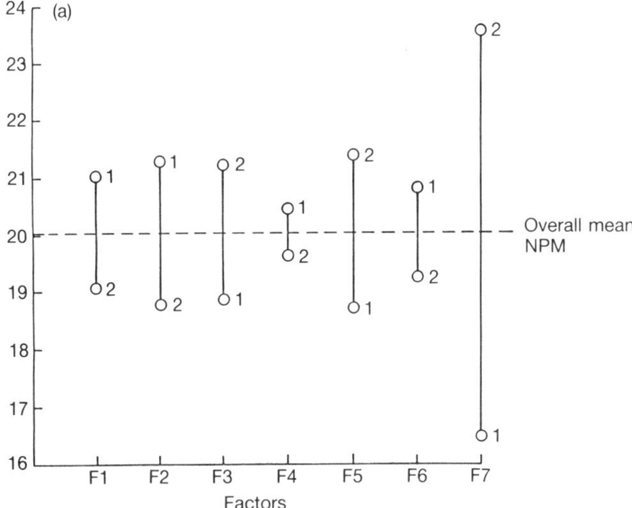

FIG. 6.17a. Factor effects on $\text{NPM(LL)} = -10\log_{10}(s^2)$.

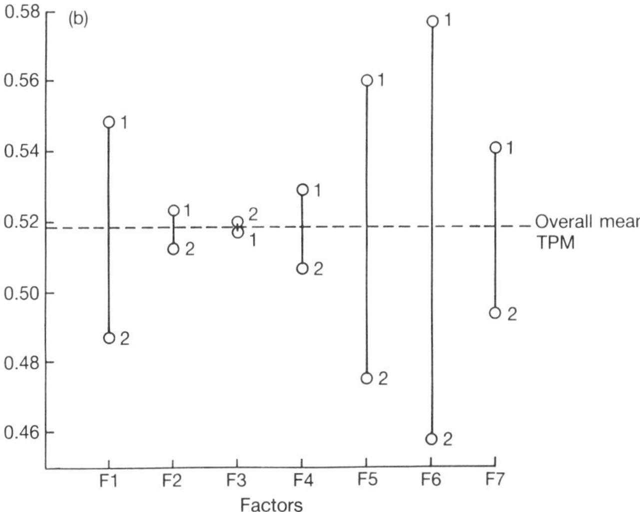

FIG. 6.17b. Factor effects on $\text{TPM(LL)} = \bar{Y}$.

('pooling' together the factors for which the sum of squares constitute less that 5% of the total sum of squares). Half normal plotting (see Section 2.6.4.2) can also be used for confirmation purposes.

Using any of the above techniques (and allowing significance levels up to 10%), only factor F7(O_2) is found to significantly affect the NPM, i.e.

TABLE 6.17c. (Pooled) ANOVA for TPM (LL-response)

Source	df	S	MS	F-ratio	$\beta = 1 - \frac{1}{F}$
F1 (T)	1	0.0075	0.0075	18.1	0.945
F5 (P100)	1	0.01457	0.01457	35.2	0.972
F6 (D100)	1	0.0283	0.0283	68.3	0.985
F7 (O_2)	1	0.00441	0.00441	10.6	0.906
Residual	3	0.001243	0.000414		
Total	7	0.0557			

the variability within the LL-response. The optimum level for this factor is the level with the highest average value of NPM, i.e. level 2. So O_2 is a variability control factor for LL with optimal level, the second i.e. the level of 1.19 sccm. Similarly for Fig. 6.17b we notice a large effect due to factors F1(T), F5(P100), F6(D100), and F7(O_2). This can be confirmed from the ANOVA Table 6.17c. (The residual sum of squares was obtained by 'pooling' together the sum of squares of the factors F2–F4).

Since factors T, P100, and D100 seem to affect only the mean response and not the variability, they can be considered as target-control factors for LL.

Remark: For cases when the desired response is 'the smaller the better' (as in the LL-case), Taguchi recommends (see Section 6.2.2.3) the following noise performance measure:

$$\text{SN} = 10 \log_{10}\left(\frac{\sum Y^2}{n}\right)$$

where $\sum Y^2$ is the sum of the n squared data values of each cell. But, an analysis of the SN can show significant effects due to F1, F5, F6, and F7. These conclusions are identical to the ones obtained with the analysis of $TPM = \bar{Y}$. This can be explained by the fact that

$$\frac{\sum Y^2}{n} = \bar{Y}^2 + \frac{n-1}{n} s^2.$$

So SN is biased by the mean to a large degree, especially when $\bar{Y}$ differs significantly between the factor levels as in our case. Therefore the use of SN for the determination of control factors is not recommended, since this can cause an undesired 'confounding' between variability-control and target-control factors (see Section 6.4). On the other hand, the use of (6.14a) and (6.14b), which were decided after the exploration for the need of a data transformation, led to much clearer results: F7 is the only factor associated with dispersion effects (variability-control factor) whereas F1, F5, and F6 are associated only with location effects (target-control factors).

6.6.4.2 Analysis of the *PER*-response data

In Section 6.6.3 we have shown that the PER-response data required the logarithmic transformation. After transforming the data, we calculate the new sample mean $\bar{Y}_T$ and sample standard deviation S_T for each cell, from which we determine the TPM and NPM measures for the PER-response (see Table 6.18a). We can now follow a similar analysis to that in Section 6.6.4.1. After calculating the level-averages for NPM and TPM for all the factors (see Table 6.18b) we can draw Figs. 6.18a and 6.18b. From Fig. 6.18a, we notice a significant effect on the NPM(PER) due to factor F7(O_2). From Fig. 6.18b we note significant effects due to

TABLE 6.18a. TPM and NPM for PER-response data

Trial	TPM(PER) = $\bar{Y}_T$	S_T(PER)	NPM(PER) = $-10\log_{10}(S_T^2)$
1	−0.2181	0.07510	22.486
2	−0.4149	0.04498	26.940
3	−0.3705	0.04514	26.908
4	−0.6474	0.08348	21.569
5	−0.2224	0.06323	23.982
6	−0.2834	0.08973	20.941
7	−0.4803	0.09309	20.622
8	−0.6906	0.05612	25.018

TABLE 6.18b. Level-averages for NPM and TPM (PER-response)

Factor	NPM(PER)		TPM(PER)	
	Level 1	Level 2	Level 1	Level 2
F1 (T)	25.30	23.50	−0.4076	−0.4152
F2 (Pre)	24.44	24.35	−0.280	−0.540
F3 (Cl_2)	23.77	25.03	−0.451	−0.372
F4 (P13)	24.35	24.44	−0.319	−0.504
F5 (P100)	23.84	24.96	−0.391	−0.432
F6 (D100)	24.94	23.85	−0.436	−0.387
F7 (O_2)	22.23	26.57	−0.402	−0.421

TABLE 6.18c. ANOVA for TPM(PER)

Source	df	S	MS	F-ratio	$\beta = 1 - \frac{1}{F}$
F2(PRE)	1	0.13658	0.13658	61.3	0.984
F3(Cl_2)	1	0.012518	0.012518	5.6	0.821
F4(P13)	1	0.068495	0.068495	30.7	0.967
Residual	4	0.008916	0.002229		
Total	7	0.22651			

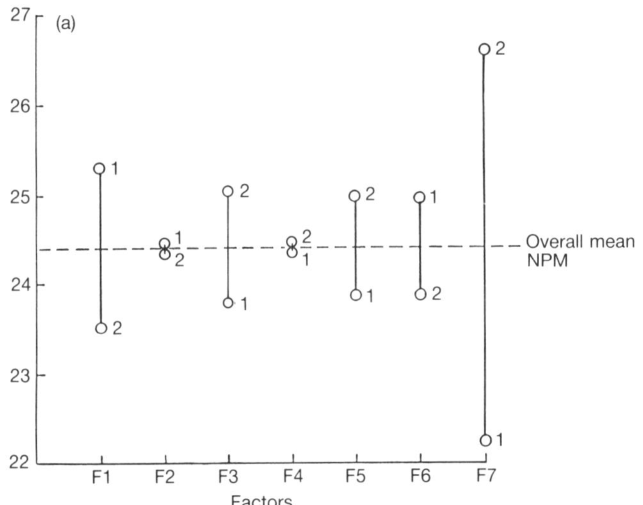

FIG. 6.18a. Factor effects on NPM(PER) = $10 \log_{10}(s_T^2)$.

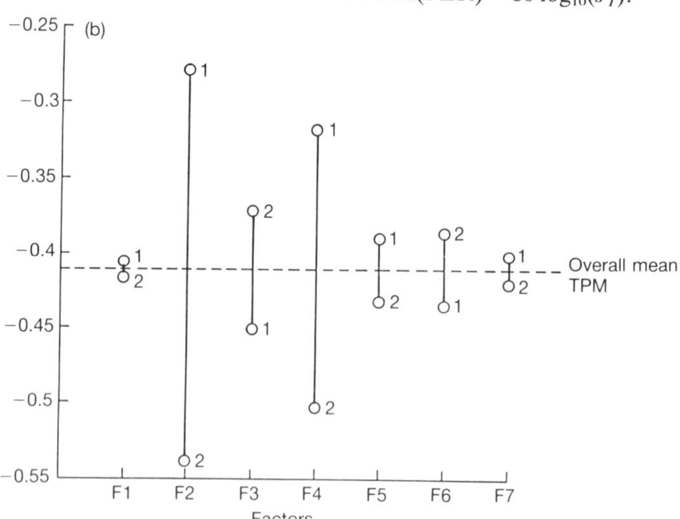

FIG. 6.18b. Factor effects on TPM(PER) = $\bar{Y}_T$.

factors F2(PRE) and F4(P13) and to a lesser degree due to F3(Cl_2). These can be confirmed by the statistical analyses of NPM and TPM (see, for example, Table 6.18c for the ANOVA of TPM).

The above analyses indicate that F7(O_2) is a variability-control factor for PER, with level 2 (1.19 sccm) as optimal, whereas F2, F3, and F4 are target-control factors.

6.6.5 Determination of 'optimal' conditions

From the analyses of Section 6.6.4 an obvious conclusion is that factor F7(O_2) is a variability-control factor for both the main responses of interest LL and PER, with optimal level (for minimum variability) the second level (1.19 sccm).

None of the other factors F1–F6 seem to affect the variability, but they do seem to affect the mean response. Their optimal conditions can be determined by studying their level-averages (for TPM).

For example, for the main response of interest LL, the 'best' level (within the experimental region) for the target-control factors are the ones with the smallest TPM value, i.e. level 2 (15 °C) for T, level 2 (25 watts) for P100 and level 2 (67%) for D100. This is because we want LL to be as low as possible, and ideally always be below 0.4. We would like to be able to predict whether, in the long run, the above settings for the controllable factors could result (with at least 95% confidence) in an LL-response below 0.4 on average. We would also like to know how far beyond the experimental range we should 'adjust' the target-control factors, so that the required targets are achieved. All these could dramatically reduce the effort required (through 'confirmatory' experiments) for the determination of the optimal factor settings.

6.6.5.1 Using regression models

Using the multiple regression technique, one can 'fit' response models to the available data, which can then be used to predict the response (see, for example, Logothetis and Haigh 1988). A measure of the prediction ability of these models, is given by their 'coefficient of determination' ($\mathbf{R}^2$) which should be as high as possible (in percentage terms). In our present case, the fitted models produced a low value for $\mathbf{R}^2$ for two out of the three responses of interest, and so it was decided not to use this prediction method.

6.6.5.2 Using the 'discount' ('beta') coefficients

This method is highly recommended by Taguchi (see Section 4.4.3) and it makes use of the level-averages and of the 'discount' (or 'beta') coefficients, which, in the case of two-level factors, are defined (for each factor) as

$$\beta = \begin{cases} 1 - \dfrac{1}{F} & \text{if} \quad F > 1 \\ 0 & \text{if} \quad F = 1, \end{cases}$$

where F is the F-ratio in the ANOVA table. We will apply this method to our case, using the level averages of Tables 6.17b and 6.18b and the F-ratios of Tables 6.17c and 6.18c (for applying the method to factors with 3 or more levels, see Section 4.4.3).

We will concentrate on the most important response, i.e. LL. We have already shown that the best levels of the significant factors within the experimental range are the second level for T, P100, D100, and O_2. The average TPM at these levels can be found in Table 6.17b and are

$$(\mathrm{T})_2 = 0.487, \quad (\mathrm{P100})_2 = 0.475, \quad (\mathrm{D100})_2 = 0.458, \quad \text{and} \quad (\mathrm{O}_2)_2 = 0.494.$$

The overall average (of all 8 experimental trials) is

$$(\mathrm{Av}) = 0.5175.$$

The discount (beta) coefficients corresponding to these factors can be found in the last column of Table 6.17c headed by 'β', i.e. $\beta(T) = 0.945$, $\beta(\mathrm{P100}) = 0.972$, $\beta(\mathrm{D100}) = 0.985$, and $\beta(\mathrm{O}_2) = 0.906$. Now, an estimated value of the process average $\hat{\mu}$, if the above factors are set at their optimal levels, is given by (see Section 4.4.3)

$$\hat{\mu} = (\mathrm{Av}) + \beta(\mathrm{T})[(\mathrm{T})_2 - (\mathrm{Av})] + \beta(\mathrm{P100})[(\mathrm{P100})_2 - (\mathrm{Av})] + \beta(\mathrm{D100})[(\mathrm{D100})_2 - (\mathrm{Av})] + \beta(\mathrm{O}_2)[(\mathrm{O}_2)_2 - (\mathrm{Av})]. \qquad (6.15)$$

Thus, in our case

$$\hat{\mu} = (0.5175) + (0.945)\{0.487 - 0.5175\} + (0.972)\{0.475 - 0.5175\} + (0.985)\{0.458 - 0.1575\} + (0.906)\{0.494 - 0.5175\} = 0.3675.$$

A 95% confidence interval for the long-run process average is given by

$$\hat{\mu} \pm t_e S_e / \sqrt{n_e}$$

where $S_e = \sqrt{MS_e}$ where MS_e is the mean sum of squares for the residual, i.e. in our case from Table 6.17c. $S_e = \sqrt{0.000414} = 0.02035$, t_e is the t-value from the statistical t-tables (see Appendix E) corresponding to a 95% two-sided interval and to 3 residual degrees of freedom, i.e. $t_e = 3.18$. The number n_e is the 'effective number of replications' defined by (see Section 4.4.2)

$$n_e = \frac{\text{Total number of experiments}}{1 + \Sigma(\beta)\,(\mathrm{df})}$$

where the sum $\sum$ is over the factors that were not disregarded when $\hat{\mu}$ was estimated (i.e. F1, F5, F6, F7); that is,

$$n_e = \frac{8}{1 + (0.945)(1) + (0.972)(1) + (0.985)(1) + (0.906)(1)} = 1.644.$$

Finally, a 95% confidence interval for the LL-process average is given by

$$0.37675 \pm (3.18)(0.02035)/\sqrt{1.664} = 0.3675 \pm 0.0502.$$

Therefore, in the long run, if the chosen factors were set at their (optimal) second level, we can predict (with 95% confidence) that the LL-response could be anything from 0.3173 to 0.4177.

Since the upper limit is over our desired limit of 0.4, we can 'adjust' one of the target-control factors to 'move' the upper limit to below 0.4. We chose to use the factor T, because it was known (from past experience) that it has a linear effect on LL. We can easily estimate an average response of 0.4565 for a temperature level of 10 °C [considering the average responses at 25 °C (level 1) and 15 °C (level 2)].

If we replace the value for $(T)_2 = 0.487$ with 0.4565 in formula (6.15) we obtain $\hat{\mu} = 0.3386$, which produces a confidence interval of 0.3386 ± 0.0502, or [0.2884, 0.3888]. These limits are well within the desired range for LL, and the decision is taken to use the following settings:

T near 10 °C, P100 near 25 watts,

D100 near 67% and O_2 *at* 1.19 sccm.

Following exactly the same procedure as above and also considering the third response RL, optimal levels for the other factors were determined as follows:

PRE at 144 millitorr, Cl_2 at 13 sccm and P13 at 80 watts.

Under the above conditions, a predicted confidence interval for the PER-response can be calculated (using Tables 6.18b and 6.18c and then retransforming back to the original scale) to be [0.486, 0.582]. These limits are well within the desired range for PER.

6.6.5.3 Using only the level averages

This is a similar procedure to the one described in Section 6.6.5.2, with the only difference that all the discount coefficients are assumed to be equal to 1. As the discount coefficients for the case of the LL- and PER-responses are close to 1, the predicted intervals are very similar.

6.6.6 Confirmation

Having tentatively determined the optimal factor settings, using only the available experimental data, some confirmatory experiments need to be conducted. The authors believe that such confirmation experiments should also be carried out in a properly designed and systematic manner: the most important factors should be allowed to systematically vary near their tentative optimum levels (perhaps according to a small experimental design) and in that way, additional valuable information can be gained while the final operating conditions can be established beyond doubt. Such properly conducted confirmation trials in our case provided no grounds for changing the optimal condititions established in Section 6.6.5.2.

Subsequent monitoring under those optimal parameter settings confirmed that optimization and robustness had been achieved. In particular

Response	Results	(Desired)
LL:	an average response of 0.25 ± 0.08	(less than 0.4)
PER:	an average response of 0.496 ± 0.02	(more than 0.4)
RL:	an average response of 0.45 ± 0.04	(less than 0.5)

These results show all responses to be well within the desired range, agree well with the predictions, and indicate a 55 per cent improvement in the mean of the most important response LL and a 50 improvement in the variability within the PER and RL response.

6.6.7 Concluding remarks

In the previous sections, we hope to have demonstrated the ability of the Taguchi approach to model and optimize successfully a multi-response dry-etching process. Statistical methods can bring order to a complex process with a limited number of experiments. The use of data transformation is a useful addition to the methods and helps in the choice of a suitable noise performance measure. Optimization followed by a confirmatory experiment led to a real improvement in response and reduction in variability. This example contains most of the ingredients of the methods.

6.7 Bibliography

A detailed outline of Taguchi's approach to experimental design can be found in Taguchi (1987). The most important points of his approach to 'parameter and tolerance design' are adequately described in Taguchi (1986); see also Kackar (1985), Burgam (1985), Barker (1986*a,b*), Byrne and S. Taguchi (1986), Phadke (1982), Taguchi and Phadke (1984), and Taguchi and Wu (1980).

Some constructive criticisms and alternatives to Taguchi's approach for design and analysis, can be found in Box (1986), Wu, Mao, and Ma (1987), Box (1988), Box, Bisgaard, and Fung (1988), Logothetis (1987*a,b*), Logothetis (1988). Critique and alternatives to Taguchi's accumulating analysis method can be found in Box and Jones (1986), Hamada and Wu (1986), and Nair (1986).

For 'Taguchi case studies' the reader is referred to the references quoted in Section 6.4 and also to Taguchi (1987), Phadke (1986), Logothetis and Haigh (1987), and to the case-studies presented in the five 'symposiums on Taguchi methods' published by the American Supplier Institute, Inc.

Suggestions on how to deal with single replicate experiments are given in Box and Meyer (1986*a,b*), Meyer (1987), and Meyer and Box (1987).

See also Song and Lawson (1988) for an alternative strategy to study the controllable and noise factors which can reduce the experimentation efforts. Phadke and Dehnad (1988) discuss the connection of Taguchi's quadratic loss function with the optimization of product and process design. An evaluation of Taguchi's signal-to-noise ratio and techniques for choosing the appropriate noise performance measure (independent of the mean), can be found in Leon *et al.* (1987), Box (1986), Box and Ramirez (1986), and Logothetis (1988).

Some papers advocating the usefulness of data-transformation are those by Curtiss (1943), Bartlett (1947), Box and Tidwell (1962), Box and Cox (1964), Perry (1987), and for the connection with Taguchi's approach see Box and Ramirez (1986), Nair and Pregibon (1986), and Logothetis (1988).

The material on variance estimation in Section 6.5.2 dates back to Bartlett and Kendall (1949). Recent work, partly stimulated by the Taguchi methodology, has been termed the study of 'dispersion effects'. A representative paper is Carroll and Rupert (1982).

7

SIMULATION AND TOLERANCE DESIGN

7.1 A general view

The availability of high-speed computers allows simulation to be performed on computer models of a component or system in order to achieve a number of different aims. This should be seen as following on from the building of mathematical and statistical models using off-line experimentation. Some parts of the system may already be sufficiently well understood to allow them to be coded into the computer with some degree of confidence. For example in electrical circuit design, parts of the circuit function may be well understood, but other parts such as the study of active devices (such as transistors) may need separate modelling to assess the effects of fabrication on their behaviour.

If the computer code is considered as a physical entity in its own right then simulations can be thought of as *computer experiments* and some of the techniques of experimental design may be applied. What constitutes noise in such situations is problematical. For given inputs x the programme should yield the same output $Y(x)$ for every computer run. However, some parts of the input may be considered as 'noise factors' and others as 'design factors'. For example, noise factors may be factors which reflect manufacturing tolerances or material variation. Design factors are those that describe the main parameters of the design: geometry, electrical characteristic, layout, strengths, and so on. But even these may be subject to some uncertainty. Thus variation around nominal values of design factors can also be considered as noise. The latter is aptly named 'tolerance' and great care needs to be taken in distinguishing in computer models tolerances on the active design factors and noise on the more passive non-design factors. An example would be the modelling of the effect of ambient temperature on a component. We may build a separate model into the programme to represent, say, the temperature characteristic of a new material but the temperature itself is not part of the design and so its effect on the behaviour is the effect due to the variation in temperature. It may be possible to perform a 'worst case' analysis to guard against extreme temperatures but this is itself an attempt to model the effect of extreme variation.

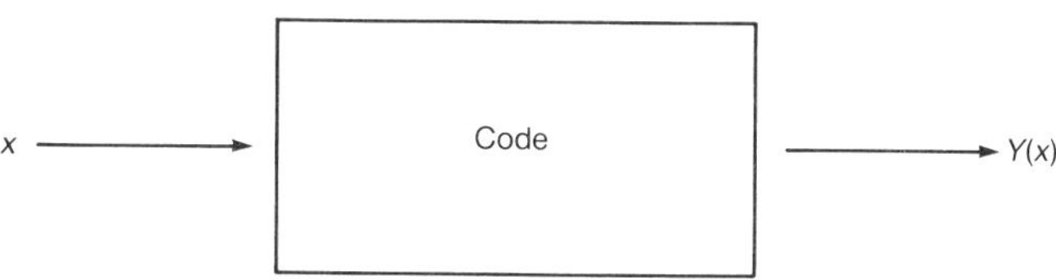

FIG. 7.1. Computer code.

Some further distinctions may need to be made between design factors which are structural in nature such as the strength characteristic of bought-in components and settings for variables which are easily tuned or adjusted.

The simple diagram of Fig. 7.1 considers the computer code as a black box with inputs and outputs. In viewing a simulation as a computer experiment there are several aspects which are easier to implement than in real physical experimentation.

(i) measuring the value of variables;
(ii) changing the level of variables;
(iii) holding variables constant;
(iv) allowing a wide range of variables;
(v) using sequential methods.

The simulation package will give for each input value x a fixed output $Y(x)$. If the input is random (or pseudo-random) then the output will be random. We should not confuse, however, random input intended to simulate the actual random behaviour of an input variable, say x_N, and random inputs generated as a technical device to elucidate some feature of $Y(x)$. The latter is called Monte Carlo methods and we give a discussion in the next section. We favour alternatives which would input *deterministic* settings of x and fit a model of some kind to the computer code. This approach is particularly valuable when the code itself is expensive to run. The approach is summarized as

(i) choose a model (submodel) for the code, $Y(x)$;
(ii) design an experiment for the model;
(iii) fit the model: $\hat{Y}(x)$;
(iv) use $\hat{Y}(x)$ (rather than $Y(x)$) to analyse $Y(x)$.

Here are some objectives for $Y(x)$ which may be easier to perform on the approximation $\hat{Y}(x)$.

(i) *Sensitivity analysis:* show how small changes in x affect small changes in Y. This can be discovered by perturbing x around nominal values and evaluating $Y(x)$ (or $\hat{Y}(x)$). In some cases one can compute derivatives: $\partial Y/\partial x_i$. (See Section 7.4).

(ii) *Evaluation of average stochastic (random) behaviour* of $Y(x)$ for stochastic x. This may simply be an average or expectation

$$\int Y(x)p(x)\,\mathrm{d}x$$

where $p(x)$ is some input density. One can compute variances and so on. (See Section 7.2.)

(iii) *Optimization:* finding the maximum of some function of Y, such as one of the integrals in (ii). One special objective is sometimes called the 'yield'

$$\mathrm{Prob}(Y > c) = \int I(x)p(x)\,\mathrm{d}x$$

where $I(x)$ is the indicator function for the set $S = \{Y > c\}$, i.e. $I(x) = 1$ on S and 0 elsewhere.

(iv) *Target attainment:* find the x for which $Y(x) = c$. This is 'root-finding' and will typically be easier to perform on $\hat{Y}$ rather than Y. One method is described in Section 7.7. We can perform target attainment while minimizing the variance of Y (or $\hat{Y}$). This is really a special case of (ii).

(v) *Estimation of long-range behaviour:* often processes can reach stability after many iterations through time. For example, one may be interested in the long-run cost of a replacement policy in reliability/renewal problems. Such long-range behaviour is often represented by a *time average.*

In this chapter we will describe some specific methods for simulation and for tolerance design and analysis.

7.2 Monte Carlo methods

This is, as explained, a method for finding integrals or expectations of Y with respect to input densities on x. We repeat that this can either be considered as (i) a purely numerical technique or (ii) a method of *post hoc* integration with respect to variation representing real noise. It should only be used when observations (runs) on the computer code are cheap. However, a number of variance reduction methods have been developed to improve the estimation procedures and three examples are given here.

Assume that $Y = f(x)$ relates a single output Y to a single input x. We must assume that $f(x)$ is known in the sense that there is a computer code to produce Y for given x. This is to be distinguished from $Y = f(x)$ being known in closed form, such as a polynomial, although this may be the case. We are interested in obtaining an estimate of the average behaviour of Y, given the random behaviour of x, i.e. an estimate of

$$\mu_Y = E(Y) = \int_a^b f(x)p(x)\,\mathrm{d}x,$$

where $[a, b]$ is some interval in which x lies. It may be $(-\infty, \infty)$. The crude estimate of μ is based on observations of Y obtained by taking a random sample from the $p(x)$ density $x_1, \ldots, x_n$, computing $Y_1 = f(x_1), \ldots, Y_n = f(x_n)$ and setting

$$\hat{\mu}_Y = \frac{1}{n}\sum_{i=1}^{n} f(x_i).$$

Note that in order to generate the values $x_1, \ldots, x_n$ there must be a pseudo-random number generator and a mechanism for translating this into random observations from the $p(x)$ density. We do not discuss these issues here.

The estimator $\hat{\mu}$ is unbiased. Indeed:

$$\begin{aligned} E(\hat{\mu}_Y) &= \frac{1}{n}\sum_{i=1}^{n} E[f(x_i)] \\ &= \frac{1}{n}\sum_{i=1}^{n} \mu_Y = \frac{n\mu_Y}{n} = \mu_Y, \end{aligned}$$

where we have used the fact that each x_i is sampled from the same distribution. Also we can compute the variance

$$\operatorname{var}(\hat{\mu}_Y) = \frac{1}{n^2}\sum_{i=1}^{n} \operatorname{var}(Y_i) = \frac{1}{n^2} n\sigma_Y^2 = \frac{1}{n}\sigma_Y^2$$

where σ_Y^2 is the variance of Y, i.e.

$$\sigma_Y^2 = \int_a^b (f(x) - \mu_Y)^2 p(x)\,\mathrm{d}x.$$

This provides a convenient benchmark which we may try to improve upon by using variance reduction methods.

7.2.1 *Importance sampling*

This goes under various names and has been rediscovered in different fields. We write μ as

$$\mu = \int_a^b f(x)\frac{p(x)}{q(x)}q(x)\,\mathrm{d}x,$$

where $q(x) > 0$ is some other density function. If we sample from $q(x)$ rather than $p(x)$ and write

$$w(x) = \frac{p(x)}{q(x)},$$

then

$$\mu = \int_a^b f^*(x)q(x)\,\mathrm{d}x,$$

where

$$f^*(x) = w(x)f(x).$$

Our new estimator is

$$\mu^* = \frac{1}{n}\sum_{i=1}^{n} f^*(x_i),$$

namely, the crude estimator, with f replaced by f^* and p by q.

The rationale for this method is that, if we use a density $q(x)$ which is close to

$$q_0(x) = \frac{f(x)p(x)}{\mu},$$

then we improve the estimate. Thus, if we use $q_0(x)$ itself, then

$$w(x) = w_0(x) = \frac{p(x)}{q_0(x)}$$

and

$$\begin{aligned} f^*(x) &= w_0(x)f(x) \\ &= \frac{p(x)}{q_0(x)}\cdot f(x) = \frac{p(x)}{f(x)}\cdot\frac{f(x)}{p(x)}\cdot\mu = \mu, \end{aligned}$$

so

$$\mu^* = \frac{1}{n}\sum_{i=1}^{n}\mu = \frac{n\mu}{n} = \mu.$$

Then automatically $\mathrm{var}(\mu^*) = 0$. A caveat is that we must have $q(x) > 0$ whenever $p(x) > 0$ and $f(x) > 0$, and this may not hold. Also it is possible to increase var(μ^*) if our guess for $q(x)$ is bad. However, μ^* is unbiased in the same way that $\hat{\mu}$ was.

A useful application of the method is to find estimators of the 'tail area', the probability $(Y > c)$ mentioned in the last section. Thus

$$p = \text{Prob}(Y > c) = \int_c^\infty f(x)p(x)\,\mathrm{d}x,$$

where $f(x)$ is the step-function (indicator function)

$$f(x)\begin{cases} = 1 & (x > c) \\ = 0 & (x \leq c) \end{cases}.$$

The caveat above says that we must take $q(x) = 0$ for $x \leq c$. Then guess at a $q(x)$ so that approximately

$$\begin{aligned} q(x) &\propto f(x)p(x) \\ &= p(x) \quad \text{for} \quad x > c. \end{aligned}$$

Essentially, then, we need a cheaper density with roughly the same shape in the tail. The crude estimator merely takes

$$\hat{p} = \frac{1}{n} \times (\text{number of } x_i > c)$$

where n is the sample size. We can compute the variance of $\hat{p}$ using the properties of the binomial distribution (see Appendix B).

$$\text{var}(\hat{p}) = p(1-p)/n.$$

A cheap estimator of this is $\hat{p}(1-\hat{p})/n$. This provides the bench mark.

Example 7.1: This is an artificial example to show the use of the method of estimating a tail probability, as above. To fix matters we take a case when p can be evaluated from tables (not usually true!). Thus let $p(x)$ be a mixture

$$p(x) = \tfrac{1}{2}p_1(x) + \tfrac{1}{2}p_2(x)$$

where $p_1(x)$ and $p_2(x)$ are the densities of normal random variables $N(-1, 4)$ and $N(1, 1)$ respectively.

Let $p = \text{Prob}(X > 2.5) = 1 - \tfrac{1}{2}[F_1(2.5) + F_2(2.5)]$ where F_1 and F_2 are the respective cumulative distribution functions; this is just

$$p = \tfrac{1}{2}[1 - \Phi(1.75)] + \tfrac{1}{2}[1 - \Phi(1.5)],$$

where Φ is the cumulative distribution function of the standard normal distribution. From Tables, this (to 5 dec. pl.) is 0.05343.

Using variance reduction we took $q(x)$ to be the density of $2.5 + |Z|$

where Z was a $N(0, 1)$ random variable. The estimated standard deviation from 100 samples each of size 100 using the variance reduction above was 0.00255. The exact standard deviation of the crude estimate $\hat{p}$ is $(p(1-p)/n)^{\frac{1}{2}} = 0.0225$, around ten times as much.

7.2.2 *Stratified sampling*

When Y is itself the sum of Y_i from different components we may write

$$Y = Y_1 + \cdots + Y_L$$

and

$$\mu_Y = \mu_1 + \cdots + \mu_L,$$

where $\mu_i = E(Y_i)$. Suppose we have the crude estimators for each μ_i based on independent sampling with sample size n_i; then we can write as an estimator for μ_Y

$$\hat{\mu}_Y = \hat{\mu}_1 + \cdots + \hat{\mu}_L.$$

This is unbiased because

$$\begin{aligned} E(\hat{\mu}_Y) &= E(\hat{\mu}_1) + \cdots + E(\hat{\mu}_L) \\ &= \mu_1 + \cdots + \mu_L = \mu_Y. \end{aligned}$$

Moreover

$$\text{var}(\hat{\mu}_Y) = \frac{\sigma_1^2}{n_1} + \cdots + \frac{\sigma_L^2}{n_L},$$

where $\sigma_i^2 = \text{var}(Y_i)$. We can think of each Y_i as coming from a separate stratum. Sample independently from the separate strata and add up the results.

Now if the total sample size is fixed, $n_1 + \cdots + n_L = n$, we can select the n_i to minimize $\text{var}(\hat{\mu}_Y)$ so long as we know the σ_i^2. Thus minimize $(\sum \sigma_i^2/n_i)$ subject to $\sum n_i = n$. Using a Lagrange multiplier λ we can seek to minimize

$$\sum \frac{\sigma_i^2}{n_i} + \lambda\left(\sum n_i - n\right),$$

giving partial derivatives

$$-\frac{\sigma_i^2}{n_i^2} + \lambda = 0;$$

solving for λ the optimum values are

$$n_i^* = \frac{n\sigma_i}{\sum \sigma_i}.$$

The minimum value of $\text{var}(\hat{\mu}_Y)$ is

$$\sum \frac{\sigma_i^2}{n_i^*} = \sum \frac{\sigma_i^2}{n\sigma_i} \sum \sigma_i = \frac{1}{n}\left(\sum \sigma_i\right)^2.$$

More generally we can consider a weighted sum

$$Y = \sum c_i Y_i \qquad \mu_Y = \sum c_i \mu_i \qquad \hat{\mu}_Y = \sum c_i \hat{\mu}_i$$

yielding

$$\text{var}(\hat{\mu}_Y) = \sum \frac{c_i^2 \sigma_i^2}{n_i}$$

$$n_i^* = \frac{nc_i\sigma_i}{\sum c_i\sigma_i}$$

and the minimum $\text{var}(\hat{\mu}_Y)$ is

$$\frac{1}{n}\left(\sum c_i\sigma_i\right)^2.$$

7.2.3 *Control variables*

(It should be noted that this terminology is merely historical and has nothing to do with the 'controllable variables' of Chapter 6).

This method is a deliberate attempt to use information on the auxiliary X to improve estimates of Y-quantities. Instead of estimating μ_Y directly from the Y-values themselves we try to estimate

$$\mu^* = E(Y^*),$$

where Y^* is a corrected variable

$$Y^* = Y - \beta X$$

and β is some real constant to be estimated.

If we assume that the mean of X is zero (which can easily be arranged by measuring X from the mean X value) then

$$E(Y^*) = E(Y) - \beta E(X) = E(Y) = \mu,$$

so that Y^* and Y have the same mean. Moreover,

$$\text{var}(Y^*) = \text{var}(Y) + \beta^2 \text{var}(X) - 2\beta \text{cov}(Y, X) < \text{var}(Y),$$

provided that

$$2\beta \text{cov}(Y, X) > \beta^2 \text{var}(X)$$

or if $\beta > 0$

$$\beta < \frac{2 \text{cov}(Y, X)}{\text{var}(X)}.$$

(See Appendix B for formulae on variances.) The best value of β to minimize var(Y^*) is

$$\beta^* = \frac{\text{cov}(Y, X)}{\text{var}(X)},$$

leading to

$$\text{var}\,(Y^*) = (1 - \rho^2)\text{var}(Y),$$

where $\rho = \text{corr}(Y, X)$. We could use the *sample* value of β^* namely

$$\frac{\frac{1}{n}\Sigma\,(Y_i - \bar{Y})(X_i - X)}{s_X^2},$$

or something similar. We call X used for this purpose a *control variable.* In statistical terminology Y^* is really a residual.

Having chosen a suitable value of β we then proceed to use a crude estimator of μ

$$\mu^* = \frac{1}{n}\sum Y_i^*.$$

The idea can be stated simply: by using a residual or estimated residual we acquire less variation in our estimator.

An even better method is to perform a regression of Y on X as follows. Suppose we observe values

$$(X_1, Y_1), (X_2, Y_2), \ldots, (X_n, Y_n).$$

Assuming a rough model

$$Y_i = \beta_0 + \beta_1 X_i,$$

we estimate β_0 and β_1 in the usual way using straight line regression (see Section 2.1.4). This gives

$$Y_i^* = Y_i - \hat{\beta}_0 - \hat{\beta}_1 X_i.$$

We then proceed to obtain μ^* as above. A refinement is to replace any sample quantity depending on the x-values by population quantities if these are known. The population x-values act as a lever to correct the simulation. Such estimates are called regression estimates. They are particularly useful when the X_i are cheap to measure but the Y_i are expensive or when the precise distribution of the X_i (or some moments) are known. A warning is that choice of the wrong model can lead to biases in estimation which exceed any gain in standard error by using the Y_i^*.

Example 7.2: This is an artificial example. Let

$$Y = \frac{X^3}{1 + X^5} + \varepsilon$$

and sample X from a uniform $[0, 1]$ distribution and ε from a normal $N(0, 0.05^2)$ distribution. From a sample of size 100 we fitted the regression model

$$\hat{Y} = -0.13864 + 0.62963X.$$

If it is assumed that the mean of X is known to be $\frac{1}{2}$ then this gives a corrected estimate for the mean of Y as 0.176175. This was checked against a sample of size 5000 which gave an (estimated) mean and standard deviation of Y as 0.1830 and 0.5301. A raw mean of 100 Y values would thus have a standard deviation of approximately 0.053 whereas the estimate was only approximately 0.007 away from the accurate value.

7.3 Resampling methods

When we have a single sample of values, $Y_1, \ldots, Y_n$, and very little is known about the distribution or population from which these values were obtained then it is usual to group the values in a histogram (see Section 2.1.3.1). Of course, it can be dangerous to use such a histogram as if it represents a true distribution. If we had taken a large sample then we would get a better estimate of the true distribution. It is safer to calculate sample quantities such as $\bar{Y}$ or s_Y^2 as estimates of the underlying population quantities μ_Y and σ_Y^2. But even this becomes problematical when standard assumptions of the normality of the underlying distribution break down. For example, if the distribution is very skewed, then s_Y^2 becomes a poor estimate of σ_Y^2. This presents a problem in saying how well our estimate $\bar{Y}$ of μ_Y is behaving.

One approach to this problem is to use the original empirical distribution of the Y_i-values by generating alternative samples and looking at $\bar{Y}$ for each sample. These alternative samples can be generated by using only the original Y_i values, no other information. Such methods are called 'resampling methods' and have become very popular in statistics in the last ten or fifteen years.

They are useful when the Y_i values are obtained after an expensive simulation and it is required to know how sensitive sample quantities such as Y are, without having to generate samples repeatedly using the entire simulation over and over again. The idea of hitching a simulation onto a single sample (or sample path) has in fact been widely used in simulation of physical systems, particularly where time is a variable.

7.3.1 *The 'bootstrap' technique*

One of the simplest techniques, known as the 'bootstrap' technique, is to sample with replacement from the Y_i values. That is to say, consider the Y_i values as being placed in a hypothetical box and sampled at random from the box n times replacing (and shaking) after each value is drawn. This will generate a new sample, possible with some of the Y_i values repeated. Repeating the whole process B times will give B alternative samples. If for each sample we calculate $\bar{Y}$ to get the 'partial estimates'

$$\bar{Y}^{(1)}, \ldots, \bar{Y}^{(B)}$$

then the histogram of these values will give some idea of the variation in $\bar{Y}$ as well as an estimate of μ_Y.

TABLE 7.1. Clearance (inches × 10^{-4})

100	100	110	100	110	95	100	120
120	110	110	95	110	110	100	95
90	85	95	85	75	80	80	85
85	75	75	80	70	75	70	75
75	70	60	65	65	65	65	65
75	75	80	70	70	80	65	65
60	105						

TABLE 7.1a. Bootstrap histogram

Mid-point	Frequency
76	1
77	1
78	1
79	13
80	24
81	48
82	91
83	139
84	169
85	179
86	127
87	91
88	57
89	28
90	20
91	2
92	1

Example 7.3: In Table 7.1 we list 50 sample values of a measurement in inches ($\times 10^{-4}$) between two machined parts which are meant to fit together. From this data 1000 samples of size 50 (with replacement) were drawn for each of which the sample mean was computed. The histogram of these means is given below as a frequency table (Table 7.1a).

The 2.5% and 97.5% values from the raw data set of 1000 means were 0.00801 and 0.00894. The 95% t-interval values for the original data set were 0.00799 and 0.00897 showing remarkable agreement with the pure simulation results.

7.3.2 The 'Jack-knife' technique

This is a resampling technique which operates by re-estimating the required parameter (μ_X or σ_X^2) several times, each time omitting a subset of the original n, say, data values. If each data point can be regarded as a subset and is omitted in turn, then $n-1$ 'partial estimates' can be generated. These newly generated 'pseudo-data-values' provide the means of calculating less biased estimates of the population parameters and confidence intervals for them. Although originally developed as a means of reducing bias in estimates, a useful feature of the jack-knife approach is that it yields standard errors of estimates (useful for calculating confidence intervals) even where this is difficult or impractical by other methods.

7.4 Moments and cumulants

In Appendix B we give a brief review of the moments and cumulants of random variables which serve to summarize aspects of the distribution. Let

$$Y = f(x_1, \ldots, x_n),$$

and assume that $x_1, \ldots, x_n$ are random. If f is known and the moments or cumulants of the x_i are known, it may be possible to compute the moments or cumulants of Y by purely algebraic methods.

A standard method which can be found in many places in the statistical and engineering literature is to first expand Y in a Taylor series around some nominal values. Suppose that the mean of each x_i is known

$$E(x_i) = \mu_i.$$

Then we may use these as the nominal values. The first-order Taylor expansion is

$$Y(x_1, \ldots, x_n) = f(\mu_1, \ldots, \mu_n) + \sum_{i=1}^{n} (x_i - \mu_i) \frac{\partial f}{\partial x_i} + \cdots.$$

If we ignore the higher-order terms then

$$E(Y) = f(\mu_1, \ldots, \mu_n) + \sum_{i=1}^{n} E(x_i - \mu_i)\frac{\partial f}{\partial x_i}$$
$$= f(\mu_1, \ldots, \mu_n),$$

since the second term is zero because $E(x_i - \mu_i) = 0$. If we assume, further, that the x_i are independent and $\text{var}(x_i) = \sigma_i^2$ then

$$\text{var}(Y) = \sum_{i=1}^{n} \text{var}\left[(x_i - \mu_i)\frac{\partial f}{\partial x_i}\right]$$
$$= \sum_{i=1}^{n} \left[\frac{\partial f}{\partial x_i}\right]^2 \sigma_i^2.$$

Thus, up to this level of approximation the variance of Y is a linear function of the variances of the x_i's. Computation of the partial derivatives may be difficult in practical examples unless f is well-defined and in somewhat closed form. If, for example, Y is the output from a simulator which does not automatically calculate the $\partial f / \partial x_i$, then some additional coding will be required.

One simple numerical technique is to compute Y at neighbouring x_i values and approximate the derivatives. Keeping all x_i fixed (at their μ_i values) except x_j, we then approximate

$$\frac{\partial f}{\partial x_j} = \frac{1}{h_j} \cdot \Delta_j(Y),$$

where

$$\Delta_j(Y) = Y(\mu_1, \ldots, \mu_{j-1}, \mu_j + \tfrac{1}{2}h_j, \mu_{j+1}, \ldots, \mu_n)$$
$$- Y(\mu_1, \ldots, \mu_{j-1}, \mu_j - \tfrac{1}{2}h_j, \mu_{j+1}, \ldots, \mu_n).$$

This leads to the approximation

$$\text{var}(Y) \cong \sum_{i=1}^{n} \left(\frac{1}{h_i} \cdot \Delta_i(Y)\right)^2 \sigma_i^2.$$

It is interesting to compare this with a simple estimator obtained by inputting a random sample of x_j:

$$x_{j1}, \ldots, x_{jm}$$

from a distribution with mean μ_j and variance σ_j^2. The estimate of σ_j^2 would be

$$s_j^2 = \frac{1}{m} \sum_{k=1}^{m} (x_{jk} - \mu_j)^2$$

(assuming the μ_j are known). This would yield a contribution to the estimate of the variance of Y of

$$\left[\frac{\partial f}{\partial x_j}\right]^2 s_j^2.$$

However, if $\partial f/\partial x_j$ has also to be approximated it at first seems unnecessary to take further X-observations to do this. We can alternatively work directly with the Y-values. If $Y_{j1}, \ldots, Y_{jm}$ are the Y-values obtained keeping all x_i except x_j fixed at their μ_i values and using random x_j we would obtain an estimate of the contribution to σ_Y^2 as

$$(\hat{\sigma}_Y^2)_j = \frac{1}{m}\sum_{k=1}^{m}[Y_{jk} - E(Y)]^2.$$

Since

$$Y_{jk} - E(Y) \cong (x_{jk} - \mu_j)\frac{\partial f}{\partial x_j}$$

we have

$$(\hat{\sigma}_Y^2)_j \simeq \frac{1}{m}\sum_{k=1}^{m}(x_{jk} - \mu_j)^2\left[\frac{\partial f}{\partial x_j}\right]^2$$

$$= s_j^2\left[\frac{\partial f}{\partial x_j}\right]^2.$$

This does not require further computations of x_j, but it does, of course, require computation of the Y values, usually the more expensive procedure. Thus it is still usually preferable to approximate $\partial f/\partial x_j$ in some way and use the s_j^2.

When Y is polynomial in X it is often possible to explicitly compute the means and variances and higher-order moments and cumulants of Y from those of X. We can go further and compute variances between several Y variables which are functions of several X-variables. We can consider again Y being obtained in this way by a higher-order Taylor expansion.

Example 7.4:

$$Y_1 = X_1 + X_2$$

$$Y_2 = X_1 X_2.$$

Compute the means and variances and covariance of Y_1 and Y_2 given that X_1 and X_2 are independent and

$$E(X_i) = \mu_i \qquad \text{var}(X_i) = \sigma_i^2 \qquad (i = 1, 2).$$

The means are straightforward

$$E(Y_1) = E(X_1) + E(X_2) = \mu_1 + \mu_2$$
$$E(Y_2) = E(X_1X_2) = E(X_1)E(X_2) = \mu_1\mu_2$$
$$\text{var}(Y_1) = \text{var}(X_1) + \text{var}(X_2) = \sigma_1^2 + \sigma_2^2$$
$$\begin{aligned}\text{var}(Y_2) &= E(X_1X_2)^2 - (E(X_1X_2))^2 \\ &= E(X_1^2X_2^2) - (E(X_1)E(X_2))^2.\end{aligned}$$

Now X_1^2 and X_2^2 are independent, so that

$$E(X_1^2X_2^2) = E(X_1^2)E(X_2^2).$$

But

$$E(X_1^2) = \sigma_1^2 + \mu_1^2$$
$$E(X_2^2) = \sigma_2^2 + \mu_2^2.$$

Therefore

$$\begin{aligned}\text{var}(Y_2) &= (\sigma_1^2 + \mu_1^2)(\sigma_2^2 + \mu_2^2) - \mu_1^2\mu_2^2 \\ &= \sigma_1^2\sigma_2^2 + \mu_1^2\sigma_2^2 + \sigma_1^2\mu_2^2.\end{aligned}$$

Moreover,

$$\begin{aligned}\text{cov}(Y_1, Y_2) &= E(Y_1Y_2) - E(Y_1)E(Y_2) \\ &= E((X_1 + X_2)(X_1X_2)) - (\mu_1 + \mu_2)(\mu_1\mu_2) \\ &= E(X_1^2X_2) + E(X_1X_2^2) - (\mu_1 + \mu_2)(\mu_1\mu_2) \\ &= E(X_1^2)E(X_2) + E(X_1)E(X_2^2) - (\mu_1 + \mu_2)(\mu_1\mu_2) \\ &= (\sigma_1^2 + \mu_1^2)\mu_2 + \mu_1(\sigma_2^2 + \mu_2^2) - (\mu_1 + \mu_2)(\mu_1\mu_2) \\ &= \sigma_1^2\mu_2 + \mu_1\sigma_2^2.\end{aligned}$$

In this example, we used the fact that if $X_1, \ldots, X_p$ are random variables then

$$E(X_1 \ldots X_p) = E(X_1) \times \ldots \times E(X_p).$$

This formula lies at the heart of most computations with independent random variables.

We give the formula in Appendix B for the mean and variance of a quadratic function of independent random variables. This can be used to explain an example frequently discussed in reference to tolerance design in the Taguchi methodology.

Let

$$Y = a_0 + \sum_i^p a_iX_i + \sum_{i,j=1}^p \sum^p a_{ij}X_iX_j$$

be a general quadratic function of random variables $X_1, \ldots, X_p$. Assume for simplicity that $X_1, \ldots, X_p$ are independent with means $\mu_1, \ldots, \mu_p$. We are interested in var(Y) and in particular finding where this quantity is a minimum. Intuition tells us it should occur where the slope of Y is zero and hence where Y has a maximum or minimum. We shall look at the case of two X-variables:

$$Y = a_0 + a_1 X_1 + a_2 X_2 + a_{11} X_1^2 + 2a_{12} X_1 X_2 + a_{22} X_2^2.$$

If X_1 and X_2 are independent with equal variance σ^2 and means μ_1 and μ_2 then

$$\begin{aligned} E(Y) &= a_0 + a_1\mu_1 + a_2\mu_2 + a_{11}(\sigma^2 + \mu_1^2) + 2a_{12}\mu_1\mu_2 + a_{22}(\sigma^2 + \mu_2^2) \\ &= a_0 + (a_{11} + a_{12})\sigma^2 + a_{11}\mu_1^2 + 2a_{12}\mu_1\mu_2 + a_{22}\mu_2^2 + a_1\mu_1 + a_2\mu_2. \end{aligned}$$

Suppose that a_{11} and $a_{22} > 0$ and that $a_{11}a_{12} > a_{12}^2$. Then this has a unique minimum at

$$\begin{aligned} \mu_1^* &= -\tfrac{1}{2}(a_{22}a_1 - a_{12}a_2)/D \\ \mu_2^* &= -\tfrac{1}{2}(-a_{12}a_1 + a_{11}a_2)/D, \end{aligned}$$

where $D = a_{11}a_{22} - a_{12}^2$, which is shown by differentiating partially with respect to μ_1 and μ_2.

Assuming in addition that X_1 and X_2 are normally distributed we can show that

$$\begin{aligned} \text{var}(Y) = {} & (a_1^2 + a_2^2)\sigma^2 + a_{11}^2(2\sigma^4 + 4\mu_1^2\sigma^2) \\ & + 4a_{12}^2[\sigma^4 + \sigma^2(\mu_1^2 + \mu_2^2)] + a_{22}^2(2\sigma^4 + 4\mu_2^2\sigma^2) \\ & + 2a_1a_{11}(2\sigma^2\mu_1) + 4a_1a_{12}(\sigma^2\mu_2) \\ & + 4a_2a_{12}(\mu_1\sigma^2) + 2a_2a_{22}(2\sigma^2\mu_2) \\ & + 4a_{11}a_{12}(2\sigma^2\mu_1\mu_2) + 4a_{12}a_{22}(2\sigma^2\mu_1\mu_2). \end{aligned}$$

Collecting terms in μ_1 and μ_2 we have

$$\begin{aligned} \text{var}(Y) = {} & \text{constant} + 4\sigma^2[\mu_1^2(a_{11}^2 + a_{12}^2) \\ & + \mu_2^2(a_{22}^2 + a_{12}^2) + 2\mu_1\mu_2(a_{11}a_{12} + a_{12}a_{22}) \\ & + \mu_1(a_1a_{11} + a_2a_{12}) + \mu_2(a_2a_{22} + a_1a_{12})]. \end{aligned}$$

This reaches a minimum at (μ_1^*, μ_2^*) confirming the intuition. This can be extended to higher dimensions.

7.5 Pseudo-random variables

Various methods of selecting sets of input values have been found to perform better than Monte Carlo in practice and to have very good theoretical properties. Imagine two input variables X_1 and X_2. It is required to evaluate some integral (or mean value) μ of Y for X_1 and X_2 uniformly distributed in the unit square with corners (0, 0), (0, 1), (1, 0), (1, 1). A pure Monte Carlo method would generate a set of (x_{1i}, x_{2i}) values ($i = 1, \ldots, n$), evaluate

$$Y_i = f(x_{1i}, x_{2i})$$

and compute

$$\hat{\mu} = \frac{1}{n} \sum Y_i.$$

This produces an estimation variance of

$$\text{var}(\hat{\mu}) = \frac{1}{n} \sigma^2,$$

where

$$\sigma^2 = \int_0^1 \int_0^1 [f(x_1, x_2) - \mu]^2 \, \mathrm{d}x_1 \, \mathrm{d}x_2.$$

There is the alternative of selecting the (x_{1i}, x_{2i}) *deterministically* according to some mechanism which spreads the points out to fill up the unit square rather better than any particular random sequence may do. We lose the statistical flavour of the problem the error being all bias in the formula

$$\text{MSE} = \text{variance} + \text{bias}^2$$

of Appendix B. The squared error or bias that we are interested in is simply

$$(\mu - \hat{\mu})^2 = \left[E(Y) - \frac{1}{n} \sum Y_i \right]^2$$

$$= \left(\iint f(x_1, x_2) \, \mathrm{d}x_1 \, \mathrm{d}x_2 - \frac{1}{n} \sum_{i=1}^{n} f(x_{1i}, x_{2i}) \right)^2$$

There are methods of judging how good a mechanism for producing such sequences is independently of what particular function $Y = f(x_1, x_2)$ is being used, the rationale being that a sequence which performs well by such a criterion will perform well for any function f which is not too variable. The sequences should not be so regular that they will be bad for functions which themselves have similar regularities, but not be too irregular that they behave much like purely random sequences.

Before describing some sequences it should be noted that in most real problems we would not be averaging with respect to uniform distributions in the unit square but, say, a bivariate normal, for example, if x_1 and x_2

are independently normally distributed. In such a case we would transform to normality by the transformation procedure described in Chapter 6.

Setting

$$\mathbf{x}_i = (x_{1i}, x_{2i}, x_{3i}, \ldots, x_{pi})$$

for p-dimensions two of the most well-known sequences are given by

$$\text{(Halton)} \quad \mathbf{x}_{i+1} = [\phi_{q_1}(i), \phi_{q_2}(i), \ldots, \phi_{q_p}(i)],$$

and

$$\text{(Hammersley)} \quad \mathbf{x}_{i+1} = \left[\frac{i}{n}, \phi_{q_1}(i), \ldots, \phi_{q_p}(i)\right],$$

where q_i are prime numbers (say, the first p prime number) and $\phi_q(i)$ is called a 'radical inverse function'. This is defined by working with 'decimals' expressed in the arithmetic to the base q where q is prime. Thus if $q = 7$ and $i = 59$, then 59 to the base 7 is 113. Write the 'decimal' (base 7) by reversing 113 to obtain 0.311

$$\phi_7(59) = \frac{3}{7} + \frac{1}{49} + \frac{1}{343} = \frac{155}{343}.$$

Let $p = 2$, $q_1 = 2$, $q_2 = 3$ and let us generate 20 points in the unit square by the Halton method. The result is given in Table 7.2.

Example 7.5: Let us compare the use of the code in Table 7.2 with pure Monte Carlo simulation for the function

$$Y = (x_1 + 1)^{\frac{1}{3}}(x_2 - 1)^2 x_1 x_2.$$

Based on a random sample of 5000 random uniformly distributed random variables, the mean of Y (equal to the integral over the unit square) is estimated to be 0.0501 and the standard deviation estimated to be 0.0473. The mean value over the 20 points of the code is 0.0508. This gives an approximate bias of 0.0035. The standard error from a random sample of size 20 is estimated at $0.0473/\sqrt{20} \simeq 0.0106$. This last value far exceeds the bias from the value based on the code.

7.6 Tolerance design

7.6.1 Tolerance philosophy

Historically, only elementary statistical methods have been used to describe engineering tolerances. We refer to prescriptions typically available in national standard or manuals on technical drawing. Such

TABLE 7.2. Halton sequence

i	Base 2	Base 3	$\phi_1(i)$	$\phi_2(i)$
1	1	1	0.50000	0.33333
2	10	2	0.25000	0.66667
3	11	10	0.75000	0.11111
4	100	11	0.12500	0.44444
5	101	12	0.62500	0.77778
6	110	20	0.37500	0.22222
7	111	21	0.87500	0.55556
8	1000	22	0.06250	0.88889
9	1001	100	0.56250	0.03704
10	1010	101	0.31250	0.37037
11	1011	102	0.81250	0.70370
12	1100	110	0.18750	0.14815
13	1101	111	0.68750	0.48148
14	1110	112	0.43750	0.81481
15	1111	120	0.93750	0.25926
16	10 000	121	0.03125	0.59259
17	10 001	122	0.53125	0.92593
18	10 010	200	0.28125	0.07407
19	10 011	201	0.78125	0.40741
20	10 100	202	0.09375	0.74074

methods specify bounds on input variables, components and so on of the form

$$c_i - \Delta_i \leq x_i \leq c_i + \Delta_i.$$

The c_i are the nominal, target, or mean values. Some attempt may be made to place a confidence or probability on these statements. Typically this will be measured in terms of standard deviations. This can be explained easily as follows. Consider X_i to be random with mean μ_i and variance σ_i^2. If the distribution is normal then

$$Z_i = (X_i - \mu_i)/\sigma_i$$

has a standard normal distribution. Any statement about X_i can be converted into a statement about Z_i (see Section 2.1.3.3). Thus

$$\mu_i - \Delta_i \leq X_i \leq \mu_i + \Delta_i$$

is equivalently

$$-\Delta_i/\sigma_i \leq Z_i \leq \Delta_i/\sigma_i.$$

If we take a 'three standard deviations' critical value for the standard normal distribution, this means computing

$$\text{Prob}(-3 \leq Z_i \leq 3) = 0.9973.$$

Thus, the chance that Z_i lies outside the range is 0.0027 or 0.27%. Converting this to a statement about X_i we merely set $\Delta_i = 3\sigma_i$, and

$$\mu_i - 3\sigma_i \leq X_i \leq \mu_i + 3\sigma_i.$$

Conversely, if we require a specific tolerance say 1/1000 or 0.1% then we have to select δ so that

$$\text{Prob}(-\delta \leq Z_i \leq \delta) = 0.999.$$

The value of δ is 4.4172.

Dealing simultaneously with many different input tolerances is often complex. A great advantage of the statistical approach is that variation in input x_i can be worked through to variation in output more easily. As a first step, finding the mean and variance of an output Y_i is the most critical. The methods described in the previous sections can be seen as a variety of ways of doing this. A principal aim is to minimize or remove sources of variability by adjusting the mean or nominal values of the x_i. We give an example to illustrate the kind of tolerance philosophy which derives from the statistical methods.

Example: Imagine a device which employs a spring to exert a force F_0. As force is proportional to the extension L we can write

$$F = kL.$$

However, springs come from a supplier with different k-values with mean μ and variance σ^2. Thus

$$E(F) = \mu L$$
$$\text{var}(F) = \sigma^2 L^2.$$

If we manufacture the device with the setting $L = L_0 = F_0/\mu$, we will achieve F_0 on the average with variance $\sigma^2 L_0^2$ and standard deviation σL_0. If this is considered too large we might ask the supplier to decrease σ by improving the process. However, if we are at liberty to choose L we may ask the supplier to work to the same tolerance (σ^2) but make a spring with a different mean $\mu' > \mu$. Then by setting $L = L_0' = F_0/\mu'$ we can achieve the same mean F value with smaller variation which is

$$\sigma L_0' = \sigma F_0/\mu' < \sigma F_0/\mu = \sigma L_0.$$

This is the 'parameter design through tolerance analysis method' (see Section 7.7.1). Any interval statements about x_i and Y are converted into statements about standard deviations provided that we have some idea of the distribution of Y and of the x_i's. However, in the absence of distributional information it generally seems safer to work with the first and the second-order theory as in the above example.

7.6.2 Tolerance regions and centring

If the output Y or outputs $Y_1, \ldots, Y_m$ are required to lie in a particular region R_Y, say, we may attempt to find the input values $x_1, \ldots, x_k$ for which this is the case. If we have an exact relationship:

$$Y = f(x_1, \ldots, x_k)$$

then we want the set of $x = (x_1, \ldots, x_k)$ values say R_x such that

$$x \text{ in } R_x \text{ gives } Y = f(x) \text{ in } R_Y.$$

We might express this briefly as

$$R_x = f^{-1}(R_Y)$$

Since f is often very complex, computation of an inverse function may be impossible or very hard. Thus we must proceed by some trial and error method to home in on R_x or some subset of R_x. Such methods are variously described as 'design centring', 'tolerance space methods', the 'method of safe domains' and so on.

When in addition the x_i are random such as arising from tolerances, measurement error, machine or process error or physical disturbances, one may be interested in p = probability(Y lies in R_Y) which is then equal to probability(x lies in R_x), assuming that Y acquires no additional error as in

$$Y = f(x_1, \ldots, x_k) + \varepsilon$$

Alternatively p itself may be the objective function and we require, say

$$p \geq c.$$

This probability is sometimes called the 'yield'. The statistical approach to estimating the yield would be to first find an estimate of the distribution of Y and then simply integrate Y over the region R_Y with respect to this density. This would be an alternative to integrating over the original space R_x.

In reality, so long as x lies in the feasible region R_x, we can fine tune the process to keep on target (that is within R_Y) while minimizing variability. Even if x is controlled with mean so that Y is near the centre of R_Y (in some sense) while minimizing variability, this will often be adequate. It will force p, the yield, to be high. Although R_Y may be

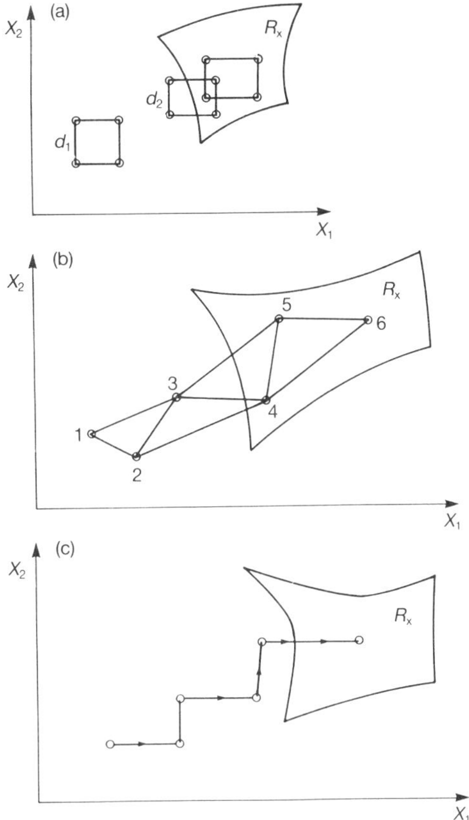

FIG. 7.2. Design centring.

simply described as

$$a_i \leq Y_i \leq b_i \qquad (i = 1, \ldots, m)$$

the inverse region R_x may be more complex. Figure 7.2a represents one approach. The region R_x is the tolerance region of good x-values. The corners of the square d_1 represent the first trial set of design values. By a statistical analysis we may ascertain a direction in which to move to a new set of design values, d_2. As soon as we have a set of configurations which lie in R_x we may experiment at their average values or proceed further into R_x.

The method is close in spirit to the EVOP method in Chapter 5 (Section 5.5.1) and to various methods of stochastic optimization not studied in this book. For example, the 'simplex' method uses triangles (tetrahedra, simplices in higher dimensions) which can be made to fit together rather better than squares as Fig. 7.2b shows.

Each new point is added by proceeding away from the worst point on the previous triangle in a direction determined by the best two points. The triangles 'flip-flop' towards the R_x region of the optimum. Another method, sometimes called the 'greedy algorithm' only allows one variable at a time maximization and typically is only successful in special applications. It is represented in Fig. 7.2c.

7.7 Taguchi's approach to tolerance design and analysis

For understanding how the 'variability in inputs' affects 'variability in the output', simulation techniques have been mentioned so far in this chapter. The Taylor expansion method has been frequently used by designers in the past, in order to calculate the functional effect of variation in the input factors on the objective characteristic. Problems in this area have arisen due mainly to the following:

(i) Assuming the existence of a *reliable* functional relationship between the input variables and the output response, a relatively large computer capability and availability is required in order for a simulation procedure (using, for example, Monte Carlo methods) to be carried out.

(ii) When the model relating input to output is known to be complicated, a Taylor-expansion formulation may be difficult.

(iii) Although one is free to decide how far to take the Taylor expansion of a function, it is sometimes hard to evaluate the error caused by omitting the higher-order terms.

Whether a functional model is available or not, Taguchi recommends the use of orthogonal experimental designs as the most important tool in the following two-phase approach.

Phase 1: Parameter design (through tolerance analysis).
This is the most important phase of product/process design work and has the objective of selecting 'optimum' levels for the system parameters, so that the response of interest is stable (i.e. of low variability) despite the use of highly variable, inexpensive components, and even if the system and environmental parameters change. At this stage, the tolerance specifications of the system parameters remain unchanged but are analysed in order to determine ways to make the product or the process insensitive to their effect.

Phase 2: Tolerance design.
This is the stage when decisions are made as to how much variability to allow in component parts. A 'rational' tightening of tolerance specification of the most crucial components is carried out to achieve further robustness and stability in the response of interest. Since setting tight tolerances might lead to additional costs, tolerance (re-)design should be used only if the required reduction in the variability cannot be achieved during the parameter design stage using existing tolerance levels.

Taguchi recommends using the 'Contribution ratios' (see Section 4.3) for a cost-effective re-specification of tolerances.

When a functional model is available, an alternative procedure using a non-linear programming approach is that recommended by Box and Fung (1986). This approach is similar to the one described in Section 7.4 and requires determination of functional partial derivatives. We will demonstrate Taguchi's suggested procedure using an actual case study which had as an objective the optimization of a filter circuit design, for which a rather complicated functional model was available.

A case-study
A filter circuit, part of the audio circuit of a mobile radio transceiver (see Fig. 7.3) is to be studied; the response of interest is the 'gain' which at a certain frequency is defined in decibels as

$$G_{\mathrm{f}} = 20 \log_{10}\{V_f\}$$

where V_f is calculated on the basis of the resistance of four resistors R_1–R_4 and of capacitance of four capacitors C_1–C_4 as follows:

$$V_f = \left\{\frac{-Kr^2}{|r^2 + r(n_3 + n_2(1+K) + n_1 n_2|}\right\}\left\{\frac{\lambda\,|r|}{|r + n_4|}\right\},$$

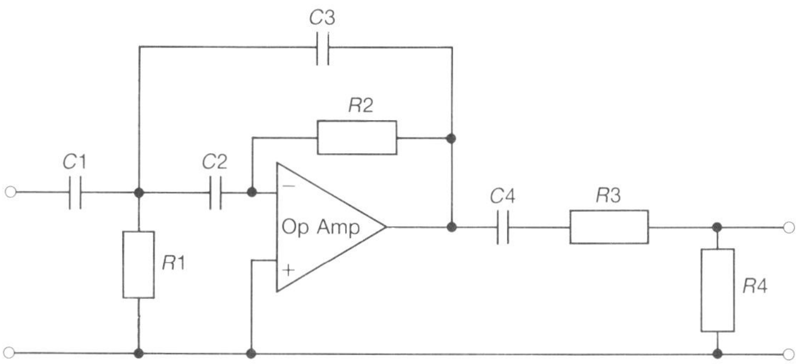

FIG. 7.3. High-pass filter circuit.

where

$$n_1 = (C_3R_1)^{-1}, \quad n_2 = (C_2R_2)^{-1}, \quad n_3 = (C_3R_2)^{-1}, \quad n_4 = [C_4(R_3 + R_4)]^{-1}$$

$$K = \frac{C_1}{C_3}, \quad \lambda = \frac{R_4}{(R_3 + R_4)}$$

and r is a complex number depending on the frequency f and is defined by

$$r = 2\pi if.$$

Note that for any complex number $z = a + ib$ we have

$$|z| = \sqrt{a^2 + b^2}.$$

For a proper performance of the circuit it is required that the gains at frequencies 300 Hz and 1000 Hz shoud consistently satisfy:

$$G_{300} - G_{1000} \leq -3 \text{ dB}$$

or equivalently

$$Y = (V_{300}/V_{1000}) \leq 0.7079.$$

Since a model relating input (resistors and capacitors) to output (Y) was available, computer simulation experiments could be performed using experimental design techniques. Taguchi's two-phase approach was followed with the objective of determining the optimal nominal levels (and, if necessary, the tolerances) of the four resistors and four capacitors, so that Y achieved the target with minimum variability.

7.7.1 Parameter design through tolerance analysis

The four resistors (R_1–R_4) and four capacitors (C_1–C_4) were studied at three nominal levels each; low cost tolerances were used at this stage. The levels are shown in Tables 7.3. The resistance values are given in kΩ and the capacitance values in μF.

The eight parameters were viewed both as 'controllable' as well as 'noise' factors; 'controllable' because the experimenter was free to choose among certain nominal values, and 'noise' because of the effect of their tolerances (the deviations from the nominal) which is the cause of variability in the circuit's performance.

Experimental design

The experimental set up is as shown in Fig. 7.4. The eight parameters when viewed as controllable factors were assigned in their nominal levels to the columns of the 'Controllable Factor Array' (CFA) which was the orthogonal array $OA_{27}(3^{13})$ (see Appendix D and Table 7.3a), i.e.

TABLE 7.3. Factor levels

Circuit parameters as controllable factors	Nominal levels			Circuit parameters as noise factors	Tolerance levels (% from nomimal)		
	1	2	3		1	2	3
R_1	10	12	15	R_1'	−2	0	+2
R_2	100	120	150	R_2'	−2	0	+2
R_3	4.7	5.6	6.8	R_3'	−2	0	+2
R_4	10	15	22	R_4'	−2	+2	
C_1	0.01	0.015	0.022	C_1'	−10	0	+10
C_2	0.01	0.015	0.022	C_2'	−10	0	+10
C_3	0.01	0.015	0.022	C_3'	−10	0	+10
C_4	0.022	0.033	0.047	C_4'	−10	0	+10

TABLE 7.3a. Controllable factor array for filter-circuit

Factors:	C_3	C_1	$C_3 \times C_1$		R_2	$C_3 \times R_2$		R_1	R_3	R_4	C_2	C_4	
Column:	1	2	3	4	5	6	7	8	9	10	11	12	13
Trial run													
1	1	1	1	1	1	1	1	1	1	1	1	1	1
2	1	1	1	1	2	2	2	2	2	2	2	2	2
3	1	1	1	1	3	3	3	3	3	3	3	3	3
4	1	2	2	2	1	1	1	2	2	2	3	3	3
⋮	⋮	⋮	⋮	⋮	⋮	⋮	⋮	⋮	⋮	⋮	⋮	⋮	⋮
27	3	3	2	1	3	2	1	2	1	3	1	3	2

TABLE 7.3b. Noise factor array for filter-circuit

Factors:	R_4'	R_1'	R_2'	R_3'	C_1'	C_2'	C_3'	C_4'
Column:	1	2	3	4	5	6	7	8
Trial run								
1	1	1	1	1	1	1	1	1
2	1	1	2	2	2	2	2	2
3	1	1	3	3	3	3	3	3
4	1	2	1	1	2	2	3	3
⋮	⋮	⋮	⋮	⋮	⋮	⋮	⋮	⋮
18	2	3	3	2	1	2	3	1

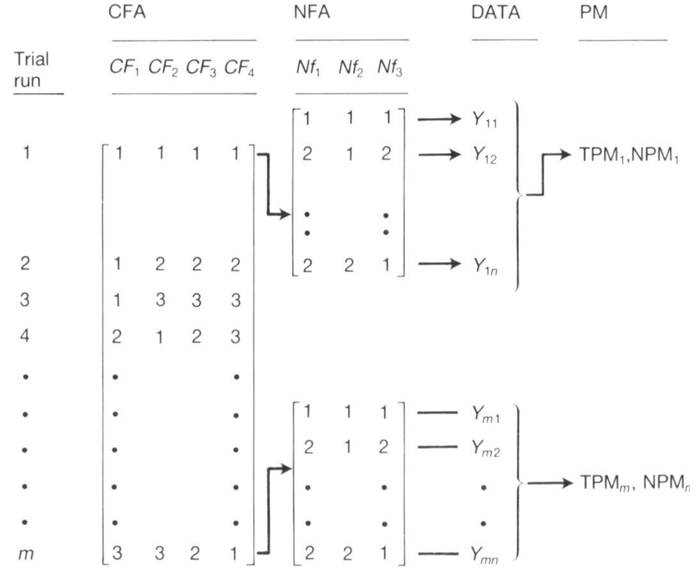

FIG. 7.4.

TABLE 7.4a. Noise factor array for trial 1 of CFA

Factors:	R'_4	R'_1	R'_2	R'_3	C'_1	C'_2	C'_3	C'_4
Columns:	1	2	3	4	5	6	7	8
Trial								
1	9.8	9.8	98	4.606	0.009	0.009	0.009	0.0198
2	9.8	9.8	100	4.7	0.010	0.010	0.010	0.022
3	9.8	9.8	102	4.794	0.011	0.011	0.011	0.0242
4	9.8	10	98	4.606	0.010	0.010	0.011	0.0242
5	9.8	10	100	4.7	0.011	0.011	0.009	0.0198
6	9.8	10	102	4.794	0.009	0.009	0.010	0.022
7	9.8	10.2	98	4.7	0.009	0.011	0.010	0.0242
8	9.8	10.2	100	4.794	0.010	0.009	0.011	0.0198
9	9.8	10.2	102	4.606	0.011	0.010	0.009	0.022
10	10.2	9.8	98	4.794	0.011	0.010	0.010	0.0198
11	10.2	9.8	100	4.606	0.009	0.011	0.011	0.022
12	10.2	9.8	102	4.7	0.010	0.009	0.009	0.0242
13	10.2	10	98	4.7	0.011	0.009	0.011	0.022
14	10.2	10	100	4.794	0.009	0.010	0.009	0.0242
15	10.2	10	102	4.606	0.010	0.011	0.010	0.0198
16	10.2	10.2	98	4.794	0.010	0.011	0.009	0.022
17	10.2	10.2	100	4.606	0.011	0.009	0.010	0.0242
18	10.2	10.2	102	4.7	0.009	0.010	0.011	0.0198

$m = 27$. The array $OA_{18}(2 \times 3^7)$ (see Appendix D) was used as the 'Noise Factor Array' (NFA), i.e. $n = 18$, where the eight parameters, viewed as noise factors, were assigned in their tolerance levels (see Table 7.3b).

For example, considering R_1: when R_1 in CFA is at nominal level 1 ($= 10\ \text{k}\Omega$) then,

level 1 of R_1' in NFA is 9.8 (nominal −2%),
level 2 of R_1' in NFA is 10 (nominal),
level 3 of R_1' in NFA is 10.2 (nominal +2%).

Similarly for the other levels of R_1 and for the levels of the other circuit parameters. Note that for R_4 there are three nominal levels in CFA and two tolerance levels in NFA, for example: when R_4 in CFA is at nominal level 3 ($= 22\ \text{k}\Omega$) then

level 1 of R_4' in NFA is 21.56 (nominal −2%)
level 2 of R_4' in NFA is 22.44 (nominal +2%).

The full outline of the CFA using the actual (nominal) values of R_1–R_4, C_1–C_4 is shown in Table 7.4 whereas Tables 7.4a and 7.4b show the full outline of the NFA for the 1st and 27th trial of CFA respectively, using the actual tolerance levels.

Results

The 18 'runs' from the NFA at each of the 27 combinations (trials) of the CFA, provided the Y-data values for the calculation of a measure reflecting the mean response (Target Performance Measure—TPM) and of a measure reflecting the variability in the response (Noise Performance Measure—NPM), see Fig. 7.4 where for our case $m = 27$ and $n = 18$. First, the sample mean $\bar{Y}$ and standard deviations for each of the 27 CFA-trials have been calculated. These are given in Table 7.5.

Analysis

Following the techniques described in Chapter 6 the appropriate TPM and NPM were determined to be

$$\text{TPM} = \bar{Y}$$

and

$$\text{NPM} = -20 \log_{10}(s).$$

An ANOVA of the TPM has shown the factors R_2, R_4, C_2, C_3, and C_4 to be highly significant for the mean response.

An ANOVA for the NPM showed only C_2 to be significantly affect the variability in the response, so C_2 can be considered as a variability-control factor with optimal level (level with maximum NPM) level 3. The factors which affect only the TPM, R_2, R_4, C_3 and C_4, can be considered

TABLE 7.4. CFA

Factors:	C_3	C_1	R_2	R_1	R_3	R_4	C_2	C_4
Columns:	1	2	5	8	9	10	11	12
Trial								
1	0.01	0.01	100	10	4.7	10	0.01	0.022
2	0.01	0.01	120	12	5.6	15	0.015	0.033
3	0.01	0.01	150	15	6.8	22	0.022	0.047
4	0.01	0.015	100	12	5.6	15	0.022	0.047
5	0.01	0.015	120	15	6.8	22	0.01	0.022
6	0.01	0.015	150	10	4.7	10	0.015	0.033
7	0.01	0.022	100	15	6.8	22	0.015	0.033
8	0.01	0.022	120	10	4.7	10	0.022	0.047
9	0.01	0.022	150	12	5.6	15	0.01	0.022
10	0.015	0.01	100	10	5.6	22	0.01	0.033
11	0.015	0.01	120	12	6.8	10	0.015	0.047
12	0.015	0.01	150	15	4.7	15	0.022	0.022
13	0.015	0.015	100	12	6.8	10	0.022	0.022
14	0.015	0.015	120	15	4.7	15	0.01	0.033
15	0.015	0.015	150	10	5.6	22	0.015	0.047
16	0.015	0.022	100	15	4.7	15	0.015	0.047
17	0.015	0.022	120	10	5.6	22	0.022	0.022
18	0.015	0.022	150	12	6.7	10	0.01	0.033
19	0.022	0.01	100	10	6.7	15	0.01	0.047
20	0.022	0.01	120	12	4.7	22	0.015	0.022
21	0.022	0.01	150	15	5.6	10	0.022	0.033
22	0.022	0.015	100	12	4.7	22	0.022	0.033
23	0.022	0.015	120	15	5.6	10	0.01	0.047
24	0.022	0.015	150	10	6.8	15	0.015	0.022
25	0.022	0.022	100	15	5.6	10	0.015	0.022
26	0.022	0.022	120	10	6.8	15	0.022	0.033
27	0.022	0.022	150	12	4.7	22	0.01	0.047

as target-control factors and can be manipulated to bring the mean response on target.

Interactions

Interactions between components in a filter circuit are sometimes significant and should be investigated. The CFA allowed the study of the interactions $C_3 \times C_1$ and $C_3 \times R_2$ which showed no significance. As a response model was available and 'computer trials' easy to perform, the experiment was re-iterated with the controllable factors re-allocated at different columns of the $OA_{27}(3^{13})$ so that other interactions could be investigated. Overall, no evidence of any significant interactions was found.

TABLE 7.4b. Noise factor array for trial 27 of CFA

Factors:	R_4'	R_1'	R_2'	R_3'	C_1'	C_2'	C_3'	C_4'
1	21.56	11.76	147	4.606	0.0198	0.009	0.0198	0.0423
2	21.56	11.76	150	4.7	0.022	0.010	0.022	0.047
3	21.56	11.76	153	4.794	0.0242	0.011	0.0242	0.0517
4	21.56	12	147	4.606	0.022	0.010	0.0242	0.0517
5	21.56	12	150	4.7	0.0242	0.011	0.0198	0.0423
6	21.56	12	153	4.794	0.0198	0.022	0.022	0.047
7	21.56	12.24	147	4.7	0.0198	0.011	0.022	0.0517
8	31.56	12.24	150	4.794	0.022	0.009	0.0242	0.0423
9	21.56	12.24	153	4.006	0.0242	0.010	0.0198	0.047
10	22.44	11.76	147	4.794	0.0242	0.010	0.022	0.0423
11	22.44	11.76	150	4.606	0.0198	0.011	0.0242	0.047
12	22.44	11.76	153	4.7	0.022	0.009	0.0198	0.0517
13	22.44	12	147	4.7	0.0242	0.009	0.0242	0.047
14	22.44	12	150	4.794	0.0198	0.010	0.0198	0.0517
15	22.44	12	153	4.606	0.022	0.011	0.022	0.0423
16	22.44	12.24	147	4.794	0.022	0.011	0.0198	0.047
17	22.44	12.24	150	4.606	0.0242	0.009	0.022	0.0517
18	22.44	12.24	153	4.7	0.0198	0.010	0.0242	0.0423

Optimal settings

Applying the techniques of Section 4.4 for 'estimating' the circuit's future performance, the following 'optimal parameter levels' were determined which would theoretically achieve the target with minimum variability:

R_1 at level 3, i.e. 15 kΩ
R_2 at level 1, i.e. 10 kΩ
R_3 at level 2, i.e. 5.6 kΩ
R_4 at level 1, i.e. 10 kΩ
C_1 at level 3, i.e. 0.022 μF
C_2 at level 3, i.e. 0.022 μF
C_3 at level 2, i.e. 0.015 μF
C_4 at level 1, i.e. 0.022 μF

Confirmation trials at the optimal setting using the associated NFA for this setting [i.e. 18 trial runs varying the existing tolerance levels according to the $OA_{18}(2 \times 3^7)$] have shown that Y had indeed achieved the target with minimum variability. In fact, in comparison with results prior to the 'parameter design stage' a twofold improvement in both the mean response and variance has been achieved, without having to tighten any existing tolerance requirements.

TABLE 7.5

Trial run (of CFA)	Mean ($\bar{Y}$)	Standard deviation (S)
1	0.21344	0.03312
2	0.69410	0.08776
3	1.12692	0.02627
4	0.74802	0.07152
5	0.46702	0.05632
6	0.67173	0.08984
7	0.53395	0.05302
8	0.71288	0.07426
9	0.41571	0.05501
10	0.52559	0.07858
11	0.99758	0.06703
12	0.82202	0.03660
13	0.67645	0.04320
14	0.65211	0.06354
15	1.19961	0.07024
16	0.70309	0.05609
17	0.88276	0.05139
18	0.64767	0.06983
19	0.79312	0.09476
20	0.98583	0.03554
21	0.85292	0.03451
22	1.02530	0.02497
23	0.81287	0.04970
24	0.95891	0.04065
25	0.56495	0.03684
26	1.00522	0.03086
27	0.97862	0.05822

7.7.2 Tolerance (re-)design

If the variability levels in the response at an 'optimal setting' determined during the parameter design stage is still unacceptable, a further improvement can be achieved through a rational reduction of certain tolerances. Of course, tightening of existing tolerances usually entails additional costs; hence this should take place *only after* the parameter design stage.

Use of Taguchi's 'contribution ratios' (see Section 4.3) can help in a cost-effective re-specification of tolerances.

Tolerance equations

Suppose it is required to reduce the variance of the Y-response to 50% of what it was *at* the 'optimal setting' determined in the 'parameter design' stage (Section 7.7.1).

To achieve this, we first need to determine which of circuit's components are the most crucial in effecting a change in variability levels. For this, we need to study the effect of the existing tolerance levels and determine how much each component 'contributes' to the total variability *at* the 'optimal setting'.

So, we will view the components as noise factors and consider a 'noise factor array' corresponding to the 'optimal setting'. Additional noise factors, not considered previously, might be included for study at this stage. For our purpose we will only consider the previously studied 'noise factors', $R_1', \ldots, R_4'$, $C_1', \ldots, C_4'$ and assign them on the $\mathrm{OA}_{18}(2 \times 3^7)$ as before.

If the distribution function describing the components-tolerances is known or if the mean μ and standard deviation σ for the tolerances can be determined, then Taguchi recommends that the values for the levels for each 'noise factor' should be set as

$$\text{Level 1: } \mu - \sqrt{\tfrac{3}{2}}\,\sigma$$

$$\text{Level 2: } \mu$$

$$\text{Level 3: } \mu + \sqrt{\tfrac{3}{2}}\,\sigma$$

Alternatively, if the tolerance range is known, then two or three levels should be chosen to cover this range. We choose to apply the latter, since the required data values had already been obtained from the confirmation trial runs of Section 7.7.1.

The analysis of variance of the 18 data values is shown in Table 7.6.

TABLE 7.6

Component	df	S	Contribution ratio ρ
R_4	1	0.00012859	0.005623
R_1	2	0.00004129	0.0018004
R_2	2	0.00036157	0.01581335
R_3	2	0.00013012	0.0056869
C_1	2	0.00265561	0.1161826
C_2	2	0.00349237	0.1527927
C_3	2	0.00551489	0.241282376
C_4	2	0.01053143	0.460766975
Residual	2	0.00000014	0.000051698
Total	17	0.02285600	1.000

Note that the contribution ratio for a factor F (in decimals) is calculated by

$$\rho_F = \frac{S_F - df_F \times MS_{Res}}{TSS}$$

where

$$MS_{Res} = S_{Res}/df_{Res}$$

(see Section 4.3).

Since it is required to reduce the existing variability by 50%, the reassignment of the tolerances can be decided using the following 'tolerance equation':

$$0.50 = \left(\frac{1}{\lambda(R_1)}\right)^2 \rho_{R_1} + \left(\frac{1}{\lambda(R_4)}\right)^2 \rho_{R_4} + \cdots + \left(\frac{1}{\lambda(C_4)}\right)^2 \rho_{C_4} + \rho_{Res}, \quad (7.1)$$

where '$1/\lambda(F)$' represents a reduction to $(1/\lambda(F))$ of the existing tolerance specification for factor F. For example, by $\lambda(C_4) = 1$ we mean that the tolerance of C_4 will remain the same, whereas $\lambda(C_4) = 3$ will mean a reduction to $\frac{1}{3}$ of the C_4-tolerance levels, i.e. $+/-10\%$ reduced to $+/-3\frac{1}{3}\% \cdots$.

Of course, there are many combinations of $\lambda(R_1), \ldots, \lambda(C_4)$ which would satisfy eqn (7.1). However the values of the contribution ratios of Table 7.6 together with cost considerations, can direct us to the most cost-effective choice. Tightening the tolerances for C_4 and C_3 will have the greatest effect, because these parameters 'contribute' the most in the ANOVA of Table 7.6.

If cost-considerations permit, by tightening the tolerances of only C_4 and C_3 by a factor of 2, the new variability target will be achieved. Indeed, by setting

$$\lambda(C_4) = 2, \qquad \lambda(C_3) = 2, \qquad \lambda(C_2) = 1, \ldots, \lambda(R_4) = 1$$

the right-hand side of (7.1) yields a value of 0.474 as required.

A model trial run at the optimal setting with the tolerances only of C_3 and C_4 tightened to $\pm 5\%$ (instead of the original $\pm 10\%$) confirmed the above prediction. In fact, new $s^2 = 0.0006413$, 'old' $s^2 = 0.0013445$. Therefore, in comparison with the 'old' variability levels prior to the 'tolerance redesign' stage, the variance in Y has dropped to 47.7% of its previous value—as 'predicted' by the tolerance equation.

It is worth considering again that, tightening of tolerance using the contribution ratios as above, should always be decided upon cost considerations. Appropriate loss functions (see, for example, Chapter 1, Section 1.3) can, of course, help in the determination of the expected quality losses or gains when one intends to follow a specific action plan.

7.8 Sources of further information

Computer simulation is a vast area and much of the best work is subject-oriented appearing in different scientific literature. The statistical literature divides roughly into (1) methods of generating pseudo-random observations from known distributions, and (2) variance reduction methods. We have completely ignored category (1) and an interested reader should consult Ripley (1987). The book by Hammersley and Handscomb (1964) is still an excellent introduction to variance reduction methods.

For an extensive survey of the use of 'uniform' sequences in numerical integration see Niederreiter (1978).

There are two kinds of tolerance theory: the statistical theory of tolerance intervals, now somewhat outdated, and the more modern methods of tolerancing in engineering. For the former the book by Guttman (1970) is one of the few works in the area. Texts on tolerancing in engineering are few and far between. Spotts (1983) contains an excellent survey of engineering problems. Statistical methods are probably most used in structural reliability and Thoft-Christensen and Baker (1982) contains a survey. The present book has ignored the large amount of literature on system reliability which makes heavy use of probability methods. Barlow and Proschan (1965) is the classical text.

A summary of the method of cumulants in tolerancing will appear in Moheiman and Wynn (1988), and there are other scattered examples of the use of such methods. A relatively new area in which statistical methods are being used is the theory of stochastic finite-statement methods (Casiati and Faravelli 1985). The general result on the minimization of the output variance mentioned at the end of Section 7.4 appears in Vuchkov and Boyadjieva (1988).

The material on design centring predates the use of Taguchi methodologies in the West (see Spence 1984). For an example of the rapidly growing application of statistical methods in circuit design and fabrication see Nassif, Stojwas, and Director (1984).

APPENDIX A

Deming's 14 points for management

1. *Create constancy of purpose for continual improvement of product and service.* Set the course today to be in business tomorrow. Provide for long-term needs rather than short-term profits. Try always to improve product quality in order to become more competitive and to stay in business.

2. *Adopt the new philosophy for economic stability.* A change is necessary to the old management methods. If you just try to meet the competitors, you will not survive in this new economic age. The customer should come first.

3. *Cease dependence on inspection to achieve quality.* Build quality into the product during the product development stage. Mass inspection cannot compensate for bad design. Do it right first time so that there is no need for rectification later.

4. *End the practice of awarding business on price tag alone.* End 'lowest tender' contracts: instead, require meaningful measures of quality for the supplies. Minimize total costs not just initial costs.

5. *Improve constantly and forever the system of production and service.* Search continually for problems to constantly improve quality and productivity and then to constantly decrease costs. Try continually to reduce the variability of products and services in order to obtain the highest quality at the lowest cost.

6. *Institute training on the job.* A continuous programme of on-the-job training needs to be instituted and this should include managers. Efficient training helps a company to make better use of all its employees. Training and education are the cornerstones of greater consistency.

7. *Adopt and institute modern methods of supervision and leadership.* Supervisors must ensure that immediate action is taken on reports of defects, poor tools and conditions detrimental to quality; efforts should be focused on helping people and machines to do a better job.

8. *Drive out fear.* Two-way communication should be encouraged so that fear is driven out throughout the organization. Ideas should be actively sought and eagerly listened to. In this way everybody may work more effectively and more productively for the company.

9. *Break down barriers between departments and individuals.* Every individual from research and development to production and sales must work as a team to foresee problems before they arise or to tackle them if they happen to arise.

10. *Eliminate the use of slogans, posters, and exhortations.* Eliminate slogans which demand zero defects and new levels of excellence without providing the methods. Arbitrary objectives without providing a 'road map' for accomplishment can be counterproductive.

11. *Eliminate work standards and numerical quotas.* Eliminate management by objective, by numbers and numerical goals. Focus on quality not quantity. The attainment of a target must not be viewed as an ultimate success because there is *always* room for further improvement.

12. *Remove barriers that rob the hourly worker of his/her right to pride of workmanship.* Abolish the annual merit rating (appraisal of performance) which destroys teamwork, fosters mediocrity, increases variability, and focuses on the short term. Evaluate individuals on long-term contribution rather than on short-term performance. Eliminating physical and mental obstacles improves communication, co-operation, and overall morale of employees.

13. *Institute a vigorous programme of education and training.* Encourage continual training to keep up with new developments, changes in product design, machinery, and techniques. Encourage self-improvement for everyone.

14. *Define top management's permanent commitment to ever-improving quality and productivity.* Put everybody in the company to work to accomplish the transformation by implementing all the preceding 13 points. All employees must understand and be committed to the new philosophy; top management will lead the way.

APPENDIX B

Basic probability and statistical theory

B.1 Introduction

Statistical analysis and statistical modelling is the science of making inferences of various kinds from data generated by a random process. The starting point is the 'random process'. The idea of randomness itself needs careful definition and we shall give the standard idea of a random variable with a *probability distribution function* or *density function*. The process or behaviour of the process depends on some unknown parameters. We use X, Y, etc., for the random process and typically θ, μ, σ, etc., for the unknown parameters. The random process produces data which are *realizations* of the random process in the way that coin-tossing is a random process and the result of a sequence of tosses a realization. From the data we produce summary *statistics* usually given Roman letters such as b, s^2, or 'hatted' Greek letters as in $\hat{\theta}$, $\hat{\mu}$. From these summary statistics we proceed to make inferences about the unknown parameters.

In Section 2.1.1 we described some important summary statistics, namely the sample mean as a measure of central tendency, and the sample standard deviation as a measure of dispersion.

Other measures of central tendency are as follows.

(i) The mode: the most commonly occurring observation.

(ii) The mid-range: the mid-point between the lowest and highest observation.

(iii) The median: the middle observation when all the observations are arranged in order of magnitude.

(iv) The geometric mean: the nth root of the product of the n observations.

Other measures of dispersion are:

(i) The range: the difference between the highest and lowest observation.

(ii) Mean deviation: the average deviation of the observations from the mean, with the deviations being taken absolutely.

When the process is 'very unknown' except for, say, its general structure, we may seek to use more *ad hoc* methods. For example if we do not know the underlying distribution function describing the process, we may approximate it using a histogram (see Section 2.1.3.1). We may find the distribution is skewed and transform it, say by a log-transformation, to look more normally distributed. A common mistake made by those introduced to statistical methods is to confuse the histogram of the data with the underlying distribution from which the data was drawn. For a larger sample the distributions may be close to each other but one is theoretical (and hopefully fixed) and the other is based on the data and will typically change if we have another sample.

If we draw another sample from the same process then the values of our summary statistics change. The behaviour of the statistics over repeated samples is important. It depends on how large the sample is and tells us how useful the statistics are in pinpointing the unknown parameters of the process or for making confidence statements about them.

It is important to bear in mind that we are at liberty to change the model *radically* in the light of the data, but it is usually wiser to update, tune, revise the model and so on, rather than make radical changes without justification. This is especially true if there are other, perhaps scientific, reasons for believing in the truth of the model. One must strike a balance between adopting an initial model which is so rigid that it will easily be rejected on a little data, and having a model which is so loose that any reasonable data set will fit it.

One situation in which rejection of a model is justified, is when the presentation of the results of an experiment triggers off thought processes or remembered technical details which should have been taken into account before the experiment. Disappointment should be replaced by the realization that this may be a successful outcome of an experiment and lead to further more focused study. For example, if a cyclic pattern is observed in the data this may point towards some rotating part causing vibrations which could then be the subject of a separate research programme. This 'hypothetico-deductive' method is an integral part of the scientific method and needs to be better understood in engineering. It could be said that even in statistics there are disputes as to what constitutes a coherent method of inference. It is doubtul, at least to the authors, whether any resolution of these disputes is possible. They reflect the tension between an objective and subjective view of the world which has bedevilled, or some might say enriched, many other subjects.

B.2 Probability

There are mathematical ways to describe certain numbers or variables which seem to change and vary in a haphazard fashion. The phase used is 'random variable'. The mathematical apparatus used to describe such entities can appear formidable but the ideas have a universal flavour and a feel for randomness can be obtained through observation and experiment. Probability is the basis for the theory and arose, so historians tell us, from games of chance, that is from observing randomness in nature or real life.

The starting point is a list of all possible outcomes of an experiment. This must include continuous sets if the outcome is a real number. This list is called the *sample space*, Ω. A subset of the sample space is an *event* usually written with capital letters: A, B, ... Probability is a number between zero and one which is attached to events and written as

$$\text{Prob(A)} \quad \text{or} \quad \text{P(A)}$$

The rules of probability are:

(i) $0 \leq \text{Prob}(A) \leq 1$.

(ii) $\text{Prob}(\Omega) = 1$.

(iii) If events A and B are disjoint (that is, cannot both happen):

$$\text{Prob}(A \text{ or } B) = \text{Prob}(A) + \text{Prob}(B).$$

Notice that the theory operates at two levels. Statements about events, $A, B, \ldots, C$ which have nothing to do with probability and then probability statements. We can use $(A \cup B)$ for (A or B) and $(A \cap B)$ for (A and B) and bring to play set theory. Here are some useful results

$$A \cup (B \cap C) = (A \cup B) \cap (A \cup C)$$

$$A \cap (B \cup C) = (A \cap B) \cup (A \cap C).$$

If $\bar{A}$, $\bar{B}$ mean 'not A', 'not B' then

$$\overline{A \cap B} = \bar{A} \cup \bar{B}$$

$$\overline{A \cup B} = \bar{A} \cap \bar{B}$$

(called De Morgan's Laws). Disjointness used in Rule (iii), can be written as

$$A \cap B = \phi,$$

where ϕ is the 'empty set'. Probability is superimposed on this set theory and has a few immediate contributions of its own to make. Of these perhaps the most important is *independence*; events A and B are independent if

$$P(A \cap B) = P(A)P(B).$$

This looks nice and symmetric but lacks intuition. For this we need conditional probability, at which point the subject starts to have a real life of its own. The conditional probability of A given B is

$$P(A/B) = \frac{P(A \cap B)}{P(B)}$$

(when $P(B) \neq 0$). This is based on the probability of A happening if we are myopic and only look at these occasions on which B also happens. In this myopic sample space, the 'when-B-happens' sample space, we need to rescale the probabilities by $1/P(B)$.

We can then happily work in this conditional sample space rather as if it were a sample space in its own right; A and B being independent, is equivalent to $P(A/B) = P(A)$. Structural reliability provides excellent examples of probability in action.

As a final formula in this section we give 'Bayes theorem' in its simple form:

$$P(E/F) = \frac{P(F/E)P(E)}{P(F/E)P(E) + P(F/\bar{E})P(\bar{E})},$$

where $P(\bar{E}) = 1 - P(E)$. This enables us to express $P(E/F)$ in terms of $P(F/E)$ and $P(E)$.

We also need to say a little about where probabilities come from. This is a large subject and there are several strands of thought each with famous names attached to it: Laplace, Bernoulli, Keynes, Von Mises, Kolmogorov, etc. The split in the subject is between the subjective view and the objective.

The subjective approach says that probabilities are internal and represent *our* assessment of whether an event will happen or not. The probability of a horse winning may be in this category. The rules are then rules of thought about probability.

The objective view is that probability is a physical phenomenon or a summary of physical phenomena. In an extreme version of this view, the probability of an event E is identified with the proportion of times that it happens in many successive repetitions of the same experiment. 'Repeatability' is a word used in engineering. This view is fraught with difficulty. To make the identification precise we would need an infinitely long run and we do not live long enough. A half-way house between these viewpoints is something like the following, which works particularly well when we are dealing with small discrete experiments: 'My view of probability is subjective (since everything is ultimately subjective) but it is based strongly on the fact that I find physical symmetries all around me (every coin or die looks alike). The rules of probability are a summary of how to deal with the effect of these symmetries on my assessment of whether any event will occur.'

This view is given a lot of support by the fact that we can simulate many random processes in a computer using pseudo-random procedures to the extent that we can barely distinguish the output from the real thing. It is no surprise to find considerable effort put into recapturing the rules of probability through careful definitions of complex computation processes.

B.3 Random variables and distributions

In describing events it is natural to use numbers rather than letters or verbal descriptions. We can go further and attach numbers to events and record them rather than the events themselves. Over repetitions of the experiment we shall see the numbers changing. We call such a quantity a random variable and denote it usually by a capital letter X, Y. The mathematical formulation then is to say that for an outcome ω in the sample space we create a number $X(\omega)$. For example, if we toss a coin three times and observe HHT as one of the eight possible outcomes and X is the number of heads, then $X = 2$. At this point we forget where the X came from and just think about X on its own. What values can it take? What is the probability it takes these values? Thus we can sit X directly on its own sample space rather than have it parasitic on some underlying Ω. For the coin-tossing experiment we can easily construct a table of probabilities:

X	0	1	2	3
Prob	$\frac{1}{8}$	$\frac{3}{4}$	$\frac{3}{4}$	$\frac{1}{8}$

The specification of the behaviour of X given by such a table is called the *distribution of X*. The X is in this case discrete, that is, it takes a finite (more generally, 'countable') set of values.

In general, if X takes values $a_1, a_2, \ldots$, then we can let

$$p_k = \text{Prob}(X = a_k).$$

Since X is not allowed to take two values at once and it must take some value, the rules of probability can be borrowed from Ω to give

(i) $0 \leq p_k \leq 1$,

(ii) $\sum p_k = 1$,

(iii) If A and B are disjoint sets of a_i

$$P(X \text{ in A or B}) = P(X \text{ in A}) + P(X \text{ in B}).$$

Continuous random variables are those that take any real numbers in some interval as their value. Here we can completely specify the distribution by defining the cumulative density function as

$$F(x) = \text{Prob}(X \leq x)$$

and its probability density function (if it exists) as

$$p(x) = \frac{\mathrm{d}F(x)}{\mathrm{d}x} = \lim_{\delta x \to 0} \frac{F(x + \delta x) - F(x)}{\delta x}$$

The numerator inside the limit is

$$F(x + \delta x) - F(x) = \text{Prob}(x < X \leq x + \delta x).$$

With a little thought it will be seen that $p(x)\delta x$ is approximately the area under the density function between x and $x + \delta x$ and that this is precisely the probability of finding X between these points (see also Section 2.1.3.3). We have to be a little careful about inequalities since 'not $\{X \leq a\}$' is $\{X > a\}$.

The random variables we find occurring naturally can often be assumed to belong to certain families. Appendix F gives many of the standard families.

Certain features of a distribution, and hence of the random variables itself, are summarized by *moments* and more general averages over the distribution. A general average is an expectation. Let $f(x)$ be a function. If X is random then $Y = f(X)$ is random. The expectation of Y is defined to be

$$E(Y) = \sum f(k)p_k \qquad \text{discrete}$$

$$= \int_{-\infty}^{\infty} f(x)p(x)\,\mathrm{d}x \quad \text{continuous.}$$

When $f(x) = x^r$ we get the 'non-central moments':

$$\mu'_r = E(X^r) = \int_{-\infty}^{\infty} x^r p(x)\,\mathrm{d}x$$

The first of these for $r = 1$ is $E(X) = \int_{-\infty}^{\infty} xp(x)\,\mathrm{d}x$ and is called the mean of the random variable (also expectation). The *central* moments are the moments about the mean, i.e. if $\mu'_1 = E(X) = \mu$ then for $r \geq 2$

$$\begin{aligned}\mu_r &= E[(X-\mu)^r] \\ &= \int_{-\infty}^{\infty} (x-\mu)^r p(x)\,\mathrm{d}x\end{aligned}$$

(Note that $\mu_1 = 0$.)

Similar expressions follow for discrete random variables. An important central moment is called the variance

$$\mathrm{var}(X) = \mu_2 = E[(X-\mu)^2]$$

A key point about expectations is that they are linear as 'operators' in that

$$E(aY + c) = aE(Y) + c.$$

From this follows a useful expansion for the variance: writing $\mu = E(X)$

$$\begin{aligned}\mathrm{var}(X) &= E[(X-\mu)^2] \\ &= E(X^2 - 2\mu X + \mu^2) \\ &= E(X^2) - 2\mu E(X) + \mu^2 \\ &= E(X^2) - 2\mu^2 + \mu \\ \mathrm{var}(X) &= E(X^2) - \mu^2 \quad \text{or} \quad \mu_2 = \mu'_2 - (\mu'_1)^2.\end{aligned}$$

We used the fact that for constant a, $E(a) = a$.

A theory can be developed using *cumulants* rather than moments. The starting point is the moment generating function:

$$\begin{aligned}\psi(t) &= \sum_{r=0}^{\infty} \left(\frac{t^r}{r!}\mu'_r\right) = E(\mathrm{e}^{tX}) \\ &= \mu'_0 + t\mu'_1 + \frac{t^2}{2!}\mu'_2 + \cdots.\end{aligned}$$

We take $\mu'_0 = 1$. Then the expansion for cumulants is

$$K(t) = \log_e \psi(t) = k_1 t + \frac{t^2}{2!}k_2 + \frac{t^3}{3!}k_3 + \cdots$$

Here are the first few formulae connecting cumulants with *non-central* or *central* moments

$$\mu_1' = k_1$$

$$\mu_2' = k_2 + k_1^2 \quad \text{or} \quad \mu_2 = k_2$$

$$\mu_3' = k_3 + 3k_2k_1 + k_1^3 \quad \text{or} \quad \mu_3 = k_3$$

$$\mu_k' = k_4 + 4k_3k_1 + 3k_2^2 + 6k_2k_1^2 + k_1^4 \quad \text{or} \quad \mu_4 = k_4 + 3k_2^2$$

and

$$k_1 = \mu_1'$$

$$k_2 = \mu_2' - (\mu_1')^2 \quad \text{or} \quad k_2 = \mu_2$$

$$k_3 = \mu_3' - 3\mu_2'\mu_1' + 2(\mu_1')^3 \quad \text{or} \quad k_3 = \mu_3$$

$$k_4 = \mu_4' - 4\mu_3'\mu_1' - 3(\mu_2')^2 + 12\mu_2'(\mu_1')^2 - 6(\mu_1')^4 \quad \text{or} \quad k_4 = \mu_4 - 3\mu_2^2.$$

One advantage of using cumulants is that for normal random variables we have $k_r = 0$ for $r > 2$. We can show this easily as follows. The normal density is

$$\frac{1}{\sqrt{2\pi}\sigma}\exp\left\{-\frac{1}{2}\left(\frac{x-\mu}{\sigma}\right)^2\right\}$$

and the moment generating function is $\exp(t\mu + \frac{1}{2}t^2\sigma^2)$ which can be obtained by integration. Taking $\log_e$ we obtain $K(t) = t\mu + \frac{1}{2}t^2\sigma^2$. Thus $k_2 = \mu_2' - (\mu_1')^2 = \sigma^2$ as expected, but higher-order terms are zero. Roughly speaking, if we find that estimates of higher-order cumulants are close to zero then this provides a check on the normality of the data.

B.4 Several random variables

On many occasions techniques are required for handling the behaviour of several random variables at once. We can extend the univariate definitions to multivariate distributions. They can be well understood by looking at the case of two random variables. The bivariate continuous joint (cumulative) distribution function of two random variables X_1 and X_2 is represented by

$$F(x_1, x_2) = \text{Prob}(X_1 \le x_1 \text{ and } X_2 \le x_2).$$

The joint probability density function is

$$p(x_1, x_2) = \frac{\partial F(x_1, x_2)}{\partial x_1 \partial x_2}$$

$$= \lim_{\delta x_1, \delta x_2 \to 0} \frac{\Delta_{12} F(x_1, x_2)}{\delta x_1 \delta x_2},$$

where

$$\begin{aligned}\Delta_{12}F(x_1, x_2) &= F(x_1+\delta x_1, x_2+\delta x_2)\\ &\quad - F(x_1+\delta x_1, x_2) - F(x_1, x_2+\delta x_2)\\ &\quad + F(x_1, x_2)\\ &= \text{Prob}(x_1 < X_1 \le x_1+\delta x_1, x_2 < X_2 \le x_2+\delta x_2).\end{aligned}$$

The separate probability density functions of X_1 and X_2 can be obtained by 'integrating out' the unwanted variable

$$p_1(x_1) = \int_{-\infty}^{\infty} p(x_1, x_2)\,\mathrm{d}x_2$$

$$p_2(x_2) = \int_{-\infty}^{\infty} p(x_1, x_2)\,\mathrm{d}x_1.$$

These are referred to as the 'marginal' probability density functions. They describe the behaviour when we 'forget about' the other variable.

The covariance between X_1 and X_2 is defined as

$$\text{cov}(X_1, X_2) = E\{[X_1 - E(X_1)][X_2 - E(X_2)]\} = E(X_1X_2) - E(X_1)E(X_2),$$

and the correlation as

$$\text{corr}(X_1X_2) = \text{cov}(X_1, X_2)/[\text{var}(X_1)\text{var}(X_2)]^{\frac{1}{2}}.$$

The following formula is useful for carrying out the kind of tolerance analysis referred to in Section 7.4. For Normally distributed random variables

$$\text{cov}(X_1X_2, X_3X_4) = \text{cov}(X_1, X_3)\text{cov}(X_2, X_4) + \text{cov}(X_1, X_4)\text{cov}(X_2, X_3).$$

A general result for a quadratic function is

$$\text{var}\left(\sum_i \sum_j a_i a_j X_i X_j\right) = \sum_{i_1}\sum_{i_2}\sum_{i_3}\sum_{i_4} a_{i_1}a_{i_2}a_{i_3}a_{i_4}\text{cov}(X_{i_1}X_{i_2}, X_{i_3}X_{i_4}).$$

If $X_1, \ldots, X_n$ are independent and identically distributed (see below), then this reduces to

$$k_4 \sum_i a_{ii}^2 + k_2^2 \sum_i \sum_j (a_{ij}^2 + a_{ij}a_{ji})$$

where k_4 and k_2 are the fourth and second cumulants, respectively.

The final idea in this section is that of independence of two (or more) random variables. This is easily stated as: X_1 and X_2 are independent if the joint density factorizes into the product of the two marginal densities, i.e. if

$$p(x_1, x_2) = p_1(x_1)p_2(x_2)$$

This says essentially that as $\delta x_1, \delta x_2 \to 0$

$$\begin{aligned}&\text{Prob}(x_1 < X_1 \le x_1+\delta x_1 \text{ and } x_2 < X_2 \le x_2+\delta x_2)\\ &\quad = \text{Prob}(x_1 < X_1 \le x_1+\delta x_1)\text{Prob}(x_2 < X_2 \le x_2+\delta x_2).\end{aligned}$$

An immediate consequence is that

$$E(X_1X_2) = E(X_1)E(X_2)$$

and hence

$$\operatorname{cov}(X_1, X_2) = 0.$$

This extends to several random variables: If $X_1, \ldots, X_n$ are independent then the joint density again factorizes into the product of marginals. As far as variances are concerned we obtain for independent random variables

$$\operatorname{var}\left(\sum a_i X_i\right) = \sum a_i^2 \operatorname{var}(X_i).$$

A purpose of the results on mean and variances is to describe the behaviour of *sample quantities* used in statistical analysis, namely quantities such as

$$\bar{X} = \frac{1}{n}\sum X_i \qquad \text{sample mean,}$$

$$s^2 = \frac{1}{n-1}\sum (X_i - \bar{X})^2 \quad \text{sample variance.}$$

The following results use most of what has gone before: if $X_1, \ldots, X_n$ are independently sampled from a distribution with mean μ and variance σ^2 then

$$E(\bar{X}) = \mu, \qquad \operatorname{var}(\bar{X}) = \frac{\sigma^2}{n}, \qquad E(s^2) = \sigma^2.$$

B.5 Estimation

As mentioned in Section B.1, the aim of evaluating sample quantities is to estimate or make statements about the unknown parameters μ, σ^2, θ, etc. We have seen examples using $\bar{X}$ as an estimate of μ and s^2 as an estimate of σ^2. The property that $E(\bar{X}) = \mu$ and $E(s^2) = \sigma^2$ is called unbiasedness and is a useful, if rather classical, property of an estimator. For a general parameter θ we define a function $\hat{\theta}$ of the observations to be unbiased if

$$E(\hat{\theta}) = \theta$$

for all θ. The 'for all θ' here means 'for all θ in the space that θ lies in'.

Example: Suppose we have a biased coin so that Prob(Heads) $= p$, Prob(tails) $= 1 - p$. We introduce 'counting' variables X_i, so that

$$i\text{th toss Heads implies } X_i = 1.$$

$$i\text{th toss Tails implies } X_i = 0.$$

Each X_i has the 'Bernoulli distribution' (see Appendix F) and it can be easily shown that $E(X_i) = p$ and $\operatorname{var}(X_i) = p(1-p)$. We can count the number of heads by

$$X = \sum X_i.$$

Indeed, X has a 'Binomial distribution' Bin (n, p) (see Appendix F). Since each binary random variable X_i has expectation p we have

$$E(X) = \sum E(X_i) = np.$$

Thus

$$\hat{p} = X/n$$

has

$$E(\hat{p}) = np/n = p$$

and is an unbiased estimator for p. The 'for all p' here refers to $0 \leq p \leq 1$. This is intuitive since $\hat{p}$ is the proportion of heads. Also

$$\begin{aligned}\operatorname{var}(\hat{p}) &= \frac{1}{n^2} \sum \operatorname{var}(X_i) \\ &= \frac{1}{n^2} np(1-p) = p(1-p)/n\end{aligned}$$

using the fact that for each X_i, $\operatorname{var}(X_i) = p(1-p)$

B.6 Risk and loss functions

We can measure how well an estimator $\hat{\theta}$ does by computing

$$E[(\hat{\theta} - \theta)^2],$$

the mean square error; that is, the expected deviation away from the true parameter value θ. But

$$\begin{aligned}E[(\hat{\theta} - \theta)^2] &= E[\hat{\theta} - E(\hat{\theta}) + E(\hat{\theta}) - \theta]^2 \\ &= E\{[\hat{\theta} - E(\hat{\theta})]^2\} \\ &\quad + E\{[\hat{\theta} - E(\hat{\theta})][E(\hat{\theta}) - \theta]\} \\ &\quad + [E(\hat{\theta}) - \theta]^2.\end{aligned}$$

The middle term is zero since

$$E\{[\hat{\theta} - E(\hat{\theta})][E(\hat{\theta}) - \theta]\} = [E(\hat{\theta}) - E(\hat{\theta})][E(\hat{\theta}) - \theta]$$

(the second bracket is a constant). Thus

$$E[(\hat{\theta} - \theta)]^2 = \operatorname{var}(\hat{\theta}) + [E(\hat{\theta}) - \theta]^2,$$

if $\hat{\theta}$ is unbiased the second term, the squared 'bias', is zero so that $\operatorname{var}(\hat{\theta})$ becomes the appropriate term to measure how good $\hat{\theta}$ is.

We repeat the important message that since the X_i are random, so is $\hat{\theta}$, being a function of the X_i, so that repeated experiments will give different values of $\hat{\theta}$. We measure how good $\hat{\theta}$ is by some function $L(\theta, \hat{\theta})$ such as $(\hat{\theta} - \theta)^2$. Since θ is unknown and $\hat{\theta}$ is random, we look at the operating characteristic of $\hat{\theta}$ as a *machine for estimation of* θ by evaluating:

$$R(\theta) = E[L(\theta, \hat{\theta})].$$

We call L the loss function and R the risk. If we are lucky, $R(\theta)$ does not depend on θ. In the coin-tossing example the risk is $p(1-p)/n$ which *does* depend on the unknown p. Also in the case of using $\bar{X}$ to estimate μ we have $\text{var}(\bar{X}) = \sigma^2/n$; so the risk (in this case the variance), though constant, still may be unknown if σ^2 is unknown. In this case s^2/n is an *estimate* of the 'risk of using $\bar{X}$ to estimate μ'.

B.7 Regression, linear models

There is one case in which the theory of estimation is particularly pleasant and it is the one used for the most part in this book.

First we need to extend the idea of unbiasedness to several parameters $\theta_0, \ldots, \theta_{m-1}$. Thus $\hat{\boldsymbol{\theta}} = (\hat{\theta}_0, \ldots, \hat{\theta}_{m-1})$ is said to be unbiased for $\boldsymbol{\theta} = (\theta_0, \ldots, \theta_{m-1})$ if

$$E(\hat{\theta}_i) = \theta_i \qquad \text{for all } \theta_i.$$

We can extend this slightly to consider arbitrary linear functions of the θ_i, namely:

$$\phi = \sum c_i\theta_i = \mathbf{c}^T\boldsymbol{\theta}.$$

If $\hat{\theta}_i$ are unbiased for θ_i ($\hat{\boldsymbol{\theta}}$ for $\boldsymbol{\theta}$) then

$$\hat{\phi} = \sum c_i\hat{\theta}_i$$

is unbiased for ϕ, since

$$E(\hat{\phi}) = \sum c_i E(\hat{\theta}_i) = \sum c_i\theta_i = \phi.$$

It will turn out that if $\hat{\theta}_i$ are least squares estimates of regression parameters θ_i then

(i) the $\hat{\theta}_i$ are unbiased;

(ii) they are *linear* in the observations;

(iii) they have minimum variance among all estimators satisfying (i) and (ii);

(iv) any $\hat{\phi}$ of the above form is itself the best linear unbiased estimator of ϕ.

This theory is best exemplified by reference to the weighing examples in Chapter 2. We give a formal development. Writing the model as

$$\underset{n\times 1}{E(\mathbf{Y})} = \underset{n\times m}{\mathbf{X}}\ \underset{m\times 1}{\boldsymbol{\theta}}$$

we can write

$$R^2 = \|\mathbf{Y} - \mathbf{X}\boldsymbol{\theta}\|^2 = (\mathbf{Y} - \mathbf{X}\boldsymbol{\theta})^{\mathrm{T}}(\mathbf{Y} - \mathbf{X}\boldsymbol{\theta}).$$

Differentiating partially with respect to each θ_i in turn and collecting the result into a vector (gradient) we have

$$\frac{\partial R^2}{\partial \boldsymbol{\theta}} = \left(\frac{\partial R^2}{\partial \theta_0}, \ldots, \frac{\partial R^2}{\partial \theta_{m-1}}\right)^T$$
$$= 2\mathbf{X}^T\mathbf{X}\boldsymbol{\theta} - 2\mathbf{X}^T\mathbf{Y}.$$

Setting this equal to zero we have

$$\mathbf{X}^T\mathbf{X}\boldsymbol{\theta} = \mathbf{X}^T\mathbf{Y}.$$

If $\mathbf{X}$ has rank m then $\mathbf{X}^T\mathbf{X}$ also has rank m (proof omitted) and is therefore invertible.

We thus obtain the *unique* least squares solutions

$$\hat{\boldsymbol{\theta}} = (\mathbf{X}^T\mathbf{X})^{-1}\mathbf{X}^T\mathbf{Y}.$$

Unbiasedness follows, since we can take expectation 'through' a matrix $\mathbf{A}$ to give

$$E(\mathbf{AY}) = \mathbf{A}E(\mathbf{Y})$$

so that

$$E(\hat{\boldsymbol{\theta}}) = (\mathbf{X}^T\mathbf{X})^{-1}\mathbf{X}^T E(\mathbf{Y})$$
$$= (\mathbf{X}^T\mathbf{X})^{-1}\mathbf{X}^T\mathbf{X}\boldsymbol{\theta}$$
$$= \boldsymbol{\theta}.$$

Moreover, if

$$\phi = \mathbf{c}^T\boldsymbol{\theta}$$

is an arbitrary linear function of the θ_i's, it can be shown that $\text{var}(\hat{\phi})$ is a minimum when and only when

$$\hat{\phi} = \mathbf{c}^T\hat{\boldsymbol{\theta}}.$$

The covariance matrix of the parameter estimates is

$$\text{cov}(\hat{\boldsymbol{\theta}}) = (\mathbf{X}^T\mathbf{X})^{-1}\mathbf{X}^T\,\text{cov}(\mathbf{Y})\mathbf{X}(\mathbf{X}^T\mathbf{X})^{-1}.$$

If we assume that $\text{cov}(\mathbf{Y}) = \sigma^2\mathbf{I}$ this reduces to

$$\text{cov}(\hat{\boldsymbol{\theta}}) = \sigma^2(\mathbf{X}^T\mathbf{X})^{-1},$$

and

$$\text{var}(\hat{\phi}) = \sigma^2\mathbf{c}^T(\mathbf{X}^T\mathbf{X})^{-1}\mathbf{c}.$$

Regression principles

Consider the matrix $\mathbf{P} = \mathbf{X}(\mathbf{X}^T\mathbf{X})^{-1}\mathbf{X}^T$, which has the following special properties:

$$\mathbf{P} = \mathbf{P}^T \quad \text{and} \quad \mathbf{P}^2 = \mathbf{P}.$$

We can use the $\mathbf{P}$ matrix to give a nice representation of the total sums of squares in terms of the sums of squares due to the regression (with the constant term) and the residual. Indeed, since

$$\mathbf{Y}^T\mathbf{Y} = \hat{\mathbf{Y}}^T\hat{\mathbf{Y}} + (\mathbf{Y} - \hat{\mathbf{Y}})^T(\mathbf{Y} - \hat{\mathbf{Y}})$$

[or, in terms of the observations,

$$\sum Y_i^2 = \sum \hat{Y}_i^2 + \sum (Y_i - \hat{Y}_i)^2]$$

using **P**

$$\begin{aligned}\mathbf{Y}^T\mathbf{Y} &= (\mathbf{PY})^T\mathbf{PY} + (\mathbf{Y} - \mathbf{PY})^T(\mathbf{Y} - \mathbf{PY}) \\ &= \mathbf{Y}^T\mathbf{P}^2\mathbf{Y} + \mathbf{Y}^T(\mathbf{I} - \mathbf{P})^T(\mathbf{I} - \mathbf{P})\mathbf{Y} \\ &= \mathbf{Y}^T\mathbf{PY} + \mathbf{Y}^T(\mathbf{I} - \mathbf{P})\mathbf{Y}\end{aligned}$$

where **I** is the identity matrix.

Mathematically, **P** is the projection matrix which projects the **Y** vector into the column space of the matrix **X**: $\hat{\mathbf{Y}} = \mathbf{PY}$ is the result. But then $(\mathbf{I} - \mathbf{P})\mathbf{Y}$ is the projection into the orthogonal subspace, all these vectors being orthogonal to the column space of **X**. Thus

$$\mathbf{Y} = \hat{\mathbf{Y}} + (\mathbf{Y} - \hat{\mathbf{Y}})$$

corresponds to

$$\mathbf{Y} = \mathbf{PY} + (\mathbf{I} - \mathbf{P})\mathbf{Y}$$

and the sums of squares are just the squared lengths of these projections. But

$$\mathbf{R} = (\mathbf{I} - \mathbf{P})\mathbf{Y}$$

is the vector of residuals. This representation makes it easier to find the covariance matrix of the residuals:

$$\begin{aligned}\text{cov}(\mathbf{R}) &= (\mathbf{I} - \mathbf{P})\text{cov}(\mathbf{Y})(\mathbf{I} - \mathbf{P})^T \\ &= (\mathbf{I} - \mathbf{P})\sigma^2\mathbf{I}(\mathbf{I} - \mathbf{P})^T \\ &= \sigma^2(\mathbf{I} - \mathbf{P})\end{aligned}$$

(using the properties $\mathbf{P}^T = \mathbf{P}$ and $\mathbf{P}^2 = \mathbf{P}$).

Thus the residuals are typically correlated but only a little if n (number of observations) is much larger than m (number of parameters) because then $\sigma^2(\mathbf{I} - \mathbf{P})$ is approximately $\sigma^2\mathbf{I}$. We may standardize the residuals by dividing by their standard deviation

$$R_i^* = \frac{R_i}{\sigma(1 - h_i)^{\frac{1}{2}}}$$

where h_i is the diagonal element of **P**. Notice also that

$$\begin{aligned}\text{cov}(\hat{\mathbf{Y}}) &= \mathbf{P}\,\text{cov}(\mathbf{Y})\mathbf{P}^T \\ &= \mathbf{P}\sigma^2\mathbf{I}\mathbf{P}^T \\ &= \sigma^2\mathbf{P}.\end{aligned}$$

P is sometimes called the 'hat' matrix for this reason and hence the h_i notation.

In most regression models we have a constant term. This makes the first column of **X** into a vector of ones. Since $(\mathbf{I} - \mathbf{P})\mathbf{X} = 0$ we have $(\mathbf{I} - \mathbf{P})j = 0$ where

j is the column of ones. But

$$\mathbf{Y}^T(\mathbf{I}-\mathbf{P})j=0$$

or

$$\mathbf{R}^T j=0,$$

giving

$$\sum R_i=0.$$

Thus, for models with constant terms, the sum of the residuals is zero. Residual plots are useful in what is called regression diagnostics, that is using them to see whether the basic assumption such as constant variance are correct. Most importantly, if we have fitted too small a model, then plots of residuals against x-values will reveal missing terms; for example, terms reflecting curvature.

Weighted least squares

When σ^2, the error variance, is non-constant then the usual least squares estimates will not be optimal in the sense of minimum variance, although they will still be unbiased, since this does not depend on the behaviour of σ^2. Suppose then that

$$\operatorname{cov}(\mathbf{Y})=\mathbf{D}=\begin{bmatrix}\sigma_1^2 & & 0\\ & \ddots & \\ 0 & & \sigma_n^2\end{bmatrix}.$$

If the σ_i^2 are known then we can convert back to the ordinary least squares case by rescaling the ith observation by $1/\sigma_i$. In matrix terms this means multiplying through by $\mathbf{D}^{-1/2}$. Thus

$$\mathbf{D}^{-1/2}\mathbf{Y}=\mathbf{D}^{-1/2}\mathbf{X}\boldsymbol{\theta}+\mathbf{D}^{-1/2}\boldsymbol{\varepsilon}.$$

But

$$\operatorname{cov}(\mathbf{D}^{-1/2}\mathbf{Y})=\mathbf{D}^{-1/2}\mathbf{D}\mathbf{D}^{-1/2}=\mathbf{I},$$

so that the transformed observations have constant variance of 1. Considering $\mathbf{D}^{-1/2}\mathbf{X}$ as a new $\mathbf{X}$-matrix we have

$$\begin{aligned}\hat{\boldsymbol{\theta}}&=(\mathbf{X}^T\mathbf{D}^{-1/2}\mathbf{D}^{-1/2}\mathbf{X})^{-1}\mathbf{X}^T\mathbf{D}^{-1/2}\mathbf{D}^{-1/2}\mathbf{Y}\\&=(\mathbf{X}^T\mathbf{D}^{-1}\mathbf{X})^{-1}\mathbf{X}^T\mathbf{D}^{-1}\mathbf{Y}.\end{aligned}$$

In fact, we only need to know $\mathbf{D}$ up to a constant unknown factor to do this. Also

$$\operatorname{cov}(\hat{\boldsymbol{\theta}})=(\mathbf{X}^T\mathbf{D}^{-1}\mathbf{X})^{-1}.$$

For some complex models such as those for variance estimation we may (i) start with the ordinary least squares, (ii) estimate $\mathbf{D}$ from the fact that under the current model,

$$\begin{aligned}\operatorname{cov}(\mathbf{R})&=(\mathbf{I}-\mathbf{P})\mathbf{D}(\mathbf{I}-\mathbf{P})^T\\&\simeq\mathbf{D},\end{aligned}$$

and (iii) use the estimate of $\mathbf{D}$ to get, hopefully, an improved estimate of $\hat{\theta}$.

Testing

The decomposition into sums of squares enables us to test linear hypotheses, that is, statements about parameters of the form $\mathbf{A}\boldsymbol{\theta} = 0$ where $\mathbf{A}$ is a matrix (or $\mathbf{A}\boldsymbol{\theta} =$ constant vector). The simplest of these is a test such as

$$\mathrm{H}: \theta_r = \theta_{r+1} = \cdots = \theta_{m-1} = 0.$$

Perhaps the most straightforward way of explaining this is to say that under a hypothesis such as H we simply have a smaller model

$$\mathbf{Y} = \mathbf{X}'\boldsymbol{\theta} + \boldsymbol{\varepsilon},$$

where $\mathbf{X}'$ only has some of the columns of the original $\mathbf{X}$ matrix. Calling $\hat{\mathbf{Y}}^T\hat{\mathbf{Y}}$ the regression sums of squares for the full model and $\hat{\mathbf{Y}}_H^T\hat{\mathbf{Y}}_H$ the regression sums of squares for the reduced model, we have

$$\mathbf{Y}^T\mathbf{Y} = \hat{\mathbf{Y}}_{\mathrm{H}}^T\hat{\mathbf{Y}}_{\mathrm{H}} + (\hat{\mathbf{Y}}^T\hat{\mathbf{Y}} - \hat{\mathbf{Y}}_{\mathrm{H}}^T\hat{\mathbf{Y}}_{\mathrm{H}}) + (\mathbf{Y} - \hat{\mathbf{Y}})^T(\mathbf{Y} - \hat{\mathbf{Y}}).$$

Under the assumption that $\mathrm{cov}(\mathbf{Y}) = \sigma^2\mathbf{I}$ and that the errors are normally distributed with zero mean, we can write

$$\frac{1}{\sigma^2}\mathbf{Y}^T\mathbf{Y} = \frac{1}{\sigma^2}\hat{\mathbf{Y}}_{\mathrm{H}}^T\hat{\mathbf{Y}}_{\mathrm{H}} + \frac{1}{\sigma^2}(\hat{\mathbf{Y}}^T\hat{\mathbf{Y}} - \hat{\mathbf{Y}}_{\mathrm{H}}^T\hat{\mathbf{Y}}_{\mathrm{H}}) + \frac{1}{\sigma^2}(\mathbf{Y} - \hat{\mathbf{Y}})^T(\mathbf{Y} - \hat{\mathbf{Y}})$$

and it is possible to show (Fisher–Cochran theorem) that the last two terms on the right-hand side of the last equation are independently χ^2 (chi-squared) distributed with degrees of freedom $m - r$ and $n - m$ respectively. (They are each independent quadratic forms in normal distributed random variables.) Letting

$$\hat{\mathbf{Y}}^T\hat{\mathbf{Y}} - \hat{\mathbf{Y}}_{\mathrm{H}}^T\hat{\mathbf{Y}}_{\mathrm{H}} = TestSS$$

$$(\mathbf{Y} - \hat{\mathbf{Y}})^T(\mathbf{Y} - \hat{\mathbf{Y}}) = RSS,$$

where *SS* stands for sums of squares, then if H is true

$$F_{\mathrm{H}} = \frac{TestSS/(m-r)}{RSS/(n-m)} = \frac{MS_{\mathrm{H}}}{MS_{Res}}$$

can be shown to have an F-distribution on $m - r$ and $n - m$ degrees of freedom written $F_{m-r,n-m}$. If F_{H} is larger than the $(1 - \alpha)\%$ critical value (say 95%) then we reject H. Note that the denominator in F is s^2, the estimator of σ^2.

The correction factor, *CF*, is referred to in various places in the text. This is the quantity $\hat{\mathbf{Y}}_{\mathrm{H}}^T\hat{\mathbf{Y}}_{\mathrm{H}}$ when all parameters are set to zero except θ_0. Thus H is

$$\theta_1 = \cdots = \theta_{m-1} = 0,$$

(but not $\theta_0 = 0$). If the reduced model is just

$$Y = \theta_0 + \varepsilon,$$

then $\hat{Y}_i = \hat{\theta}_0 = \bar{Y}$ and $\hat{\mathbf{Y}}$ is a vector composed entirely of $\bar{Y}$. Thus

$$\hat{\mathbf{Y}}_{\mathrm{H}}^T\hat{\mathbf{Y}}_{\mathrm{H}} = n(\bar{Y})^2$$

and

$$\begin{aligned}TestSS &= \hat{\mathbf{Y}}^T\hat{\mathbf{Y}} - \hat{\mathbf{Y}}_{\mathrm{H}}^T\hat{\mathbf{Y}}_{\mathrm{H}} \\ &= \hat{\mathbf{Y}}^T\hat{\mathbf{Y}} - n(\bar{Y})^2 \\ &= \sum \hat{Y}_i^2 - n(\bar{Y})^2.\end{aligned}$$

This last term can also be written

$$\sum (\hat{Y}_i - \bar{Y})^2.$$

(Note: this last expression should only be used for models with a constant term.)

The test for a single parameter, say $\mathrm{H}: \theta_q = 0$, can be written

$$t = \hat{\theta}_q / \hat{\sigma}_q$$

where $\hat{\sigma}_q^2 = \widehat{\mathrm{var}}(\hat{\theta}_q)$, which is the qth diagonal element of $\mathrm{cov}(\hat{\boldsymbol{\theta}}) = s^2(\mathbf{X}^T\mathbf{X})^{-1}$. We reject H if t is beyond the critical value on the t-distribution with $n - m$ degrees of freedom. It is important to note, although we omit the proof, that $F = t^2$ is precisely the F-statistic we would use if we did the F-test rather than the t-test. The tests are the same. Many regression packages give the t-value for each parameter *in the presence of other parameters*, and the 'standard error' $\hat{\sigma}_q$, namely the estimated standard deviations. This enables one to see at a glance which are the significant parameters. However, if many such tests are performed simultaneously, care must be exercised to control the overall rejection rate.

B.8 Transforming the variables

If $\log_e Y$ is a normally distributed random variable with mean μ and variance σ^2, then Y is said to have a lognormal distribution. Transforming random variables to obtain new random variables requires a little additional analysis. Suppose

$$Y = f(X)$$

where f is a strictly increasing function, then

$$\begin{aligned}F_Y(y) &= \mathrm{Prob}(Y \le y) \\ &= \mathrm{Prob}[f(X) \le y] \\ &= \mathrm{Prob}[X \le f^{-1}(y)]. \\ &= F_X[f^{-1}(y)].\end{aligned}$$

That is, the cumulative distribution function of X evaluated at $x = f^{-1}(y)$.

Differentiating

$$\begin{aligned} p_Y(y) &= \frac{\mathrm{d}F_Y(y)}{\mathrm{d}y} \\ &= \frac{\mathrm{d}}{\mathrm{d}y} F_X(x) \\ &= \frac{\mathrm{d}}{\mathrm{d}x} F_X(x) \frac{\mathrm{d}x}{\mathrm{d}y} \\ &= p_X(x) \frac{\mathrm{d}x}{\mathrm{d}y} \\ &= p_X[f^{-1}(y)]\left(\frac{\mathrm{d}y}{\mathrm{d}x}\right)^{-1}. \end{aligned}$$

The way to remember this is, $p_Y\,\mathrm{d}y = p_X\,\mathrm{d}x$ showing that the area under the two different density functions remains constant. Applying this to the lognormal distribution we have $\log_e Y = X$ or $Y = \mathrm{e}^X$. Thus $y = f(x) = \mathrm{e}^x$ and $\mathrm{d}y/\mathrm{d}x = y$. Finally

$$p_Y(y) = \frac{1}{\sqrt{2\pi}\sigma}\frac{1}{y}\exp\left(-\frac{1}{2}\left(\frac{\log y - \mu}{\sigma}\right)^2\right)$$

from which we find

$$E(Y) = \exp(\mu + \tfrac{1}{2}\sigma^2)$$

and

$$\mathrm{var}(Y) = \exp(2\mu + \sigma^2)(\mathrm{e}^{\sigma^2} - 1).$$

The same kind of technique can be applied to find the distribution of $Y = \log_e(X)$ when X has a chi-squared distribution with q degrees of freedom. This is useful in modelling the variance (see Section 6.5).

B.9 Orthogonal polynomials

When one deals with a quantitative multi-level factor, there is often a need to evaluate the effect of the linear, quadratic, cubic, etc., component of that factor. When the factor has equidistant levels, the use of orthogonal polynomials due to P. L. Chebyshev, is recommended.

For example, if A is a quantitative factor with k equidistant levels such that

$$A(i) = A(1) + (i - 1)h, \qquad i = 1, \ldots, k$$

(i.e. the levels $A(i)$ are at equal intervals of spacing h), let us define $\bar{A}$ to be the mean of the level values of the k levels for A, i.e.

$$\bar{A} = A(1) + \frac{(k-1)}{2}h.$$

Then one can represent the population mean μ of the experimental values for factor A by an orthogonal polynomial as follows

$$\mu = b_0 + b_1(A - \bar{A}) + b_2\left[(A - \bar{A})^2 - \frac{(k^2-1)}{12}h^2\right] + b_3\left[(A - \bar{A})^3 - \frac{(3k^2-7)}{20}(A - \bar{A})h^2\right] + \ldots .$$

Estimates of the coefficients $b_0, b_1, \ldots,$ can be obtained by solving, through the method of least squares, the following k equations:

$$b_0 + b_1(A(1) - \bar{A}) + b_2\left[(A(1) - \bar{A})^2 - \frac{(k^2-1)}{12}h^2\right] + \cdots = \frac{A_1}{r}$$

$$\vdots$$

$$b_0 + b_1(A(k) - \bar{A}) + b_2\left[(A(k) - \bar{A})^2 - \frac{(k^2-1)}{12}h^2\right] + \cdots = \frac{A_k}{r}$$

where A_i, is the sum of all observations, say r of them, corresponding to level $A(i)$, $i = 1, \ldots, k$. The above equations are so constructed that, if $p \neq q$, the sum of products of the coefficients of b_p and b_q is zero. One can thus obtain, for example,

$$\hat{b}_0 = \frac{A_1 + \cdots + A_k}{k \,.\, r} \qquad \text{(grand mean of all observations)}$$

$$\hat{b}_1 = \frac{[A(1) - \bar{A}]A_1 + \cdots + [A(k) - \bar{A}]A_k}{[A(1) - \bar{A}]^2 r + \cdots + [A(k) - \bar{A}]^2 r},$$

etc. What is of interest, is the variation of $\hat{b}_1$, $\hat{b}_2$, etc., because this provides estimates of the variability due respectively to the linear component, the quadratic component, etc., of factor A. For example, the variance of $\hat{b}_1$ is equivalent to the sum of squares due to the linear component of A, the variance of $\hat{b}_2$ is equivalent to the sum of squares due to the quadratic component of A, etc.

An easy way of obtaining these sum of squares is by applying the formula

$$\text{var}(\hat{b}_j) = \frac{(W_1A_1 + \cdots + W_kA_k)^2}{Sr},$$

where $S = \sum_{i=1}^{k} W_i^2$, and the coefficients W_i, $i = 1, \ldots, k$ for each $\hat{b}_j$ [linear ($j = 1$), quadratic ($j = 2$), etc.] can be found in Table T4 of Appendix E.

For example, if A is a 3-level factor then

coefficient	$\hat{b}_1$	$\hat{b}_2$
W_1	−1	1
W_2	0	−2
W_3	1	1
S	2	6

and so

$$\operatorname{var}(\hat{b}_1) = S_{A_L} = \frac{(-A_1 + A_3)^2}{2r}$$

and

$$\operatorname{var}(\hat{b}_2) = S_{A_Q} = \frac{(A_1 - 2A_2 + A_3)^2}{6r}$$

Clearly, the above sum of squares correspond to 1 degree of freedom each, and the sum of squares S_A for the (total) main effect of A is given by

$$S_A = S_{A_L} + S_{A_Q} \qquad (\text{df} = 2).$$

The above formula for $\operatorname{var}(\hat{b}_j)$ can be applied even for slightly unequal-spaced levels, provided the deviation from what would be considered as 'equidistant' is within about 20% compared with the spacing.

If the number of replications per level is not the same for all levels, say r_i for level $A(i)$, provided that each r_i comes within roughly a factor of 2 of the mean number of replications, the above formula can still be applied, by using the harmonic mean (see Section 4.5.1) of the r_i's for each level.

B.10 A brief guide to some of the more commonly used statistical computer packages

B.10.1 Minitab

An easy to use interactive package most useful for preliminary inspection and plotting of data and for regression analysis. Some data manipulation facilities are available as well as elementary time series and analysis of variance. Includes a useful 'help' command. Not suitable for large data sets.

B.10.2 GLIM

Although primarily designed for fitting generalized linear models (of the form $g(E(Y_i)) = \sum_j \beta_j x_{ij}$) GLIM has a command language of considerable power including facilities for defining macros, loops, and branching. Also available interactively, GLIM has achieved considerable popularity although the uninitiated may find the output confusing. Also not suitable for large data sets.

B.10.3 SPSS-X

The new version of the extremely popular SPSS package. A simple control language, good documentation, and powerful data manipulation facilities are the main reasons for this popularity. The range of statistical techniques though initially poor has been greatly enhanced and the new version contains many improvements.

B.10.4 BMDP

Although initially written for medical applications, this is another package of general applicability. It consists of a set of independent programs each covering a particular field. This has one advantage for the occasional user that he/she need only read the relevant program specification.

B.10.5 SAS

This is a highly regarded general purpose package. Facilities for data manipulation are unequalled and these are combined with a good range of statistical procedures including multivariate analysis, regression, and analysis of variance.

B.10.6 Genstat

Another general package, Genstat has a powerful command language allowing the user to program complex algorithms as well as call procedures already available. The price of this versatility is that Genstat can seem difficult to the lay user. A good introductory guide has been written, however.

B.10.7 Statgraphics

Probably the best menu-driven package, Statgraphics has a wide range of facilities including some design generation algorithms. High-resolution graphics, mathematical functions and an optional APL interface are other features.

Note: Most of the above packages are now available on micros or workstations.

APPENDIX C

Constructing orthogonal arrays

C.1 Using Latin squares

The simplest way of constructing orthogonal arrays is by superimposing 'orthogonal Latin squares' (see Chapter 3, Sections 3.2.3 and 3.2.4) on each other. In general:

(C1) We can construct an orthogonal array $\mathrm{OA}[q;r]$ where

$$\mathrm{OA}[q;r] = \mathrm{OA}_{q^r}(q^{(q^r-1)/(q-1)})$$

for any $r = 2, 3, \ldots,$ by simply using $(q-1)$ mutually orthogonal $q \times q$ Latin squares. For example, if

$$L_1(3) = \begin{bmatrix} 1 & 2 & 3 \\ 2 & 3 & 1 \\ 3 & 1 & 2 \end{bmatrix}$$

and

$$L_2(3) = \begin{bmatrix} 1 & 2 & 3 \\ 3 & 1 & 2 \\ 2 & 3 & 1 \end{bmatrix}$$

are two 3×3 Latin squares, orthogonal to each other, then by superimposing $L_2(3)$ on $L_1(3)$ as Table C1 shows, a Graeco-Latin square (see Section 3.2.3) is obtained, and this can be expressed in array form as Table C2 shows.

TABLE C1

		Columns		
		1	2	3
	1	1,1	2,2	3,3
Rows	2	2,3	3,1	1,2
	3	3,2	1,3	2,1

TABLE C2

	R	C		
	1	1	1	1
	1	2	2	2
	1	3	3	3
	2	1	2	3
OA[3; 2] =	2	2	3	1
	2	3	1	2
	3	1	3	2
	3	2	1	3
	3	3	2	1

$= OA_9(3^4)$

The letters R and C represent the Rows and Columns of the 'mixed' array of Table C1. The array of Table C2 can be used to study four three-level factors, assuming no interactions, or two three-level factors (first two columns) and their interaction (last two columns). So, using two 3×3 orthogonal Latin squares $L_1(3)$ and $L_2(3)$, we have constructed the $OA_N(3^n)$, where

$$N = 9 = 3^2 \quad \text{and} \quad n = 4 = \frac{3^2 - 1}{3 - 1},$$

i.e. in this case (C1) is true with $q = 3$ and $r = 2$. [Note that $L_2(3)$ was obtained by simply interchanging the second and third row of $L_1(3)$.]

Similarly for $q = 4$ and $r = 2$, the

$$OA_{4^2}[4^{(4^2-1)/(4-1)}] = OA_{16}(4^5)$$

can be easily obtained by using the three mutually orthogonal 4×4 Latin squares (superimposed on each other):

$$L_1(4) = \begin{bmatrix} 1 & 2 & 3 & 4 \\ 2 & 1 & 4 & 3 \\ 3 & 4 & 1 & 2 \\ 4 & 3 & 2 & 1 \end{bmatrix} \quad L_2(4) = \begin{bmatrix} 1 & 2 & 3 & 4 \\ 3 & 4 & 1 & 2 \\ 4 & 3 & 2 & 1 \\ 2 & 1 & 4 & 3 \end{bmatrix} \quad L_3(4) = \begin{bmatrix} 1 & 2 & 3 & 4 \\ 4 & 3 & 2 & 1 \\ 2 & 1 & 4 & 3 \\ 3 & 4 & 1 & 2 \end{bmatrix}$$

as Table C3 shows. In this case $q = 4$, $r = 2$.

Before we generalize for $r > 2$, we examine the special case $q = 2$ where condition (C1) still applies. Indeed,

$$L(2) = \begin{bmatrix} 1 & 2 \\ 2 & 1 \end{bmatrix}$$

can be considered a 2×2 Latin square and is the only square needed for the construction of

$$OA_{2^2}(2^{(2^2-1)/(2-1)}) = OA_4(2^3),$$

as Table C4 shows.

Table C3

$\text{OA}[4; 2] =$

R	C			
1	1	1	1	1
1	2	2	2	2
1	3	3	3	3
1	4	4	4	4
2	1	2	3	4
2	2	1	4	3
2	3	4	1	2
2	4	3	2	1
3	1	3	4	2
3	2	4	3	1
3	3	1	2	4
3	4	2	1	3
4	1	4	2	3
4	2	3	1	4
4	3	2	4	1
4	4	1	3	2

$= \text{OA}_{16}(4^5)$

Table C4

$$\text{OA}[2; 2] = \begin{bmatrix} 1 & 1 & 1 \\ 1 & 2 & 2 \\ 2 & 1 & 2 \\ 2 & 2 & 1 \end{bmatrix} = \text{OA}_4(2^3)$$

Remark: A complete orthogonal system of Latin squares of order q is a set of $q-1$ pairwise orthogonal $q \times q$ Latin squares. This is also known as a Hyper-Graeco-Latin square. Assuming the existence of a complete orthogonal system of Latin squares of order q, we can construct $\text{OA}[q; 2]$ and consequently (see the Generalization below) the $\text{OA}[q; r]$ $r \geq 3$. The question, of course, arises as to when such a system exists. This is, in fact, one of the famous unsolved problems of discrete mathematics. It can be shown that there does not exist even one pair of orthogonal 6×6 Latin squares. The question as to the existence of a complete orthogonal system of Latin squares of order 10 is unresolved. However, if q is a prime power such a system exists. Ways to construct such a system of Latin squares can be found in A. Dey (1985). The construction in general of an $\text{OA}[p; 2]$ when p is a prime number is described in Section C.2 below.

Generalization

For any $q \geq 2$, let us call the array

$$\text{OA}[q, r = 2] = \text{OA}_{q^2}(q^{(q^2-1)/(q-1)}) = \text{OA}_{q^2}(q^{q+1})$$

a 'reference-array $\{q\}$'. For example, the array of Table C2 is a reference array $\{3\}$, etc.

(C2) Every Reference array q can be considered as a full-factorial design for the study of two q-level factors.

This is obvious in the arrays constructed so far: for each ref array $\{q\}$, two q-level factors can be assigned at the first two columns (e.g. under 'R' and 'C'), with the remaining columns 'being assigned' their two-way interaction. Note that the number of remaining columns is $\{(q^2-1)/(q-1)\}-2=(q-1)$ and this is always the number of interaction columns required for the interaction between two q-level factors. Of course, if more than two factors are present, under the assumption of no interactions, up to $(q+1)=(q^2-1)/(q-1)$ q-level factors can be studied with a reference array $\{q\}$.

(C3) For any $q \geq 2$, we can construct sequentially for $r=3, 4, \ldots,$ an OA$[q; r]$, assuming the existence of a reference array $\{q\}$.

We will demonstrate (C3) by first referring to the special case when $q=3$, $r=3$. Since the reference array $\{3\}$ OA[3; 2] exists (Table C2), we can construct the

$$\text{OA}[3;3]=\text{OA}_{3^3}(3^{(3^3-1)/(3-1)})=\text{OA}_{27}(3^{13})$$

as follows.

(1) Expand each column of OA[3; 2] to three times its size by repeating (three times) each entry, e.g. replace 1 with $(1,1,1)^T$, 2 with $(2,2,2)^T$ etc; where by $\mathbf{z}^T$ we represent the transpose of the vector $\mathbf{z}$, e.g.

$$(1,1,1)^T=\begin{pmatrix}1\\1\\1\end{pmatrix},$$

etc. This 'expansion' provides the first four columns in the $\text{OA}_{27}(3^{13})$; see Table C5.

(2) Add a fifth column by simply repeating $(1,2,3)^T$ nine times; see Table C5.

(3) The remaining eight columns are found by constructing the 'interaction columns' between each of the first four columns with column 5. Note that there are four interactions to be considered, namely

$$(1\times 5),\ (2\times 5),\ (3\times 5),\quad \text{and}\quad (4\times 5),$$

and for each interaction, two additional columns need to be constructed; this can be done easily by using the reference array $\{3\}$ of Table C2. Indeed, the last two $(q-1=3-1)$ columns of this array indicate how the two interaction columns corresponding to the R × C interactions, i.e. the interaction between the factors assigned to columns 1 and 2 of OA[3; 2] (see Table C2) should be constructed.

For example, the interaction (1×5), i.e. the interaction between the first column of the new array,

$$(1,1,1,\ \ 1,1,1,\ \ 1,1,1,\ \ 2,2,2,\ \ 2,2,2,\ \ 2,2,2,\ \ 3,3,3,\ \ 3,3,3,\ \ 3,3,3)^T$$

TABLE C5

Columns	1	2	3	4	5	6	7	8	9	10	11	12	13
						(1 × 5)		(2 × 5)		(3 × 5)		(4 × 5)	
	1	1	1	1	1	1	1	1	1	1	1	1	1
	1	1	1	1	2	2	2	2	2	2	2	2	2
	1	1	1	1	3	3	3	3	3	3	3	3	3
	1	2	2	2	1	1	1	2	3	2	3	2	3
	1	2	2	2	2	2	2	3	1	3	1	3	1
	1	2	2	2	3	3	3	1	2	1	2	1	2
	1	3	3	3	1	1	1	3	2	3	2	3	2
	1	3	3	3	2	2	2	1	3	1	3	1	3
	1	3	3	3	3	3	3	2	1	2	1	2	1
	2	1	2	3	1	2	3	1	1	2	3	3	2
	2	1	2	3	2	3	1	2	2	3	1	1	3
	2	1	2	3	3	1	2	3	3	1	2	2	1
	2	2	3	1	1	2	3	2	3	3	2	1	1
OA[3; 3] =	2	2	3	1	2	3	1	3	1	1	3	2	2
	2	2	3	1	3	1	2	1	2	2	1	3	3
	2	3	1	2	1	2	3	3	2	1	1	2	3
	2	3	1	2	2	3	1	1	3	2	2	3	1
	2	3	1	2	3	1	2	2	1	3	3	1	2
	3	1	3	2	1	3	2	1	1	3	2	2	3
	3	1	3	2	2	1	3	2	2	1	3	3	1
	3	1	3	2	3	2	1	3	3	2	1	1	2
	3	2	1	3	1	3	2	2	3	1	1	3	2
	3	2	1	3	2	1	3	3	1	2	2	1	3
	3	2	1	3	3	2	1	1	2	3	3	2	1
	3	3	2	1	1	3	2	3	2	2	3	1	1
	3	3	2	1	2	1	3	1	3	3	1	2	2
	3	3	2	1	3	2	1	2	1	1	2	3	3

and the fifth column of the new array,

$$(1, 2, 3, \quad 1, 2, 3, \quad 1, 2, 3, \quad 1, 2, 3, \quad 1, 2, 3, \quad 1, 2, 3, \quad 1, 2, 3, \quad 1, 2, 3, \quad 1, 2, 3)^T$$

produces the two columns (columns 6 and 7 in Table C5)

$$(1, 2, 3, \quad 1, 2, 3, \quad 1, 2, 3, \quad 2, 3, 1, \quad 2, 3, 1, \quad 2, 3, 1, \quad 3, 1, 2, \quad 3, 1, 2, \quad 3, 1, 2,)^T,$$

and

$$(1, 2, 3, \quad 1, 2, 3, \quad 1, 2, 3, \quad 3, 1, 2, \quad 3, 1, 2, \quad 3, 1, 2, \quad 2, 3, 1, \quad 2, 3, 1, \quad 2, 3, 1)^T.$$

This is because of the following correspondence in the column entries (see Table

C2):

Col 1	Col 2	Interaction columns
1	1	$\to$ 1 1
1	2	$\to$ 2 2
⋮	⋮	⋮ ⋮
3	3	$\to$ 2 1

The four interactions will thus lead to the construction of eight (4×2) additional columns. The complete $OA[3; 3] = OA_{27}(3^{13})$ is shown in Table C5.

Having constructed $OA[3; 3]$ we can now construct $OA[3, 4] = OA_{34}(3^{40})$ as follows:

(1) expand the 13 columns of $OA[3; 3]$ by repeating each entry three times;

(2) construct the 14th column by repeating $(1, 2, 3)^T$ 27 times;

(3) construct the interaction columns corresponding to the interaction between each of the first 13 columns and the 14th column (of the new array) using the reference array $\{3\}$ as before. This will result in the construction of 26 (13×2) additional columns making 40 columns all together.

This procedure can be generalized for any $q \geq 2$ as follows: To construct $OA[q; r+1]$ assuming the existence of $OA[q; r]$:

1. Expand the $(q^r - 1)/(q - 1)$ columns of $OA[q; r]$ by repeating each column entry q times. This will provide the first $(q^r - 1)/(q - 1)$ columns of $OA[q; r+1]$.
2. Add another column by repeating $(1, 2, \ldots, q)^T q^r$ times. This will provide the $\left(\frac{q^r - 1}{q + 1} + 1\right)$-th column of $OA[q; r+1]$.
3. The remaining columns needed are the 'interaction-columns' between each of the columns constructed in step 1 and the column constructed in step 2. There are $(q^r - 1)/(q - 1)$ interactions to be considered and for each one there are $q - 1$ columns to be constructed. So there are altogether

$$\frac{(q^r - 1)}{(q - 1)}(q - 1) = q^r - 1$$

interaction-columns to be constructed. These can be constructed on the basis of the reference array $\{q\}$. The complete $OA[q; r+1]$ will thus have

$$\frac{q^r - 1}{q - 1} + 1 + q^r - 1 = \frac{q^{r+1} - 1}{q - 1} \text{ columns,}$$

leading to the array

$$OA_{q^{r+1}}(q^{(q^{r+1} - 1)/(q - 1)}),$$

which requires q^{r+1} trials for the study of $(q^{r+1} - 1)/(q - 1)$ q-level effects.

C.2 *P*-Level designs when *p* is a prime number

For any prime number p there exists an orthogonal array with p^2 rows and $(p+1)$ p-level columns; i.e., an

$$\mathrm{OA}_{p^2}(p^{p+1}).$$

This can be constructed as follows. First consider the p-tuples

$$e_1 = (0, 1, 2, \ldots, p-1)$$

and

$$e_2 = (1, 2, 3, \ldots, p).$$

For $i = 2, \ldots, p$, construct the p-tuples $e_i = (e_i(1), e_i(2), e_i(3), \ldots, e_i(p))$ as follows:

$$e_i = e_2 + (i-2)e_1 = e_{i-1} + e_1$$

by adding the corresponding entries in the p-tuples and taking the answer mod p, i.e. for each k

$$p + k = k.$$

For example,

$$e_3 = e_2 + e_1(\mathrm{mod}\, p) = (1, 3, 5, \ldots, p-1).$$

Similarly

$$\begin{aligned} e_4 &= e_2 + 2e_1 = e_3 + e_1 = (1, 4, 7, \ldots, p-2) \\ &\vdots \\ e_p &= e_2 + (p-2)e_1 = e_{p-1} + e_1 = (1, p, p-1, \ldots, 2). \end{aligned}$$

Then we construct an array of $(p+1)$ columns and p^2 rows as Table C6 shows.

If we divide the array of Table C6 into p sections we note the following. In the ith section, the first column consists of a repetition (p times) of level i. Columns 2 to $(p+1)$ in the first section are all identical to the column $(1, 2, 3, \ldots, p)^T$. The first row of section i, $i = 2, \ldots, p$, from column 2 to $(p+1)$, consists of the elements of the p-tuple $e_i = [e_i(1), e_i(2), \ldots, e_i(p)]$, where $e_i(j)$ is the jth element of e_i. The entries of the remaining rows of Section i, from column 2 to $(p+1)$, are formed by adding 1 to the previous column entry (of the previous row), taking the answer mod p.

Examples:

1. Since 3 is a prime we can construct $\mathrm{OA}_{3^2}(3^4)$: we have

$$e_1 = (0, 1, 2), \qquad e_2 = [e_2(1), e_2(2), e_2(3)] = (1, 2, 3)$$

and

$$e_3 = [e_3(1), e_3(2), e_3(3)] = e_2 + e_1(\mathrm{mod}\, 3) = (1, 3, 2).$$

TABLE C6: $OA_{p^2}(p^{p+1})$

Row	Col. 1	Col. 2	Col. 3	Col. 4	... Col. $(p+1)$	
1	1	1	1	1	1	
2	1	2	2	2	2	
3	1	3	3	3	3	
4	1	4	4	4	4	
⋮						
p	1	p	p	p	p	
$p+1$	2	$e_2(1)$	$e_2(2)$	$e_2(3)$	$e_2(p)$	
$p+2$	2	$e_2(1)+1$	$e_2(2)+1$	$e_2(3)+1$	$e_2(p)+1$	(mod p)
$p+3$	2	$e_2(1)+2$	$e_2(2)+2$	$e_2(3)+2$	$e_2(p)+2$	(mod p)
⋮						
$2p$	2	$e_2(1)+(p-1)$	$e_2(2)+(p-1)$	$e_2(3)+(p-1)$	$e_2(p)+(p-1)$	(mod p)
$2p+1$	3	$e_3(1)$	$e_3(2)$	$e_3(3)$	$e_3(p)$	
$2p+2$	3	$e_3(1)+1$	$e_3(2)+1$	e_3+1	$e_3(p)+1$	(mod p)
⋮						
$3p$	3	$e_3(1)+(p-1)$	$e_3(2)+(p-1)$	$e_3(3)+(p-1)$	$e_3(p)+(p-1)$	(mod p)
⋮						
$p(p-1)+1$	p	$e_p(1)$	$e_p(2)$	$e_p(3)$	$e_p(p)$	
$p(p-1)+2$	p	$e_p(1)+1$	$e_p(2)+1$	$e_p(3)+1$	$e_p(p)+1$	(mod p)
⋮						⋮
p^2	p	$e_p(1)+(p-1)$	$e_p(2)+(p-1)$	$e_p(3)+(p-1)$	$e_p(p)+(p-1)$	(mod p)

Hence

	Trial	Col 1	Col 2	Col 3	Col 4	
Section 1	1	1	1	1	1	
	2	1	2	2	2	
	3	1	3	3	3	
Section 2	4	2	$e_2(1)=1$	$e_2(2)=2$	$e_2(3)=3$	
	5	2	$e_2(1)+1=2$	$e_2(2)=1+3$	$e_2(3)+1=1$	(mod 3)
	6	2	$e_2(1)+2=3$	$e_2(2)+2=1$	$e_2(3)+2=2$	(mod 3)
Section 3	7	3	$e_3(1)=1$	$e_3(2)=3$	$e_3(3)=2$	(mod 3)
	8	3	$e_3(1)+1=2$	$e_3(2)+1=1$	$e_3(3)+1=3$	(mod 3)
	9	3	$e_3(1)+2=3$	$e_3(2)+2=2$	$e_3(3)+2=1$	(mod 3)

2. To construct $OA_{5^2}(5^6)$ we have

$$e_1=(0,1,2,3,4),\quad e_2=(1,2,3,4,5),$$
$$e_3=(1,3,5,2,4),\quad e_4=(1,4,2,5,3),$$
$$e_5=(1,5,4,3,2).$$

Applying the format of Table C6, $OA_{25}(5^6)$ can be easily obtained (see Table C7). This array can also be found in Appendix D. We can similarly obtain $OA_{7^2}(7^8)$, $OA_{11^2}(11^{12})$, etc.

TABLE C7. $OA_{5^2}(5^6)$

Trial							
1	1	1	1	1	1	1	
2	1	2	2	2	2	2	
3	1	3	3	3	3	3	
4	1	4	4	4	4	4	
5	1	5	5	5	5	5	
6	2	(1	2	3	4	5)	$\rightarrow e_2$
7	2	2	3	4	5	1	
8	2	3	4	5	1	2	
9	2	4	5	1	2	3	
10	2	5	1	2	3	4	
11	3	(1	3	5	2	4)	$\rightarrow e_3$
12	3	2	4	1	3	5	
13	3	3	5	2	4	1	
14	3	4	1	3	5	2	
15	3	5	2	4	1	3	
16	4	(1	4	2	5	3)	$\rightarrow e_4$
17	4	2	5	3	1	4	
18	4	3	1	4	2	5	
19	4	4	2	5	3	1	
20	4	5	3	1	4	2	
21	5	(1	5	4	3	2)	$\rightarrow e_5$
22	5	2	1	5	4	3	
23	5	3	2	1	5	4	
24	5	4	3	2	1	5	
25	5	5	4	3	2	1	

C.3 Plackett and Burman (P–B) designs

These are designs suitable for studying $k = (N-1)/(L-1)$ factors each with L levels with the expense of N trials (see also Section 3.3.4, Plackett and Burman (1946)), i.e. designs of the form $OA_N(L^K)$. When $L = 2$, (P–B) designs exist for N being a multiple of 4, from $N = 4$ up to 100 (except for the isolated case of $N = 92$). The designs are constructed with the help of a 'generating' vector, which is first written down as a column. A second column is obtained by moving down the elements of the previous column once, and placing the last element in first position. This procedure is repeated until altogether $(N-1)/(L-1)$ columns are obtained. Finally a row of elements, all representing the first factor level, is added to complete the design.

For example, the $OA_4(2^3)$ can be constructed considering that $N = 4$, $L = 2$,

$K = (4-1)/(2-1) = 3$, and using the generating vector

$$(1 \quad 2 \quad 2)$$

as

$$\begin{array}{ccc} 1 & 2 & 2 \\ 2 & 1 & 2 \\ 2 & 2 & 1 \\ \hline 1 & 1 & 1 \end{array}$$

The generating vectors for various values of N and L are given below. The factor levels are represented by the numerals $1, 2, \ldots, L$. There is *always* a final row of ones to be added. In the designs for $L = 2$ and $N = 28$, 52, 76, 100, some 'generating' square blocks are supplied; these blocks are permuted cyclically amongst themselves. For example, for $N = 28$, if A, B, and C are three supplied 9×9 square blocks, these are written down cyclically as

$$\begin{array}{ccc} A & B & C \\ B & C & A \\ C & A & B \end{array}$$

and the resulting 27 rows are followed by a row of ones. Note that in the cases of $N = 52$, 76, and 100, the first row and first column are also given; these have alternatively '1' and '2' throughout, apart from the corner element (which is a '2').

Table of generating vectors and blocks

(i) $(L=2)$

$N=4$: (122)
$N=8$: (2221211)
$N=12$: (22122211121)
$N=16$: (222212122112111)
$N=20$: (2211222212121111221)
$N=24$: (22222121221122112121111)
$N=28$:

212222111	121112112	221212212
221222111	112211211	122221221
122222111	211121121	212122122
111212222	112121112	212221212
111221222	211112211	221122221
111122222	121211121	122212122
222111212	112112121	212212221
222111221	211211112	221221122
222111122	121121211	122122211

$N=32$: (1111212122212211122222112212112)
$N=36$: (12122211122222122211211112121221121)
$N=40$: Double the design for $N=20$
$N=44$: (2211212112221222221112122211111211122121221)
$N=48$: (22222122221121212221121122122111212122111121111)
$N=52$:

2	2121212121	2121212121	2121212121	2121212121	2121212121
2	1211111111	2211112222	2222112211	2211221122	2222221111
1	2212121212	2112122121	2121122112	2112211221	2121211212
2	1112111111	2222111122	1122221122	2222112211	1122222211
1	1222121212	2121121221	1221211221	2121122112	1221212112
2	1111121111	2222221111	2211222211	1122221122	1111222222
1	1212221212	2121211212	2112212112	1221211221	1212212121
2	1111111211	1122222211	1122112222	2211222211	2211112222
1	1212122212	1221212112	1221122121	2112212112	2112122121
2	1111111112	1111222222	2211221122	1122112222	2222111122
1	1212121222	1212212121	2112211221	1221122121	2121121221

$N=56$:Double the design for $N=28$
$N=60$: (22122212121121122212222112222211111221111211122122121211121)
$N=64$: Double the design for $N=33$
$N=68$:

(2211212112211122221212222221121112122212211111121211112221122121221)
N = 72:

(22222221222121122122211122121221211122212112121122211121122121112111111)

N = 76:

2	21	21	21	21	21	21	21	21	21	21	21	21	21	21	21	21	21	21	21	21
2	12	11	22	11	11	22	22	11	22	11	11	11	11	22	22	22	11	22	22	22
1	22	12	21	12	22	21	21	12	21	12	12	12	12	21	21	21	12	21	21	21

21	21	21	21	21	21	21	21	21	21	21	21	21	21	21	21	21
22	11	22	22	22	11	11	11	11	22	11	22	22	11	11	22	11
21	12	21	21	21	12	12	12	12	21	12	21	21	12	12	21	12

$N = 80$:
(22212211222212112122222212211112211121212121222112222112111111212212
11112211211)

$N = 84$:
(22122112122221112211121212222222121122212211211122121111111212122211
222111121221121)

$N = 88$: Double the design for $N = 44$

$N = 92$: See note at the end of this section

$N = 96$: Double the design for $N = 48$

$N = 100$:

2	21212121212121	21212121212121	21212121212121
2	12111111111111	22112222112211	22111122221122
1	22121212121212	21122121122112	21121221211221
2	11121111111111	11221122221122	22221111222211
1	12221212121212	12211221211221	21211212212112
2	11111211111111	22112211222211	11222211112222
1	12122212121212	21122112212112	12212112122121
2	11111112111111	11221122112222	22112222111122
1	12121222121212	12211221122121	21122121121221
2	11111111121111	22112211221122	22221122221111
1	12121212221212	21122112211221	21211221211212
2	11111111111211	22221122112211	11222211222211
1	12121212122212	21211221122112	12212112212112
2	11111111111112	11222211221122	11112222112222
1	12121212121222	12212112211221	12122121122121

21212121212121	21212121212121	21212121212121
22222211111122	22221111112222	22221122221111
21212112121221	21211212122121	21211221211212
22222222111111	22222211111122	11222211222211
21212121121212	21212112121221	12212112212112
11222222221111	22222222111111	11112222112222
12212121211212	21212121121212	12122121122121
11112222222211	11222222221111	22111122221122
12122121212112	12212121211212	21121221211221
11111122222222	11112222222211	22221111222211
12121221212121	12122121212112	21211212212112
22111111222222	11111122222222	11222211112222
21121212212121	12121221212121	12212112122121
22221111112222	22111111222222	22112222111122
21211212122121	21121212212121	21122121121221

21212121212121
22112211222211
21122112212112
11221122112222
12211221122121
22112211221122
21122112211221
22221122112211
21211221122112
11222211221122
12212112211221
22112222112211
21122121122112
11221122221122
12211221211221

(ii) $L=3$

$N=9$: (12331322)

$N=27$: (11212322312221131323321333)

$N=81$: (12222 31232 23231 31332 21312 21123 33132 11311 13333 21323 32321 21223 31213 31132 22123 11211)

(iii) $L=5$

$N=25$: (15223 21433 53125 54513 4424)

$N=125$: (13332 15225 24245 23132 21355 42513 11555 31433

43232 43515 33154 42341 51144 45125 52535 32541 45514
22352 14112 22413 44345 45342 12441 23354 3121)

(iv) $L = 7$

$N = 49$: (12373 32716 43446 31524 22541 76266 72135 65536 14757 745)

Note: For the special case $L = 2$, $N = 92$, see Baumert, Golumb, and Hall (1962). For the cases $L = 2$ and N is a multiple of 4 between 100 and 200, see Raghavarao (1971), Turyn (1973), and Hedayat and Wallis (1978).

C.4 Other constructing methods

(i) The method of 'blocking' as described in Chapter 2, (Section 2.4) can be used to create highly fractional designs starting from an existing full-factorial array.

(ii) Applying the methods described in Chapter 3, Section 3.3.5, multi-level factorial designs can be constructed starting from an existing two-level or three-level orthogonal array. For example, by the following correspondence scheme

$$\begin{array}{ccc} 1 & 1 & 1 \rightarrow 1 \\ 1 & 2 & 2 \rightarrow 2 \\ 2 & 1 & 2 \rightarrow 3 \\ 2 & 2 & 1 \rightarrow 4 \end{array}$$

one can construct an $OA_8(4 \times 2^4)$ from $OA_8(2^7)$, etc.; see Section 3.3.5.1. Of course, one can also act conversely: for example, assuming that $OA_{16}(4^5)$ has been obtained by the method described in Section C.1 (see Table C3), one can construct

$$OA_{16}(4^4 \times 2^3) \quad \text{or} \quad OA_{16}(2^{15}), \text{ etc.,}$$

by using the above correspondence scheme in the inverse way.

(iii) Applying the methods described in Chapter 3, Section 3.3.6, the number of column levels in an orthogonal array can be reduced to accommodate factors with a small number of factor levels. For example, a four-level column can accommodate a three-level factor by the scheme (see Section 3.3.6.1a)

$$1 \rightarrow 1, \quad 2 \rightarrow 2, \quad 3 \rightarrow 2, \quad 4 \rightarrow 3.$$

It is apparent that the 'new' design will be neither completely balanced nor orthogonal. However, the desired statistical property of orthogonal arrays, namely that 'the estimated effect of a factor assigned to a new column created in this way will remain uncorrelated with all other estimated effects' can still be retained, provided that, for any q-level factor A and p-level factor B assigned to the 'new' array,

$$n_{ij} = \frac{(n_i n_j)}{n} \qquad i = 1, \ldots, q; j = 1, \ldots, p, \tag{C4}$$

where n_i is the number of times the ith level of A occurs, n_j is the number of times the jth level of B occurs, and n_{ij} is the number of times the ith level of A occurs together with the jth level of B. This is in accord with the 'proportional frequencies' criterion which states that: a necessary and sufficient condition that the estimates of two main effects are uncorrelated, is that the levels of one factor occur with each of the levels of the other factor with proportional frequencies, i.e. (C4) is satisfied. The proof can be found in Dey (1985). If this criterion is not satisfied, appropriate analysis techniques should be followed to take care of any introduced bias (see, for example, Section 3.3.6.3).

APPENDIX D

Taguchi's recommended designs and interaction matrices

This Appendix contains the most commonly used orthogonal arrays and their associated triangular interaction tables; these arrays are strongly recommended by Taguchi (see also Taguchi and Konishi 1987). Note that in some cases such as for $OA_4(2^3)$, $OA_9(3^4)$, $OA_{16}(4^5)$, and $OA_{25}(5^6)$, the interaction columns between any two columns are simply the other remaining columns in the array.

Although the interaction matrices provide only the columns for two-way interactions (see Sections 3.3.2.3 and 3.3.2.4), indications on how to study other types of interactions are provided at the bottom of the arrays where low-key letters are used. For example, in a two-level array,

$$\begin{pmatrix} a \\ b \\ c \end{pmatrix}$$

at the bottom of a column means that, the interaction $a \times b \times c$ can be assigned to this column. However, if a, b, c are factors with more than two levels (say, three levels), then

$$\begin{pmatrix} a \\ b \\ c \end{pmatrix}$$

means the interaction of factor a with the *linear* components of b and c, i.e. $a \times b_L \times c_L$, assuming b and c are quantitative factors. The quadratic components are indicated with a square, the cubic components with a power of three, etc. For example,

$$\begin{pmatrix} b \\ c^2 \\ d \end{pmatrix}$$

indicates the interaction of b with the *quadratic* component of c and the *linear* component of d [see, for example, $OA_{81}(3^{40})$]. Also

$$\begin{pmatrix} a \\ b^3 \end{pmatrix}$$

corresponds to the interaction of a with the cubic component of b [see, for example, $OA_{25}(5^6)$] and so on.

Indications on how to group the columns for 'split-unit designs' (see Section 3.3.5.4) are also provided at the bottom of the arrays.

$OA_4(2^3)$

Tr.	Col: 1	2	3
1	1	1	1
2	1	2	2
3	2	1	2
4	2	2	1
	a	*b*	*a* *b*
Group	1	2	

Interactions between two columns

(Col.) \ Col:	1	2	3
	(1)	3	2
		(2)	1

$OA_8(2^7)$

Tr.	Col: 1	2	3	4	5	6	7
1	1	1	1	1	1	1	1
2	1	1	1	2	2	2	2
3	1	2	2	1	1	2	2
4	1	2	2	2	2	1	1
5	2	1	2	1	2	1	2
6	2	1	2	2	1	2	1
7	2	2	1	1	2	2	1
8	2	2	1	2	1	1	2
	a	*b*	*a* *b*	*c*	*a* *c*	*b* *c*	*a* *b* *c*
Group	1	2		3			

Interactions between two columns

(Col.) \ Col:	1	2	3	4	5	6	7
	(1)	3	2	5	4	7	6
		(2)	1	6	7	4	5
			(3)	7	6	5	4
				(4)	1	2	3
					(5)	3	2
						(6)	1

$OA_{12}(2^{11})$

Tr.	Col: 1	2	3	4	5	6	7	8	9	10	11
1	1	1	1	1	1	1	1	1	1	1	1
2	1	1	1	1	1	2	2	2	2	2	2
3	1	1	2	2	2	1	1	1	2	2	2
4	1	2	1	2	2	1	2	2	1	1	2
5	1	2	2	1	2	2	1	2	1	2	1
6	1	2	2	2	1	2	2	1	2	1	1
7	2	1	2	2	1	1	2	2	1	2	1
8	2	1	2	1	2	2	2	1	1	1	2
9	2	1	1	2	2	2	1	2	2	1	1
10	2	2	2	1	1	1	1	2	2	1	2
11	2	2	1	2	1	2	1	1	1	2	2
12	2	2	1	1	2	1	2	1	2	2	1
Group	1 (cols 1–2)		2 (cols 3–11)								

$OA_{12}(2^{11})$ has the property that no interactions can be estimated orthogonally.

$OA_{16}(2^{15})$

Tr.	Col: 1	2	3	4	5	6	7	8	9	10	11	12	13	14	15
1	1	1	1	1	1	1	1	1	1	1	1	1	1	1	1
2	1	1	1	1	1	1	1	2	2	2	2	2	2	2	2
3	1	1	1	2	2	2	2	1	1	1	1	2	2	2	2
4	1	1	1	2	2	2	2	2	2	2	2	1	1	1	1
5	1	2	2	1	1	2	2	1	1	2	2	1	1	2	2
6	1	2	2	1	1	2	2	2	2	1	1	2	2	1	1
7	1	2	2	2	2	1	1	1	1	2	2	2	2	1	1
8	1	2	2	2	2	1	1	2	2	1	1	1	1	2	2
9	2	1	2	1	2	1	2	1	2	1	2	1	2	1	2
10	2	1	2	1	2	1	2	2	1	2	1	2	1	2	1
11	2	1	2	2	1	2	1	1	2	1	2	2	1	2	1
12	2	1	2	2	1	2	1	2	1	2	1	1	2	1	2
13	2	2	1	1	2	2	1	1	2	2	1	1	2	2	1
14	2	2	1	1	2	2	1	2	1	1	2	2	1	1	2
15	2	2	1	2	1	1	2	1	2	2	1	2	1	1	2
16	2	2	1	2	1	1	2	2	1	1	2	1	2	2	1
	a	*b*	*a* *b*	*c*	*a* *c*	*b* *c*	*a* *b* *c*	*d*	*a* *d*	*b* *d*	*a* *b* *d*	*c* *d*	*a* *c* *d*	*b* *c* *d*	*a* *b* *c* *d*
Group	1 (col 1)	2 (cols 2–3)		3 (cols 4–7)				4 (cols 8–15)							

$OA_{16}(2^{15})$ interactions between two columns

(Col.) \ Col:	1	2	3	4	5	6	7	8	9	10	11	12	13	14	15
	(1)	3	2	5	4	7	6	9	8	11	10	13	12	15	14
		(2)	1	6	7	4	5	10	11	8	9	14	15	12	13
			(3)	7	6	5	4	11	10	9	8	15	14	13	12
				(4)	1	2	3	12	13	14	15	8	9	10	11
					(5)	3	2	13	12	15	14	9	8	11	10
						(6)	1	14	15	12	13	10	11	8	9
							(7)	15	14	13	12	11	10	9	8
								(8)	1	2	3	4	5	6	7
									(9)	3	2	5	4	7	6
										(10)	1	6	7	4	5
											(11)	7	6	5	4
												(12)	1	2	3
													(13)	3	2
														(14)	1

$OA_{32}(2^{31})$

Tr.	Col: 1	2	3	4	5	6	7	8	9	10	11	12	13	14	15	16	17	18	19	20	21	22	23	24	25	26	27	28	29	30	31
1	1	1	1	1	1	1	1	1	1	1	1	1	1	1	1	1	1	1	1	1	1	1	1	1	1	1	1	1	1	1	1
2	1	1	1	1	1	1	1	1	1	1	1	1	1	1	1	2	2	2	2	2	2	2	2	2	2	2	2	2	2	2	2
3	1	1	1	1	1	1	1	2	2	2	2	2	2	2	2	1	1	1	1	1	1	1	1	2	2	2	2	2	2	2	2
4	1	1	1	1	1	1	1	2	2	2	2	2	2	2	2	2	2	2	2	2	2	2	2	1	1	1	1	1	1	1	1
5	1	1	1	2	2	2	2	1	1	1	1	2	2	2	2	1	1	1	1	2	2	2	2	1	1	1	1	2	2	2	2
6	1	1	1	2	2	2	2	1	1	1	1	2	2	2	2	2	2	2	2	1	1	1	1	2	2	2	2	1	1	1	1
7	1	1	1	2	2	2	2	2	2	2	2	1	1	1	1	1	1	1	1	2	2	2	2	2	2	2	2	1	1	1	1
8	1	1	1	2	2	2	2	2	2	2	2	1	1	1	1	2	2	2	2	1	1	1	1	1	1	1	1	2	2	2	2
9	1	2	2	1	1	2	2	1	1	2	2	1	1	2	2	1	1	2	2	1	1	2	2	1	1	2	2	1	1	2	2
10	1	2	2	1	1	2	2	1	1	2	2	1	1	2	2	2	2	1	1	2	2	1	1	2	2	1	1	2	2	1	1
11	1	2	2	1	1	2	2	2	2	1	1	2	2	1	1	1	1	2	2	1	1	2	2	2	2	1	1	2	2	1	1
12	1	2	2	1	1	2	2	2	2	1	1	2	2	1	1	2	2	1	1	2	2	1	1	1	1	2	2	1	1	2	2
13	1	2	2	2	2	1	1	1	1	2	2	2	2	1	1	1	1	2	2	2	2	1	1	1	1	2	2	2	2	1	1
14	1	2	2	2	2	1	1	1	1	2	2	2	2	1	1	2	2	1	1	1	1	2	2	2	2	1	1	1	1	2	2
15	1	2	2	2	2	1	1	2	2	1	1	1	1	2	2	1	1	2	2	2	2	1	1	2	2	1	1	1	1	2	2
16	1	2	2	2	2	1	1	2	2	1	1	1	1	2	2	2	2	1	1	1	1	2	2	1	1	2	2	2	2	1	1
17	2	1	2	1	2	1	2	1	2	1	2	1	2	1	2	1	2	1	2	1	2	1	2	1	2	1	2	1	2	1	2

(continued)

$OA_{32}(2^{31})$ (*continued*)

Tr.	Col: 1	2	3	4	5	6	7	8	9	10	11	12	13	14	15	16	17	18	19	20	21	22	23	24	25	26	27	28	29	30	31
18	2	1	2	1	2	1	2	1	2	1	2	1	2	1	2	2	1	2	1	2	1	2	1	2	1	2	1	2	1	2	1
19	2	1	2	1	2	1	2	2	1	2	1	2	1	2	1	1	2	1	2	1	2	1	2	2	1	2	1	2	1	2	1
20	2	1	2	1	2	1	2	2	1	2	1	2	1	2	1	2	1	2	1	2	1	2	1	1	2	1	2	1	2	1	2
21	2	1	2	2	1	2	1	1	2	1	2	2	1	2	1	1	2	1	2	2	1	2	1	1	2	1	2	2	1	2	1
22	2	1	2	2	1	2	1	1	2	1	2	2	1	2	1	2	1	2	1	1	2	1	2	2	1	2	1	1	2	1	2
23	2	1	2	2	1	2	1	2	1	2	1	1	2	1	2	1	2	1	2	2	1	2	1	2	1	2	1	1	2	1	2
24	2	1	2	2	1	2	1	2	1	2	1	1	2	1	2	2	1	2	1	1	2	1	2	1	2	1	2	2	1	2	1
25	2	2	1	1	2	2	1	1	2	2	1	1	2	2	1	1	2	2	1	1	2	2	1	1	2	2	1	1	2	2	1
26	2	2	1	1	2	2	1	1	2	2	1	1	2	2	1	2	1	1	2	2	1	1	2	2	1	1	2	2	1	1	2
27	2	2	1	1	2	2	1	2	1	1	2	2	1	1	2	1	2	2	1	1	2	2	1	2	1	1	2	2	1	1	2
28	2	2	1	1	2	2	1	2	1	1	2	2	1	1	2	2	1	1	2	2	1	1	2	1	2	2	1	1	2	2	1
29	2	2	1	2	1	1	2	1	2	2	1	2	1	1	2	1	2	2	1	2	1	1	2	1	2	2	1	2	1	1	2
30	2	2	1	2	1	1	2	1	2	2	1	2	1	1	2	2	1	1	2	1	2	2	1	2	1	1	2	1	2	2	1
31	2	2	1	2	1	1	2	2	1	1	2	1	2	2	1	1	2	2	1	2	1	1	2	2	1	1	2	1	2	2	1
32	2	2	1	2	1	1	2	2	1	1	2	1	2	2	1	2	1	1	2	1	2	2	1	1	2	2	1	2	1	1	2
	a	*b*	*ab*	*c*	*ac*	*bc*	*abc*	*d*	*ad*	*bd*	*abd*	*cd*	*acd*	*bcd*	*abcd*	*e*	*ae*	*be*	*abe*	*ce*	*ace*	*bce*	*abce*	*de*	*ade*	*bde*	*abde*	*cde*	*acde*	*bcde*	*abcde*
Group	1	2		3				4								5															

OA$_{32}$(2^{31}) (*continued*)

Interactions between two columns

(Col.)\Col:	1	2	3	4	5	6	7	8	9	10	11	12	13	14	15	16	17	18	19	20	21	22	23	24	25	26	27	28	29	30	31
	(1)	3	2	5	4	7	6	9	8	11	10	13	12	15	14	17	16	19	18	21	20	23	22	25	24	27	26	29	28	31	30
		(2)	1	6	7	4	5	10	11	8	9	14	15	12	13	18	19	16	17	22	23	20	21	26	27	24	25	30	31	28	29
			(3)	7	6	5	4	11	10	9	8	15	14	13	12	19	18	17	16	23	22	21	20	27	26	25	24	31	30	29	28
				(4)	1	2	3	12	13	14	15	8	9	10	11	20	21	22	23	16	17	18	19	28	29	30	31	24	25	26	27
					(5)	3	2	13	12	15	14	9	8	11	10	21	20	23	22	17	16	19	18	29	28	31	30	25	24	27	26
						(6)	1	14	15	12	13	10	11	8	9	22	23	20	21	18	19	16	17	30	31	28	29	26	27	24	25
							(7)	15	14	13	12	11	10	9	8	23	22	21	20	19	18	17	16	31	30	29	28	27	26	25	24
								(8)	1	2	3	4	5	6	7	24	25	26	27	28	29	30	31	16	17	18	19	20	21	22	23
									(9)	3	2	5	4	7	6	25	24	27	26	29	28	31	30	17	16	19	18	21	20	23	22
										(10)	1	6	7	4	5	26	27	24	25	30	31	28	29	18	19	16	17	22	23	20	21
											(11)	7	6	5	4	27	26	25	24	31	30	29	28	19	18	17	16	23	22	21	20
												(12)	1	2	3	28	29	30	31	24	25	26	27	20	21	22	23	16	17	18	19
													(13)	3	2	29	28	31	30	25	24	27	26	21	20	23	22	17	16	19	18
														(14)	1	30	31	28	29	26	27	24	25	22	23	20	21	18	19	16	17
															(15)	31	30	29	28	27	26	25	24	23	22	21	20	19	18	17	16
																(16)	1	2	3	4	5	6	7	8	9	10	11	12	13	14	15
																	(17)	3	2	5	4	7	6	9	8	11	10	13	12	15	14
																		(18)	1	6	7	4	5	10	11	8	9	14	15	12	13
																			(19)	7	6	5	4	11	10	9	8	15	14	13	12
																				(20)	1	2	3	12	13	14	15	8	9	10	11
																					(21)	3	2	13	12	15	14	9	8	11	10
																						(22)	1	14	15	12	13	10	11	8	9
																							(23)	15	14	13	12	11	10	9	8
																								(24)	1	2	3	4	5	6	7
																									(25)	3	2	5	4	7	6
																										(26)	1	6	7	4	5
																											(27)	7	6	5	4
																												(28)	1	2	3
																													(29)	3	2
																														(30)	1

$OA_{64}(2^{63})$

Tr.	Col: 1	2	3	4	5	6	7	8	9	10	11	12	13	14	15	16	17	18	19	20	21	22	23	24	25	26	27	28	29	30	31
1	1	1	1	1	1	1	1	1	1	1	1	1	1	1	1	1	1	1	1	1	1	1	1	1	1	1	1	1	1	1	1
2	1	1	1	1	1	1	1	1	1	1	1	1	1	1	1	1	1	1	1	1	1	1	1	1	1	1	1	1	1	1	1
3	1	1	1	1	1	1	1	1	1	1	1	1	1	1	1	2	2	2	2	2	2	2	2	2	2	2	2	2	2	2	2
4	1	1	1	1	1	1	1	1	1	1	1	1	1	1	1	2	2	2	2	2	2	2	2	2	2	2	2	2	2	2	2
5	1	1	1	1	1	1	1	2	2	2	2	2	2	2	2	1	1	1	1	1	1	1	1	2	2	2	2	2	2	2	2
6	1	1	1	1	1	1	1	2	2	2	2	2	2	2	2	1	1	1	1	1	1	1	1	2	2	2	2	2	2	2	2
7	1	1	1	1	1	1	1	2	2	2	2	2	2	2	2	2	2	2	2	2	2	2	2	1	1	1	1	1	1	1	1
8	1	1	1	1	1	1	1	2	2	2	2	2	2	2	2	2	2	2	2	2	2	2	2	1	1	1	1	1	1	1	1
9	1	1	1	2	2	2	2	1	1	1	1	2	2	2	2	1	1	1	1	2	2	2	2	1	1	1	1	2	2	2	2
10	1	1	1	2	2	2	2	1	1	1	1	2	2	2	2	1	1	1	1	2	2	2	2	1	1	1	1	2	2	2	2
11	1	1	1	2	2	2	2	1	1	1	1	2	2	2	2	2	2	2	2	1	1	1	1	2	2	2	2	1	1	1	1
12	1	1	1	2	2	2	2	1	1	1	1	2	2	2	2	2	2	2	2	1	1	1	1	2	2	2	2	1	1	1	1
13	1	1	1	2	2	2	2	2	2	2	2	1	1	1	1	1	1	1	1	2	2	2	2	2	2	2	2	1	1	1	1
14	1	1	1	2	2	2	2	2	2	2	2	1	1	1	1	1	1	1	1	2	2	2	2	2	2	2	2	1	1	1	1
15	1	1	1	2	2	2	2	2	2	2	2	1	1	1	1	2	2	2	2	1	1	1	1	1	1	1	1	2	2	2	2
16	1	1	1	2	2	2	2	2	2	2	2	1	1	1	1	2	2	2	2	1	1	1	1	1	1	1	1	2	2	2	2
17	1	2	2	1	1	2	2	1	1	2	2	1	1	2	2	1	1	2	2	1	1	2	2	1	1	2	2	1	1	2	2
18	1	2	2	1	1	2	2	1	1	2	2	1	1	2	2	1	1	2	2	1	1	2	2	1	1	2	2	1	1	2	2
19	1	2	2	1	1	2	2	1	1	2	2	1	1	2	2	2	2	1	1	2	2	1	1	2	2	1	1	2	2	1	1
20	1	2	2	1	1	2	2	1	1	2	2	1	1	2	2	2	2	1	1	2	2	1	1	2	2	1	1	2	2	1	1
21	1	2	2	1	1	2	2	2	2	1	1	2	2	1	1	1	1	2	2	1	1	2	2	2	2	1	1	2	2	1	1
22	1	2	2	1	1	2	2	2	2	1	1	2	2	1	1	1	1	2	2	1	1	2	2	2	2	1	1	2	2	1	1
23	1	2	2	1	1	2	2	2	2	1	1	2	2	1	1	2	2	1	1	2	2	1	1	1	1	2	2	1	1	2	2
24	1	2	2	1	1	2	2	2	2	1	1	2	2	1	1	2	2	1	1	2	2	1	1	1	1	2	2	1	1	2	2
25	1	2	2	2	2	1	1	1	1	2	2	2	2	1	1	1	1	2	2	2	2	1	1	1	1	2	2	2	2	1	1
26	1	2	2	2	2	1	1	1	1	2	2	2	2	1	1	1	1	2	2	2	2	1	1	1	1	2	2	2	2	1	1
27	1	2	2	2	2	1	1	1	1	2	2	2	2	1	1	2	2	1	1	1	1	2	2	2	2	1	1	1	1	2	2
28	1	2	2	2	2	1	1	1	1	2	2	2	2	1	1	2	2	1	1	1	1	2	2	2	2	1	1	1	1	2	2
29	1	2	2	2	2	1	1	2	2	1	1	1	1	2	2	1	1	2	2	2	2	1	1	2	2	1	1	1	1	2	2
30	1	2	2	2	2	1	1	2	2	1	1	1	1	2	2	1	1	2	2	2	2	1	1	2	2	1	1	1	1	2	2
31	1	2	2	2	2	1	1	2	2	1	1	1	1	2	2	2	2	1	1	1	1	2	2	1	1	2	2	2	2	1	1
32	1	2	2	2	2	1	1	2	2	1	1	1	1	2	2	2	2	1	1	1	1	2	2	1	1	2	2	2	2	1	1
33	2	1	2	1	2	1	2	1	2	1	2	1	2	1	2	1	2	1	2	1	2	1	2	1	2	1	2	1	2	1	2
34	2	1	2	1	2	1	2	1	2	1	2	1	2	1	2	1	2	1	2	1	2	1	2	1	2	1	2	1	2	1	2
35	2	1	2	1	2	1	2	1	2	1	2	1	2	1	2	2	1	2	1	2	1	2	1	2	1	2	1	2	1	2	1
36	2	1	2	1	2	1	2	1	2	1	2	1	2	1	2	2	1	2	1	2	1	2	1	2	1	2	1	2	1	2	1
37	2	1	2	1	2	1	2	2	1	2	1	2	1	2	1	1	2	1	2	1	2	1	2	2	1	2	1	2	1	2	1
38	2	1	2	1	2	1	2	2	1	2	1	2	1	2	1	1	2	1	2	1	2	1	2	2	1	2	1	2	1	2	1
39	2	1	2	1	2	1	2	2	1	2	1	2	1	2	1	2	1	2	1	2	1	2	1	1	2	1	2	1	2	1	2
40	2	1	2	1	2	1	2	2	1	2	1	2	1	2	1	2	1	2	1	2	1	2	1	1	2	1	2	1	2	1	2
41	2	1	2	2	1	2	1	1	2	1	2	2	1	2	1	1	2	1	2	2	1	2	1	1	2	1	2	2	1	2	1
42	2	1	2	2	1	2	1	1	2	1	2	2	1	2	1	1	2	1	2	2	1	2	1	1	2	1	2	2	1	2	1
43	2	1	2	2	1	2	1	1	2	1	2	2	1	2	1	2	1	2	1	1	2	1	2	2	1	2	1	1	2	1	2
44	2	1	2	2	1	2	1	1	2	1	2	2	1	2	1	2	1	2	1	1	2	1	2	2	1	2	1	1	2	1	2
45	2	1	2	2	1	2	1	2	1	2	1	1	2	1	2	1	2	1	2	2	1	2	1	2	1	2	1	1	2	1	2
46	2	1	2	2	1	2	1	2	1	2	1	1	2	1	2	1	2	1	2	2	1	2	1	2	1	2	1	1	2	1	2
47	2	1	2	2	1	2	1	2	1	2	1	1	2	1	2	2	1	2	1	1	2	1	2	1	2	1	2	2	1	2	1
48	2	1	2	2	1	2	1	2	1	2	1	1	2	1	2	2	1	2	1	1	2	1	2	1	2	1	2	2	1	2	1

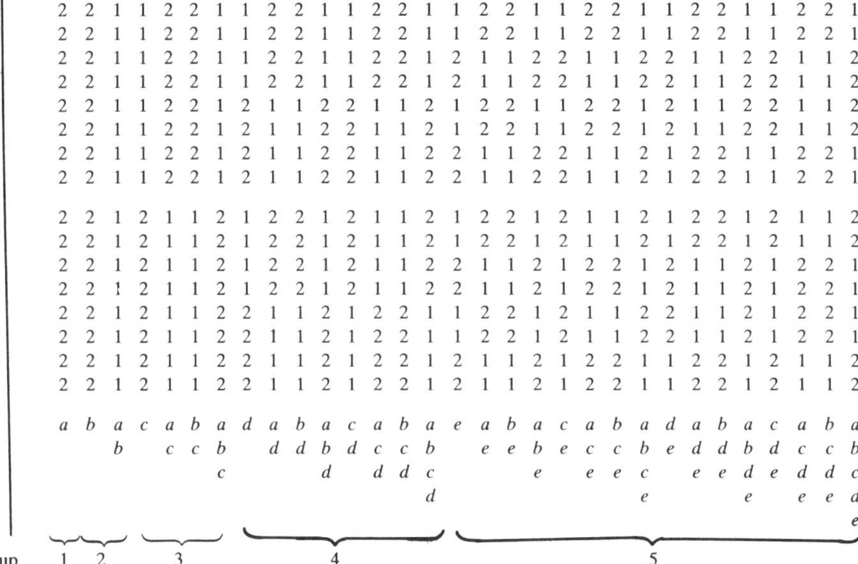

	2	2	1	1	2	2	1	1	2	2	1	1	2	2	1	1	2	2	1	1	2	2	1	1	2	2	1	1	2	2	1
	2	2	1	1	2	2	1	1	2	2	1	1	2	2	1	1	2	2	1	1	2	2	1	1	2	2	1	1	2	2	1
	2	2	1	1	2	2	1	1	2	2	1	1	2	2	1	2	1	1	2	2	1	1	2	2	1	1	2	2	1	1	2
	2	2	1	1	2	2	1	1	2	2	1	1	2	2	1	2	1	1	2	2	1	1	2	2	1	1	2	2	1	1	2
	2	2	1	1	2	2	1	2	1	1	2	2	1	1	2	1	2	2	1	1	2	2	1	2	1	1	2	2	1	1	2
	2	2	1	1	2	2	1	2	1	1	2	2	1	1	2	1	2	2	1	1	2	2	1	2	1	1	2	2	1	1	2
	2	2	1	1	2	2	1	2	1	1	2	2	1	1	2	2	1	1	2	2	1	1	2	1	2	2	1	1	2	2	1
	2	2	1	1	2	2	1	2	1	1	2	2	1	1	2	2	1	1	2	2	1	1	2	1	2	2	1	1	2	2	1
	2	2	1	2	1	1	2	1	2	2	1	2	1	1	2	1	2	2	1	2	1	1	2	1	2	2	1	2	1	1	2
	2	2	1	2	1	1	2	1	2	2	1	2	1	1	2	1	2	2	1	2	1	1	2	1	2	2	1	2	1	1	2
	2	2	1	2	1	1	2	1	2	2	1	2	1	1	2	2	1	1	2	1	2	2	1	2	1	1	2	1	2	2	1
	2	2	1	2	1	1	2	1	2	2	1	2	1	1	2	2	1	1	2	1	2	2	1	2	1	1	2	1	2	2	1
	2	2	1	2	1	1	2	2	1	1	2	1	2	2	1	1	2	2	1	2	1	1	2	2	1	1	2	1	2	2	1
	2	2	1	2	1	1	2	2	1	1	2	1	2	2	1	1	2	2	1	2	1	1	2	2	1	1	2	1	2	2	1
	2	2	1	2	1	1	2	2	1	1	2	1	2	2	1	2	1	1	2	1	2	2	1	1	2	2	1	2	1	1	2
	2	2	1	2	1	1	2	2	1	1	2	1	2	2	1	2	1	1	2	1	2	2	1	1	2	2	1	2	1	1	2
	a	*b*	*ab*	*c*	*ac*	*bc*	*abc*	*d*	*ad*	*bd*	*abd*	*cd*	*acd*	*bcd*	*abcd*	*e*	*ae*	*be*	*abe*	*ce*	*ace*	*bce*	*abce*	*de*	*ade*	*bde*	*abde*	*cde*	*acde*	*bcde*	*abcde*
oup	1	2		3				4								5															

$OA_{64}(2^{63})$ (*continued*)

32	33	34	35	36	37	38	39	40	41	42	43	44	45	46	47	48	49	50	51	52	53	54	55	56	57	58	59	60	61	62	63
1	1	1	1	1	1	1	1	1	1	1	1	1	1	1	1	1	1	1	1	1	1	1	1	1	1	1	1	1	1	1	1
2	2	2	2	2	2	2	2	2	2	2	2	2	2	2	2	2	2	2	2	2	2	2	2	2	2	2	2	2	2	2	2
1	1	1	1	1	1	1	1	1	1	1	1	1	1	1	1	2	2	2	2	2	2	2	2	2	2	2	2	2	2	2	2
2	2	2	2	2	2	2	2	2	2	2	2	2	2	2	2	1	1	1	1	1	1	1	1	1	1	1	1	1	1	1	1
1	1	1	1	1	1	1	1	2	2	2	2	2	2	2	2	1	1	1	1	1	1	1	1	2	2	2	2	2	2	2	2
2	2	2	2	2	2	2	2	1	1	1	1	1	1	1	1	2	2	2	2	2	2	2	2	1	1	1	1	1	1	1	1
1	1	1	1	1	1	1	1	2	2	2	2	2	2	2	2	2	2	2	2	2	2	2	2	1	1	1	1	1	1	1	1
2	2	2	2	2	2	2	2	1	1	1	1	1	1	1	1	1	1	1	1	1	1	1	1	2	2	2	2	2	2	2	2
1	1	1	1	2	2	2	2	1	1	1	1	2	2	2	2	1	1	1	1	2	2	2	2	1	1	1	1	2	2	2	2
2	2	2	2	1	1	1	1	2	2	2	2	1	1	1	1	2	2	2	2	1	1	1	1	2	2	2	2	1	1	1	1
1	1	1	1	2	2	2	2	1	1	1	1	2	2	2	2	2	2	2	2	1	1	1	1	2	2	2	2	1	1	1	1
2	2	2	2	1	1	1	1	2	2	2	2	1	1	1	1	1	1	1	1	2	2	2	2	1	1	1	1	2	2	2	2
1	1	1	1	2	2	2	2	2	2	2	2	1	1	1	1	1	1	1	1	2	2	2	2	2	2	2	2	1	1	1	1
2	2	2	2	1	1	1	1	1	1	1	1	2	2	2	2	2	2	2	2	1	1	1	1	1	1	1	1	2	2	2	2
1	1	1	1	2	2	2	2	2	2	2	2	1	1	1	1	2	2	2	2	1	1	1	1	1	1	1	1	2	2	2	2
2	2	2	2	1	1	1	1	1	1	1	1	2	2	2	2	1	1	1	1	2	2	2	2	2	2	2	2	1	1	1	1
1	1	2	2	1	1	2	2	1	1	2	2	1	1	2	2	1	1	2	2	1	1	2	2	1	1	2	2	1	1	2	2
2	2	1	1	2	2	1	1	2	2	1	1	2	2	1	1	2	2	1	1	2	2	1	1	2	2	1	1	2	2	1	1
1	1	2	2	1	1	2	2	1	1	2	2	1	1	2	2	2	2	1	1	2	2	1	1	2	2	1	1	2	2	1	1
2	2	1	1	2	2	1	1	2	2	1	1	2	2	1	1	1	1	2	2	1	1	2	2	1	1	2	2	1	1	2	2
1	1	2	2	1	1	2	2	2	2	1	1	2	2	1	1	1	1	2	2	1	1	2	2	2	2	1	1	2	2	1	1
2	2	1	1	2	2	1	1	1	1	2	2	1	1	2	2	2	2	1	1	2	2	1	1	1	1	2	2	1	1	2	2
1	1	2	2	1	1	2	2	2	2	1	1	2	2	1	1	2	2	1	1	2	2	1	1	1	1	2	2	1	1	2	2
2	2	1	1	2	2	1	1	1	1	2	2	1	1	2	2	1	1	2	2	1	1	2	2	2	2	1	1	2	2	1	1

(*continued*)

$OA_{64}(2^{63})$ (*continued*)

32	33	34	35	36	37	38	39	40	41	42	43	44	45	46	47	48	49	50	51	52	53	54	55	56	57	58	59	60	61	62	63
1	1	2	2	2	2	1	1	1	1	2	2	2	2	1	1	1	1	2	2	2	2	1	1	1	1	2	2	2	2	1	1
2	2	1	1	1	1	2	2	2	2	1	1	1	1	2	2	2	2	1	1	1	1	2	2	2	2	1	1	1	1	2	2
1	1	2	2	2	2	1	1	1	1	2	2	2	2	1	1	2	2	1	1	1	1	2	2	2	2	1	1	1	1	2	2
2	2	1	1	1	1	2	2	2	2	1	1	1	1	2	2	1	1	2	2	2	2	1	1	1	1	2	2	2	2	1	1
1	1	2	2	2	2	1	1	2	2	1	1	1	1	2	2	1	1	2	2	2	2	1	1	2	2	1	1	1	1	2	2
2	2	1	1	1	1	2	2	1	1	2	2	2	2	1	1	2	2	1	1	1	1	2	2	1	1	2	2	2	2	1	1
1	1	2	2	2	2	1	1	2	2	1	1	1	1	2	2	2	2	1	1	1	1	2	2	1	1	2	2	2	2	1	1
2	2	1	1	1	1	2	2	1	1	2	2	2	2	1	1	1	1	2	2	2	2	1	1	2	2	1	1	1	1	2	2
1	2	1	2	1	2	1	2	1	2	1	2	1	2	1	2	1	2	1	2	1	2	1	2	1	2	1	2	1	2	1	2
2	1	2	1	2	1	2	1	2	1	2	1	2	1	2	1	2	1	2	1	2	1	2	1	2	1	2	1	2	1	2	1
1	2	1	2	1	2	1	2	1	2	1	2	1	2	1	2	2	1	2	1	2	1	2	1	2	1	2	1	2	1	2	1
2	1	2	1	2	1	2	1	2	1	2	1	2	1	2	1	1	2	1	2	1	2	1	2	1	2	1	2	1	2	1	2
1	2	1	2	1	2	1	2	2	1	2	1	2	1	2	1	1	2	1	2	1	2	1	2	2	1	2	1	2	1	2	1
2	1	2	1	2	1	2	1	1	2	1	2	1	2	1	2	2	1	2	1	2	1	2	1	1	2	1	2	1	2	1	2
1	2	1	2	1	2	1	2	2	1	2	1	2	1	2	1	2	1	2	1	2	1	2	1	1	2	1	2	1	2	1	2
2	1	2	1	2	1	2	1	1	2	1	2	1	2	1	2	1	2	1	2	1	2	1	2	2	1	2	1	2	1	2	1
1	2	1	2	2	1	2	1	1	2	1	2	2	1	2	1	1	2	1	2	2	1	2	1	1	2	1	2	2	1	2	1
2	1	2	1	1	2	1	2	2	1	2	1	1	2	1	2	2	1	2	1	1	2	1	2	2	1	2	1	1	2	1	2
1	2	1	2	2	1	2	1	1	2	1	2	2	1	2	1	2	1	2	1	1	2	1	2	2	1	2	1	1	2	1	2
2	1	2	1	1	2	1	2	2	1	2	1	1	2	1	2	1	2	1	2	2	1	2	1	1	2	1	2	2	1	2	1
1	2	1	2	2	1	2	1	2	1	2	1	1	2	1	2	1	2	1	2	2	1	2	1	2	1	2	1	1	2	1	2
2	1	2	1	1	2	1	2	1	2	1	2	2	1	2	1	2	1	2	1	1	2	1	2	1	2	1	2	2	1	2	1
1	2	1	2	2	1	2	1	2	1	2	1	1	2	1	2	2	1	2	1	1	2	1	2	1	2	1	2	2	1	2	1
2	1	2	1	1	2	1	2	1	2	1	2	2	1	2	1	1	2	1	2	2	1	2	1	2	1	2	1	1	2	1	2
1	2	2	1	1	2	2	1	1	2	2	1	1	2	2	1	1	2	2	1	1	2	2	1	1	2	2	1	1	2	2	1
2	1	1	2	2	1	1	2	2	1	1	2	2	1	1	2	2	1	1	2	2	1	1	2	2	1	1	2	2	1	1	2
1	2	2	1	1	2	2	1	1	2	2	1	1	2	2	1	2	1	1	2	2	1	1	2	2	1	1	2	2	1	1	2
2	1	1	2	2	1	1	2	2	1	1	2	2	1	1	2	1	2	2	1	1	2	2	1	1	2	2	1	1	2	2	1
1	2	2	1	1	2	2	1	2	1	1	2	2	1	1	2	1	2	2	1	1	2	2	1	2	1	1	2	2	1	1	1
2	1	1	2	2	1	1	2	1	2	2	1	1	2	2	1	2	1	1	2	2	1	1	2	1	2	2	1	1	2	2	2
1	2	2	1	1	2	2	1	2	1	1	2	2	1	1	2	2	1	1	2	2	1	1	2	1	2	2	1	1	2	2	1
2	1	1	2	2	1	1	2	1	2	2	1	1	2	2	1	1	2	2	1	1	2	2	1	2	1	1	2	2	1	1	2
1	2	2	1	2	1	1	2	1	2	2	1	2	1	1	2	1	2	2	1	2	1	1	2	1	2	2	1	2	1	1	2
2	1	1	2	1	2	2	1	2	1	1	2	1	2	2	1	2	1	1	2	1	2	2	1	2	1	1	2	1	2	2	1
1	2	2	1	2	1	1	2	1	2	2	1	2	1	1	2	2	1	1	2	1	2	2	1	2	1	1	2	1	2	2	1
2	1	1	2	1	2	2	1	2	1	1	2	1	2	2	1	1	2	2	1	2	1	1	2	1	2	2	1	2	1	1	2
1	2	2	1	2	1	1	2	2	1	1	2	1	2	2	1	1	2	2	1	2	1	1	2	2	1	1	2	1	2	2	1
2	1	1	2	1	2	2	1	1	2	2	1	2	1	1	2	2	1	1	2	1	2	2	1	1	2	2	1	2	1	1	2
1	2	2	1	2	1	1	2	2	1	1	2	1	2	2	1	2	1	1	2	1	2	2	1	1	2	2	1	2	1	1	2
2	1	1	2	1	2	2	1	1	2	2	1	2	1	1	2	1	2	2	1	2	1	1	2	2	1	1	2	1	2	2	1
f	*a*	*b*	*a*	*c*	*a*	*b*	*a*	*d*	*a*	*b*	*a*	*c*	*a*	*b*	*a*	*e*	*a*	*b*	*a*	*c*	*a*	*b*	*a*	*d*	*a*	*b*	*a*	*c*	*a*	*b*	*a*
	f	*f*	*b*	*f*	*c*	*c*	*b*	*f*	*d*	*d*	*b*	*d*	*c*	*c*	*b*	*f*	*e*	*e*	*b*	*e*	*c*	*c*	*b*	*e*	*d*	*d*	*b*	*d*	*c*	*c*	*b*
			f		*f*	*f*	*c*		*f*	*f*	*d*	*f*	*d*	*d*	*c*		*f*	*f*	*e*	*f*	*e*	*e*	*c*	*f*	*e*	*e*	*d*	*e*	*d*	*d*	*c*
							f				*f*		*f*	*f*	*d*				*f*		*f*	*f*	*e*		*f*	*f*	*e*	*f*	*e*	*e*	*d*
															f								*f*				*f*		*f*	*f*	*e*
																															f

6

$OA_{64}(2^{63})$ (*continued*)
Interactions between two columns

(Col.) \ Col: 1	2	3	4	5	6	7	8	9	10	11	12	13	14	15	16	17	18	19	20	21	22	23	24	25	26	27	28	29	30	31
(1)	3	2	5	4	7	6	9	8	11	10	13	12	15	14	17	16	19	18	21	20	23	22	25	24	27	26	29	28	31	30
	(2)	1	6	7	4	5	10	11	8	9	14	15	12	13	18	19	16	17	22	23	20	21	26	27	24	25	30	31	28	29
		(3)	7	6	5	4	11	10	9	8	15	14	13	12	19	18	17	16	23	22	21	20	27	26	25	24	31	30	29	28
			(4)	1	2	3	12	13	14	15	8	9	10	11	20	21	22	23	16	17	18	19	28	29	30	31	24	25	26	27
				(5)	3	2	13	12	15	14	9	8	11	10	21	20	23	22	17	16	19	18	29	28	31	30	25	24	27	26
					(6)	1	4	15	12	13	10	11	8	9	22	23	20	21	18	19	16	17	30	31	28	29	26	27	24	25
						(7)	15	14	13	12	11	10	9	8	23	22	21	20	19	18	17	16	31	30	29	28	27	26	25	24
							(8)	1	2	3	4	5	6	7	24	25	26	27	28	20	30	31	16	17	18	19	20	21	22	23
								(9)	3	2	5	4	7	6	25	24	27	26	29	28	31	30	17	16	19	18	21	20	23	22
									(10)	1	6	7	4	5	26	27	24	25	30	31	28	29	18	19	16	17	22	23	20	21
										(11)	7	6	5	4	27	26	25	24	31	30	29	28	19	18	17	16	13	22	21	20
											(12)	1	2	3	28	29	30	31	24	25	26	27	20	21	22	23	16	17	18	19
												(13)	3	2	29	28	31	30	25	24	27	26	21	20	23	22	17	16	19	18
													(14)	1	30	31	28	29	26	27	24	25	22	23	20	21	18	19	16	17
														(15)	31	30	29	28	27	26	25	24	23	22	21	20	19	18	17	16
															(16)	1	2	3	4	5	6	7	8	9	10	11	12	13	14	15
																(17)	3	2	5	4	7	6	9	8	11	10	13	12	15	14
																	(18)	1	6	7	4	5	10	11	8	9	14	15	12	13
																		(19)	7	6	5	4	11	10	9	8	15	14	13	12
																			(20)	1	2	3	12	13	14	15	8	9	10	11
																				(21)	3	2	13	12	15	14	9	8	11	10
																					(22)	1	14	15	12	13	10	11	8	9
																						(23)	15	14	13	12	11	10	9	8
																							(24)	1	2	3	4	5	6	7
																								(25)	3	2	5	4	7	6
																									(26)	1	6	7	4	5
																										(27)	7	6	5	4
																											(28)	1	2	1
																												(29)	3	2
																													(30)	1
																														(31)

$OA_{64}(2^{63})$ (*continued*)

Interactions between two columns

(Col:) \ Col:	32	33	34	35	36	37	38	39	40	41	42	43	44	45	46	47	48	49	50	51	52	53	54	55	56	57	58	59	60	61	62	63
	33	32	35	34	37	36	39	38	41	40	43	42	45	44	47	46	49	48	51	50	53	52	55	54	57	56	59	58	61	60	63	62
	34	35	32	33	38	39	36	37	42	43	40	41	46	47	44	45	50	51	48	49	54	55	52	53	58	59	56	57	62	63	60	61
	35	34	33	32	39	38	37	36	43	42	41	40	47	46	45	44	51	50	49	48	55	54	53	52	59	58	57	56	63	62	61	60
	36	37	38	39	32	33	34	35	44	45	46	47	40	41	42	43	52	53	54	55	48	49	50	51	60	61	62	63	56	57	58	59
	37	36	39	38	33	32	35	34	45	44	47	46	41	40	43	42	53	52	55	54	49	48	51	50	61	60	63	62	57	56	59	58
	38	39	36	37	34	35	32	33	46	47	44	45	42	43	40	41	54	55	52	53	50	51	48	49	62	63	60	61	58	59	56	57
	39	38	37	36	35	34	33	32	47	46	45	44	43	42	41	40	55	54	53	52	51	50	49	48	63	62	61	60	59	58	57	56
	40	41	42	43	44	45	46	47	32	33	34	35	36	37	38	39	56	57	58	59	60	61	62	63	48	49	50	51	52	53	54	55
	41	40	43	42	45	44	47	46	33	32	35	34	37	36	39	38	57	56	59	58	61	60	63	62	49	48	51	50	53	52	55	54
	42	43	40	41	46	47	44	45	34	35	32	33	38	39	36	37	58	59	56	57	62	63	60	61	50	51	48	49	54	55	52	53
	43	42	41	40	47	46	45	44	35	34	33	32	39	38	37	36	59	58	57	56	63	62	61	60	51	50	49	48	55	54	53	52
	44	45	46	47	40	41	42	43	36	37	38	39	32	33	34	35	60	61	62	63	56	57	58	59	52	53	54	55	48	49	50	51
	45	44	47	46	41	40	43	42	37	36	39	38	33	32	35	34	61	60	63	62	57	56	59	58	53	52	55	54	49	48	51	50
	46	47	44	45	42	43	40	41	38	39	36	37	34	35	32	33	62	63	60	61	58	59	56	57	54	55	52	53	50	51	48	49
	47	46	45	44	43	42	41	40	39	38	37	36	35	34	33	32	63	62	61	60	59	58	57	57	55	54	53	52	51	50	49	48
	48	49	50	51	52	53	54	55	56	57	58	59	60	61	62	63	32	33	34	35	36	37	38	39	40	41	42	43	44	45	46	47
	49	48	51	50	53	52	55	54	57	56	59	58	61	60	63	62	33	32	35	34	37	36	39	38	41	40	43	42	45	44	47	46
	50	51	48	49	54	55	52	53	58	59	56	57	62	63	60	61	34	35	32	33	38	39	36	37	42	43	40	41	46	47	44	45
	51	50	49	48	55	54	53	52	59	58	57	56	63	62	61	60	35	34	33	32	39	38	37	36	43	42	41	40	47	46	45	44
	52	53	54	55	48	49	50	51	60	61	62	63	56	57	58	59	36	37	38	39	32	33	34	35	44	45	46	47	40	41	42	43
	53	52	55	54	49	48	51	50	61	60	63	62	57	56	59	58	37	36	39	38	33	32	35	34	45	44	47	46	41	40	43	42
	54	55	52	53	50	51	48	49	62	63	60	61	58	59	56	57	38	39	36	37	34	35	32	33	46	47	44	45	42	43	40	41
	55	54	53	52	51	50	49	48	63	62	61	60	59	58	57	56	39	38	37	36	35	34	33	32	47	46	45	44	43	42	41	40
	56	57	58	59	60	61	62	63	48	49	50	51	52	53	54	55	40	41	42	43	44	45	46	47	32	33	34	35	36	37	38	39
	57	56	59	58	61	60	63	62	49	48	51	50	53	52	55	54	41	40	43	42	45	44	47	46	33	32	35	34	37	36	39	38

58	59	56	57	62	63	60	61	50	51	48	49	54	55	52	53	42	43	40	41	46	47	44	45	34	35	32	33	38	39	36	37
59	58	57	56	63	62	61	60	51	50	49	48	55	54	53	52	43	42	41	40	47	46	45	44	35	34	33	32	39	38	37	36
60	61	62	63	56	57	58	59	52	53	54	55	48	49	50	51	44	45	46	47	40	41	42	43	36	37	38	39	32	33	34	35
61	60	63	62	57	56	59	58	53	52	55	54	49	48	51	50	45	44	47	46	41	40	43	42	37	36	39	38	33	32	35	34
62	63	60	61	58	59	56	57	54	55	52	53	50	51	48	49	46	47	44	45	42	43	40	41	38	39	36	37	34	35	32	33
63	62	61	60	59	58	57	56	55	54	53	52	51	50	49	48	47	46	45	44	43	42	41	40	39	38	37	36	35	34	33	32
(32)	1	2	3	4	5	6	7	8	9	10	11	12	13	14	15	16	17	18	19	20	21	22	23	24	25	26	27	28	29	30	31
	(33)	3	2	5	4	7	6	9	8	11	10	13	12	15	14	17	16	19	18	21	20	23	22	25	24	27	26	29	28	31	30
		(34)	1	6	7	4	5	10	11	8	9	14	15	12	13	18	19	16	17	22	23	20	21	26	27	24	25	30	31	28	29
			(35)	7	6	5	4	11	10	9	8	15	14	13	12	19	18	17	16	23	22	21	20	27	26	25	24	31	30	29	28
				(36)	1	2	3	12	13	14	15	8	9	10	11	20	21	22	23	16	17	18	19	28	29	30	31	24	25	26	27
					(37)	3	2	13	12	15	14	9	8	11	10	21	20	23	22	17	16	19	18	29	28	31	30	25	24	27	26
						(38)	1	14	15	12	13	10	11	8	9	22	23	20	21	18	19	16	17	30	31	28	29	26	27	24	25
							(39)	15	14	13	12	11	10	9	8	23	22	21	20	19	18	17	16	31	30	29	28	27	26	25	24
								(40)	1	2	3	4	5	6	7	24	25	26	27	28	29	30	31	16	17	18	19	20	21	22	23
									(41)	3	2	5	4	7	6	25	24	27	26	29	28	31	30	17	16	19	18	21	20	23	22
										(42)	1	6	7	4	5	26	27	24	25	30	31	28	29	18	19	16	17	22	23	20	21
											(43)	7	6	5	4	27	26	25	24	31	30	29	28	19	18	17	16	23	22	21	20
												(44)	1	2	3	28	29	30	31	24	25	26	27	20	21	22	23	16	17	18	19
													(45)	3	2	29	28	31	30	25	24	27	26	21	20	23	22	17	16	19	18
														(46)	1	30	31	28	29	26	27	24	25	22	23	20	21	18	19	16	17
															(47)	31	30	29	28	27	26	25	24	23	22	21	20	19	18	17	16
																(48)	1	2	3	4	5	6	7	8	9	10	11	12	13	14	15
																	(49)	3	2	5	4	7	6	9	8	11	10	13	12	15	14
																		(50)	1	6	7	4	5	10	11	8	9	14	15	12	13
																			(51)	7	6	5	4	11	10	9	8	15	14	13	12
																				(52)	1	2	3	12	13	14	15	8	9	10	11
																					(53)	3	2	13	12	15	14	9	8	11	10
																						(54)	1	14	15	12	13	10	11	8	9
																							(55)	15	14	13	12	11	10	9	8
																								(56)	1	2	3	4	5	6	7
																									(57)	3	2	5	4	7	6
																										(58)	1	6	7	4	5
																											(59)	7	6	5	4
																												(60)	1	2	3
																													(61)	3	2
																														(62)	1

$OA_9(3^4)$

Tr.	Col: 1	2	3	4
1	1	1	1	1
2	1	2	2	2
3	1	3	3	3
4	2	1	2	3
5	2	2	3	1
6	2	3	1	2
7	3	1	3	2
8	3	2	1	3
9	3	3	2	1
	a	b	a b	a b^2
Group	1	2		

Interactions between two columns

(Col.)\Col:	1	2	3	4
	(1)	3 4	2 4	2 3
		(2)	1 4	1 3
			(3)	1 2

$OA_{18}(2^1 \times 3^7)$

Tr.	Col: 1	2	3	4	5	6	7	8
1	1	1	1	1	1	1	1	1
2	1	1	2	2	2	2	2	2
3	1	1	3	3	3	3	3	3
4	1	2	1	1	2	2	3	3
5	1	2	2	2	3	3	1	1
6	1	2	3	3	1	1	2	2
7	1	3	1	2	1	3	2	3
8	1	3	2	3	2	1	3	1
9	1	3	3	1	3	2	1	2
10	2	1	1	3	3	2	2	1
11	2	1	2	1	1	3	3	2
12	2	1	3	2	2	1	1	3
13	2	2	1	2	3	1	3	2
14	2	2	2	3	1	2	1	3
15	2	2	3	1	2	3	2	1
16	2	3	1	3	2	3	1	2
17	2	3	2	1	3	1	2	3
18	2	3	3	2	1	2	3	1
Group	1	2	3					

Note: The only interaction information that can be obtained from OA_{18} is that between the first two columns.

$OA_{27}(3^{13})$

Tr.	Col: 1	2	3	4	5	6	7	8	9	10	11	12	13
1	1	1	1	1	1	1	1	1	1	1	1	1	1
2	1	1	1	1	2	2	2	2	2	2	2	2	2
3	1	1	1	1	3	3	3	3	3	3	3	3	3
4	1	2	2	2	1	1	1	2	2	2	3	3	3
5	1	2	2	2	2	2	2	3	3	3	1	1	1
6	1	2	2	2	3	3	3	1	1	1	2	2	2
7	1	3	3	3	1	1	1	3	3	3	2	2	2
8	1	3	3	3	2	2	2	1	1	1	3	3	3
9	1	3	3	3	3	3	3	2	2	2	1	1	1
10	2	1	2	3	1	2	3	1	2	3	1	2	3
11	2	1	2	3	2	3	1	2	3	1	2	3	1
12	2	1	2	3	3	1	2	3	1	2	3	1	2
13	2	2	3	1	1	2	3	2	3	1	3	1	2
14	2	2	3	1	2	3	1	3	1	2	1	2	3
15	2	2	3	1	3	1	2	1	2	3	2	3	1
16	2	3	1	2	1	2	3	3	1	2	2	3	1
17	2	3	1	2	2	3	1	1	2	3	3	1	2
18	2	3	1	2	3	1	2	2	3	1	1	2	3
19	3	1	3	2	1	3	2	1	3	2	1	3	2
20	3	1	3	2	2	1	3	2	1	3	2	1	3
21	3	1	3	2	3	2	1	3	2	1	3	2	1
22	3	2	1	3	1	3	2	2	1	3	3	2	1
23	3	2	1	3	2	1	3	3	2	1	1	3	2
24	3	2	1	3	3	2	1	1	3	2	2	1	3
25	3	3	2	1	1	3	2	3	2	1	2	1	3
26	3	3	2	1	2	1	3	1	3	2	3	2	1
27	3	3	2	1	3	2	1	2	1	3	1	3	2
	a	b	a	a	c	a	a	b	a	a	b	a	a
			b	b^2		c	c^2	c	b	b^2	c^2	b^2	b
									c	c^2		c	c^2
Group	1	2			3								

$OA_{27}(3^{13})$ (*continued*)

Interactions between two columns

(Col.) \ Col:	1	2	3	4	5	6	7	8	9	10	11	12	13
(1)		3	2	2	6	5	5	9	8	8	12	11	11
		4	4	3	7	7	6	10	10	9	13	13	12
(2)			1	1	8	9	10	5	6	7	5	6	7
			4	3	11	12	13	11	12	13	8	9	10
(3)				1	9	10	8	7	5	6	6	7	5
				2	13	11	12	12	13	11	10	8	9
(4)					10	8	9	6	7	5	7	5	6
					12	13	11	13	11	12	9	10	8
(5)						1	1	2	3	4	2	4	3
						7	6	11	13	12	8	10	9
(6)							1	4	2	3	3	2	4
							5	13	12	11	10	9	8
(7)								3	4	2	4	3	2
								12	11	13	9	8	10
(8)									1	1	2	3	4
									10	9	5	7	6
(5)										1	4	2	3
										8	7	6	5
(10)											3	4	2
											6	5	7
(11)												1	1
												13	12
(12)													1
													11

$OA_{81}(3^{40})$

Tr.	Col. 1	2	3	4	5	6	7	8	9	10	11	12	13	14	15	16	17	18	19	20	21	22	23	24	25	26	27	28	29	30	31	32	33	34	35	36	37	38	39	40
1	1	1	1	1	1	1	1	1	1	1	1	1	1	1	1	1	1	1	1	1	1	1	1	1	1	1	1	1	1	1	1	1	1	1	1	1	1	1	1	1
2	1	1	1	1	1	1	1	1	1	1	1	1	1	2	2	2	2	2	2	2	2	2	2	2	2	2	2	2	2	2	2	2	2	2	2	2	2	2	2	2
3	1	1	1	1	1	1	1	1	1	1	1	1	1	3	3	3	3	3	3	3	3	3	3	3	3	3	3	3	3	3	3	3	3	3	3	3	3	3	3	3
4	1	1	1	1	2	2	2	2	2	2	2	2	2	1	1	1	1	1	1	1	1	1	2	2	2	2	2	2	2	2	2	3	3	3	3	3	3	3	3	3
5	1	1	1	1	2	2	2	2	2	2	2	2	2	2	2	2	2	2	2	2	2	2	3	3	3	3	3	3	3	3	3	1	1	1	1	1	1	1	1	1
6	1	1	1	1	2	2	2	2	2	2	2	2	2	3	3	3	3	3	3	3	3	3	1	1	1	1	1	1	1	1	1	2	2	2	2	2	2	2	2	2
7	1	1	1	1	3	3	3	3	3	3	3	3	3	1	1	1	1	1	1	1	1	3	3	3	3	3	3	3	3	3	3	2	2	2	2	2	2	2	2	2
8	1	1	1	1	3	3	3	3	3	3	3	3	3	2	2	2	2	2	2	2	2	2	1	1	1	1	1	1	1	1	1	3	3	3	3	3	3	3	3	3
9	1	1	1	1	3	3	3	3	3	3	3	3	3	3	3	3	3	3	3	3	3	3	2	2	2	2	2	2	2	2	2	1	1	1	1	1	1	1	1	1
10	1	2	2	2	1	1	1	2	2	2	3	3	3	1	1	1	2	2	2	3	3	3	1	1	1	2	2	2	3	3	3	1	1	1	2	2	2	2	3	3
11	1	2	2	2	1	1	1	2	2	2	3	3	3	2	2	2	3	3	3	1	1	1	2	2	2	3	3	3	1	1	1	2	2	2	3	3	3	1	1	1
12	1	2	2	2	1	1	1	2	2	2	3	3	3	3	3	3	1	1	1	2	2	2	3	3	3	1	1	1	2	2	2	3	3	3	1	1	1	2	2	2
13	1	2	2	2	2	2	2	3	3	3	1	1	1	1	1	1	2	2	2	3	3	3	2	2	2	3	3	3	1	1	1	3	3	3	1	1	1	2	2	2
14	1	2	2	2	2	2	2	3	3	3	1	1	1	2	2	2	3	3	3	1	1	1	3	3	3	1	1	1	2	2	2	1	1	1	2	2	2	3	3	3
15	1	2	2	2	2	2	2	3	3	3	1	1	1	3	3	3	1	1	1	2	2	2	1	1	1	2	2	2	3	3	3	2	2	2	3	3	3	1	1	1
16	1	2	2	2	3	3	3	1	1	1	2	2	2	1	1	1	2	2	2	2	2	2	2	2	2	1	1	1	2	2	2	2	2	2	3	3	3	1	1	1
17	1	2	2	2	3	3	3	1	1	1	2	2	2	2	2	2	3	3	3	1	1	1	1	1	1	2	2	2	3	3	3	3	3	3	1	1	1	2	2	2
18	1	2	2	2	3	3	3	1	1	1	2	2	2	3	3	3	1	1	1	2	2	2	2	2	2	3	3	3	1	1	1	1	1	1	3	3	3	3	3	3
19	1	3	3	3	1	1	1	3	3	3	2	2	2	1	1	1	3	3	3	2	2	2	1	1	1	3	3	3	2	2	2	1	1	1	3	3	3	2	2	2
20	1	3	3	3	1	1	1	3	3	3	2	2	2	2	2	2	1	1	1	3	3	3	2	2	2	1	1	1	3	3	3	2	2	2	1	1	1	3	3	3
21	1	3	3	3	1	1	1	3	3	3	2	2	2	3	3	3	2	2	2	1	1	1	3	3	3	2	2	2	1	1	1	3	3	3	2	2	2	1	1	1
22	1	3	3	3	2	2	2	1	1	1	3	3	3	1	1	1	3	3	3	2	2	2	2	2	2	1	1	1	3	3	3	3	3	3	2	2	2	1	1	1
23	1	3	3	3	2	2	2	1	1	1	3	3	3	2	2	2	1	1	1	3	3	3	3	3	3	2	2	2	1	1	1	1	1	1	3	3	3	2	2	2
24	1	3	3	3	2	2	2	1	1	1	3	3	3	3	3	3	2	2	2	1	1	1	1	1	1	3	3	3	2	2	2	2	2	2	1	1	1	3	3	3
25	1	3	3	3	3	3	3	2	2	2	1	1	1	1	1	1	3	3	3	2	2	2	3	3	3	2	2	2	1	1	1	2	2	2	1	1	1	3	3	3
26	1	3	3	3	3	3	3	2	2	2	1	1	1	2	2	2	1	1	1	3	3	3	1	1	1	3	3	3	2	2	2	3	3	3	2	2	2	1	1	1
27	1	3	3	3	3	3	3	2	2	2	1	1	1	3	3	3	2	2	2	1	1	1	2	2	2	1	1	1	3	3	3	1	1	1	3	3	3	2	2	2
28	2	1	2	3	1	2	3	1	2	3	1	2	3	1	2	3	1	2	3	1	2	3	1	2	3	1	2	3	1	2	3	1	2	3	1	2	3	1	2	3
29	2	1	2	3	1	2	3	1	2	3	1	2	3	2	3	1	2	3	1	2	3	1	2	3	1	2	3	1	2	3	1	2	3	1	2	3	1	2	3	1
30	2	1	2	3	1	2	3	1	2	3	1	2	3	3	1	2	3	1	2	3	1	2	3	1	2	3	1	2	3	1	2	3	1	2	3	1	2	3	1	2
31	2	1	2	3	2	3	1	2	3	1	2	3	1	1	2	3	1	2	3	1	2	3	2	3	1	2	3	1	2	3	1	3	1	2	3	1	2	3	1	2
32	2	1	2	3	2	3	1	2	3	1	2	3	1	2	3	1	2	3	1	2	3	1	3	1	2	3	1	2	3	1	2	1	2	3	1	2	3	1	2	3

33	2	1	2	3	2	3	1	2	3	1	2	3	1	3	1	2	3	1	2	3	1	2	1	2	3	1	2	3	1	2	3	2	3	1	2	3	1	2	3	1
34	2	1	2	3	3	1	2	3	1	2	3	1	2	1	2	3	1	2	3	1	2	3	3	1	2	3	1	2	3	1	2	2	3	1	2	3	1	2	3	1
35	2	1	2	3	3	1	2	3	1	2	3	1	2	2	3	1	2	3	1	2	3	1	1	2	3	1	2	3	1	2	3	3	1	2	3	1	2	3	1	2
36	2	1	2	3	3	1	2	3	1	2	3	1	2	3	1	2	3	1	2	3	1	2	2	3	1	2	3	1	2	3	1	1	2	3	1	2	3	1	2	3
37	2	2	3	1	1	2	3	2	3	1	3	1	2	1	2	3	2	3	1	3	1	2	1	2	3	2	3	1	3	1	2	1	2	3	2	3	1	3	1	2
38	2	2	3	1	1	2	3	2	3	1	3	1	2	2	3	1	3	1	2	1	2	3	2	3	1	3	1	2	1	2	3	2	3	1	3	1	2	1	2	3
39	2	2	3	1	1	2	3	2	3	1	3	1	2	3	1	2	1	2	3	2	3	1	3	1	2	1	2	3	2	3	1	3	1	2	1	2	3	2	3	1
40	2	2	3	1	2	3	1	3	1	2	1	2	3	1	2	3	2	3	1	3	1	2	2	3	1	3	1	2	1	2	3	3	1	2	1	2	3	2	3	1
41	2	2	3	1	2	3	1	3	1	2	1	2	3	2	3	1	3	1	2	1	2	3	3	1	2	1	2	3	2	3	1	1	2	3	2	3	1	3	1	2
42	2	2	3	1	2	3	1	3	1	2	1	2	3	3	1	2	1	2	3	2	3	1	1	2	3	2	3	1	3	1	2	2	3	1	3	1	2	1	2	3
43	2	2	3	1	3	1	2	1	2	3	2	3	1	1	2	3	2	3	1	3	1	2	3	1	2	1	2	3	2	3	1	2	3	1	3	1	2	1	2	3
44	2	2	3	1	3	1	2	1	2	3	2	3	1	2	3	1	3	1	2	1	2	3	1	2	3	2	3	1	3	1	2	3	1	2	1	2	3	2	3	1
45	2	2	3	1	3	1	2	1	2	3	2	3	1	3	1	2	1	2	3	2	3	1	2	3	1	3	1	2	1	2	3	1	2	3	2	3	1	3	1	2
46	2	3	1	2	1	2	3	3	1	2	2	3	1	1	2	3	3	1	2	2	3	1	1	2	3	3	1	2	2	3	1	1	2	3	3	1	2	2	3	1
47	2	3	1	2	1	2	3	3	1	2	2	3	1	2	3	1	1	2	3	3	1	2	2	3	1	1	2	3	3	1	2	2	3	1	1	2	3	3	1	2
48	2	3	1	2	1	2	3	3	1	2	2	3	1	3	1	2	2	3	1	1	2	3	3	1	2	2	3	1	1	2	3	3	1	2	2	3	1	1	2	3
49	2	3	1	2	2	3	1	1	2	3	3	1	2	1	2	3	3	1	2	2	3	1	2	3	1	1	2	3	3	1	2	3	1	2	2	3	1	1	2	3
50	2	3	1	2	2	3	1	1	2	3	3	1	2	2	3	1	1	2	3	3	1	2	3	1	2	2	3	1	1	2	3	1	2	3	3	1	2	2	3	1
51	2	3	1	2	2	3	1	1	2	3	3	1	2	3	1	2	2	3	1	1	2	3	1	2	3	3	1	2	2	3	1	2	3	1	1	2	3	3	1	2
52	2	3	1	2	3	1	2	2	3	1	1	2	3	1	2	3	3	1	2	2	3	1	3	1	2	2	3	1	1	2	3	2	3	1	1	2	3	3	1	2
53	2	3	1	2	3	1	2	2	3	1	1	2	3	2	3	1	1	2	3	3	1	2	1	2	3	3	1	2	2	3	1	3	1	2	2	3	1	1	2	3
54	2	3	1	2	3	1	2	2	3	1	1	2	3	3	1	2	2	3	1	1	2	3	2	3	1	1	2	3	3	1	2	1	2	3	3	1	2	2	3	1
55	3	1	3	2	1	3	2	1	3	2	1	3	2	1	3	2	1	3	2	1	3	2	1	3	2	1	3	2	1	3	2	1	3	2	1	3	2	1	3	2
56	3	1	3	2	1	3	2	1	3	2	1	3	2	2	1	3	2	1	3	2	1	3	2	1	3	2	1	3	2	1	3	2	1	3	2	1	3	2	1	3
57	3	1	3	2	1	3	2	1	3	2	1	3	2	3	2	1	3	2	1	3	2	1	3	2	1	3	2	1	3	2	1	3	2	1	3	2	1	3	2	1
58	3	1	3	2	2	1	3	2	1	3	2	1	3	1	3	2	1	3	2	1	3	2	2	1	3	2	1	3	2	1	3	3	2	1	3	2	1	3	2	1
59	3	1	3	2	2	1	3	2	1	3	2	1	3	2	1	3	2	1	3	2	1	3	3	2	1	3	2	1	3	2	1	1	3	2	1	3	2	1	3	2
60	3	1	3	2	2	1	3	2	1	3	2	1	3	3	2	1	3	2	1	3	2	1	1	3	2	1	3	2	1	3	2	2	1	3	2	1	3	2	1	3
61	3	1	3	2	3	2	1	3	2	1	3	2	1	1	3	2	1	3	2	1	3	2	3	2	1	3	2	1	3	2	1	2	1	3	2	1	3	2	1	3
62	3	1	3	2	3	2	1	3	2	1	3	2	1	2	1	3	2	1	3	2	1	3	1	3	2	1	3	2	1	3	2	3	2	1	3	2	1	3	2	1
63	3	1	3	2	3	2	1	3	2	1	3	2	1	3	2	1	3	2	1	3	2	1	2	1	3	2	1	3	2	1	3	1	3	2	1	3	2	1	3	2
64	3	2	1	3	1	3	2	2	1	3	3	2	1	1	3	2	2	1	3	3	2	1	1	3	2	2	1	3	3	2	1	1	3	2	2	1	3	3	2	1
65	3	2	1	3	1	3	2	2	1	3	3	2	1	2	1	3	3	2	1	1	3	2	2	1	3	3	2	1	1	3	2	2	1	3	3	2	1	1	3	2
66	3	2	1	3	1	3	2	2	1	3	3	2	1	3	2	1	1	3	2	2	1	3	3	2	1	1	3	2	2	1	3	3	2	1	1	3	2	2	1	3
67	3	2	1	3	2	1	3	3	2	1	1	3	2	1	3	2	2	1	3	3	2	1	2	1	3	3	2	1	1	3	2	3	2	1	1	3	2	2	1	3
68	3	2	1	3	2	1	3	3	2	1	1	3	2	2	1	3	3	2	1	1	3	2	3	2	1	1	3	2	2	1	3	1	3	2	2	1	3	3	2	1
69	3	2	1	3	2	1	3	3	2	1	1	3	2	3	2	1	1	3	2	2	1	3	1	3	2	2	1	3	3	2	1	2	1	3	3	2	1	1	3	2

$OA_{81}(3^{40})$ (*continued*)

Tr.	Col. 1	2	3	4	5	6	7	8	9	10	11	12	13	14	15	16	17	18	19	20	21	22	23	24	25	26	27	28	29	30	31	32	33	34	35	36	37	38	39	40
70	3	2	1	3	3	2	1	1	3	2	2	1	3	1	3	2	2	1	3	3	2	1	3	2	1	1	3	2	2	1	3	2	1	3	3	2	1	1	3	2
71	3	2	1	3	3	2	1	1	3	2	2	1	3	2	1	3	3	2	1	1	3	2	1	3	2	2	1	3	3	2	1	3	2	1	1	3	2	2	1	3
72	3	2	1	3	3	2	1	1	3	2	2	1	3	3	2	1	1	3	2	2	1	3	2	1	3	3	2	1	1	3	2	1	3	2	2	1	3	3	2	1
73	3	3	2	1	1	3	2	3	2	1	2	1	3	1	3	2	3	2	1	2	1	3	1	3	2	3	2	1	2	1	3	1	3	2	3	2	1	2	1	3
74	3	3	2	1	1	3	2	3	2	1	2	1	3	2	1	3	1	3	2	3	2	1	2	1	3	1	3	2	3	2	1	2	1	3	1	3	2	3	2	1
75	3	3	2	1	1	3	2	3	2	1	2	1	3	3	2	1	2	1	3	1	3	2	3	2	1	2	1	3	1	3	2	3	2	1	2	1	3	1	3	2
76	3	3	2	1	2	1	3	1	3	2	3	2	1	1	3	2	3	2	1	2	1	3	2	1	3	1	3	2	3	2	1	3	2	1	2	1	3	1	3	2
77	3	3	2	1	2	1	3	1	3	2	3	2	1	2	1	3	1	3	2	3	2	1	3	2	1	2	1	3	1	3	2	1	3	2	3	2	1	2	1	3
78	3	3	2	1	2	1	3	1	3	2	3	2	1	3	2	1	2	1	3	1	3	2	1	3	2	3	2	1	2	1	3	2	1	3	1	3	2	3	2	1
79	3	3	2	1	3	2	1	2	1	3	1	3	2	1	3	2	3	2	1	2	1	3	3	2	1	2	1	3	1	3	2	2	1	3	1	3	2	3	2	1
80	3	3	2	1	3	2	1	2	1	3	1	3	2	2	1	3	1	3	2	3	2	1	1	3	2	3	2	1	2	1	3	3	2	1	2	1	3	1	3	2
81	3	3	2	1	3	2	1	2	1	3	1	3	2	3	2	1	2	1	3	1	3	2	2	1	3	1	3	2	3	2	1	1	3	2	3	2	1	2	1	3
	a	b	ab	ab^2	c	ac	ac^2	bc	abc	ab^2c^2	bc^2	ab^2c	abc^2	d	ad	ad^2	bd	abd	ab^2d^2	bd^2	ab^2d	abd^2	cd	acd	ac^2d^2	bcd	$abcd$	$ab^2c^2d^2$	bc^2d^2	ab^2cd	abc^2d^2	cd^2	ac^2d	acd^2	bc^2d	abc^2d	ab^2cd^2	bcd^2	ab^2c^2d	$abcd^2$
Group	1		2						3																		4													

$OA_{81}(3^{40})$ (*continued*)

Interactions between two columns

(Col.) \ Col:	14	15	16	17	18	19	20	21	22	23	24	25	26	27	28	29	30	31	32	33	34	35	36	37	38	39	40
(1)	15	14	14	18	17	17	21	20	20	24	23	23	27	26	26	30	29	29	33	32	32	36	35	35	39	38	38
	16	16	15	19	19	18	22	22	21	25	25	24	28	28	27	31	31	30	34	34	33	37	37	36	40	40	39
(2)	17	18	19	14	15	16	14	15	16	26	27	28	23	24	25	23	24	25	35	36	37	32	33	34	32	33	34
	20	21	22	20	21	22	17	18	19	29	30	31	29	30	31	26	27	28	38	39	40	38	39	40	35	36	37
(3)	18	19	17	16	14	15	15	16	14	27	28	26	25	23	24	24	25	23	36	37	35	34	32	33	33	34	32
	22	20	21	21	22	20	19	17	18	31	29	30	30	31	29	28	26	27	40	38	39	39	40	38	37	35	36
(4)	19	17	18	15	16	14	16	14	15	28	26	27	24	25	23	25	23	24	37	35	36	33	34	32	34	32	33
	21	22	20	22	20	21	18	19	17	30	31	29	31	29	30	27	28	26	39	40	38	40	38	39	36	37	35
(5)	23	24	25	26	27	28	29	30	31	14	15	16	17	18	19	20	21	22	14	15	16	17	18	19	20	21	22
	32	33	34	35	36	37	38	39	40	32	33	34	35	36	37	38	39	40	23	24	25	26	27	28	29	30	31
(6)	24	25	23	27	28	26	30	31	29	16	14	15	19	17	18	22	20	21	15	16	14	18	19	17	21	22	20
	34	32	33	37	35	36	40	38	29	33	34	32	36	37	35	39	40	38	25	23	24	28	26	27	31	29	30
(7)	25	23	24	28	26	27	31	29	30	15	16	14	18	19	17	21	22	20	16	14	15	19	17	18	22	20	21
	33	34	32	36	37	35	39	40	38	34	32	33	37	35	36	40	38	39	24	25	23	27	28	26	30	31	29
(8)	26	27	28	29	30	31	23	24	25	20	21	22	14	15	16	17	18	19	17	18	19	20	21	22	14	15	16
	38	39	40	32	33	34	35	36	37	35	36	37	38	39	40	32	33	34	29	30	31	23	24	25	26	27	28
(9)	27	28	26	30	31	29	24	25	23	22	20	21	16	14	15	19	17	18	18	19	17	21	22	20	15	16	14
	40	38	39	34	32	33	37	35	36	36	37	35	39	40	38	33	34	32	31	29	30	25	23	24	28	26	27
(10)	28	26	27	31	29	30	25	23	24	21	22	20	15	16	14	18	19	17	19	17	18	22	20	21	16	14	15
	39	40	38	33	34	32	36	37	35	37	35	36	40	38	39	34	32	33	30	31	29	24	25	23	27	28	26
(11)	29	30	31	23	24	25	26	27	28	17	18	19	20	21	22	14	15	16	20	21	22	14	15	16	17	18	19
	35	36	37	38	39	40	32	33	34	38	39	40	32	33	34	35	36	37	26	27	28	29	30	31	23	24	25
(12)	30	31	29	24	25	23	27	28	26	19	17	18	22	20	21	16	14	15	21	22	20	15	16	14	18	19	17
	37	35	36	40	38	39	34	32	33	39	40	38	33	34	32	36	37	35	28	26	27	31	29	30	25	23	24
(13)	31	29	30	25	23	24	28	26	27	18	19	17	21	22	20	15	16	14	22	20	21	16	14	15	19	17	18
	36	37	35	39	40	38	33	34	32	40	38	39	34	32	33	37	35	36	27	28	26	30	31	29	24	25	23
(14)		1	1	2	3	4	2	4	3	5	6	7	8	9	10	11	12	13	5	7	6	11	13	12	8	10	9
		16	15	20	22	21	17	19	18	32	34	33	38	40	39	35	37	36	23	25	24	29	31	30	26	28	27
(15)			1	4	2	3	3	2	4	7	5	6	10	8	9	13	11	12	6	5	7	12	11	13	9	8	10
			14	22	21	20	19	18	17	34	33	32	40	39	38	37	36	35	25	24	23	31	30	29	28	27	26
				3	4	2	4	3	2	6	7	5	9	10	8	12	13	11	7	6	5	13	12	11	10	9	8

$OA_{81}(3^{40})$ (*continued*)
Interaction between two columns

(Col.)	Col: 14	15	16	17	18	19	20	21	22	23	24	25	26	27	28	29	30	31	32	33	34	35	36	37	38	39	40
			(16)	21	20	22	18	17	19	33	32	34	39	38	40	36	35	37	24	23	25	30	29	31	27	26	28
				(17)	1	1	2	3	4	11	12	13	5	6	7	8	9	10	8	10	9	5	7	6	11	13	12
					19	18	14	16	15	38	40	39	35	37	36	32	34	33	29	31	30	26	28	27	23	25	24
					(18)	1	4	2	3	13	11	12	7	5	6	10	8	9	9	8	10	6	5	7	12	11	13
						17	16	15	14	40	39	38	37	36	35	34	33	32	31	30	29	28	27	26	25	24	23
						(19)	3	4	2	12	13	11	6	7	5	9	10	8	10	9	8	7	6	5	13	12	11
							15	14	16	39	38	40	36	35	37	33	32	34	30	29	31	27	26	28	24	23	25
							(20)	1	1	8	9	10	11	12	13	5	6	7	11	13	12	8	10	9	5	7	6
								22	21	35	37	36	22	34	33	38	40	39	26	28	27	23	25	24	29	31	30
								(21)	1	10	8	9	13	11	12	7	5	6	12	11	13	9	8	10	6	5	7
									20	37	36	35	34	33	32	40	39	38	28	27	26	25	24	23	31	30	29
									(22)	9	10	8	12	13	11	6	7	5	13	12	11	10	9	8	7	6	5
										36	35	37	33	32	34	39	38	40	27	26	28	24	23	25	30	29	31
										(23)	1	1	2	3	4	2	4	3	5	6	7	8	9	10	11	12	13
											25	24	29	31	30	26	28	27	14	16	15	20	22	21	17	19	18
											(24)	1	4	2	3	3	2	4	7	5	6	10	8	9	13	11	12
												23	31	30	29	28	27	26	16	15	14	22	21	20	19	18	17
												(25)	3	4	2	4	3	2	6	7	5	9	10	8	12	13	11
													30	29	31	27	26	28	15	14	16	21	20	22	18	17	19
													(26)	1	1	2	3	4	11	12	13	5	6	7	8	9	10
														28	27	23	25	24	20	22	21	17	19	18	14	16	15
														(27)	1	4	2	3	13	11	12	7	5	6	10	8	9
															26	25	24	23	22	21	20	19	18	17	16	15	14
															(28)	3	4	2	12	13	11	6	7	5	9	10	8
																24	23	25	21	20	22	18	17	19	15	14	16
																(29)	1	1	8	9	10	11	12	13	5	6	7
																	31	30	17	19	18	14	16	15	20	22	21
																	(30)	1	10	8	9	13	11	12	7	5	6
																		29	19	18	17	16	15	14	22	21	20
																		(31)	9	10	8	12	13	11	6	7	5

18	17	19	15	14	16	21	20	22
(32)	1	1	2	3	4	2	4	3
	34	33	38	40	39	35	37	36
	(33)	1	4	2	3	3	2	4
		32	40	39	38	37	36	35
		(34)	3	4	2	4	3	2
			39	38	40	36	35	37
			(35)	1	1	2	3	4
				37	36	32	34	33
				(36)	1	4	2	3
					35	34	33	32
					(37)	3	4	2
						33	32	34
						(38)	1	1
							40	39
							(39)	1
								38

Note: For interactions among Columns 1–13, see the triangular table for array $OA_{27}(3^{13})$

$OA_{54}(2^1 \times 3^{25})$

Tr.	Col: 1	2	3	4	5	6	7	8	9	10	11	12	13	14	15	16	17	18	19	20	21	22	23	24	25	2
1	1	1	1	1	1	1	1	1	1	1	1	1	1	1	1	1	1	1	1	1	1	1	1	1	1	1
2	1	1	1	1	1	1	1	1	2	2	2	2	2	2	2	2	2	2	2	2	2	2	2	2	2	2
3	1	1	1	1	1	1	1	1	3	3	3	3	3	2	3	3	3	3	3	3	3	3	3	3	3	3
4	1	1	2	2	2	2	2	2	1	1	1	1	1	1	2	3	2	3	2	3	2	3	2	3	2	3
5	1	1	2	2	2	2	2	2	2	2	2	2	2	2	3	1	3	1	3	1	3	1	3	1	3	1
6	1	1	2	2	2	2	2	2	3	3	3	3	3	3	1	2	1	2	1	2	1	2	1	2	1	2
7	1	1	3	3	3	3	3	3	1	1	1	1	1	1	3	2	3	2	3	2	3	2	3	2	3	2
8	1	1	3	3	3	3	3	3	2	2	2	2	2	2	1	3	1	3	1	3	1	3	1	3	1	3
9	1	1	3	3	3	3	3	3	3	3	3	3	3	3	2	1	2	1	2	1	2	1	2	1	2	1
10	1	2	1	1	2	2	3	3	1	1	2	2	3	3	1	1	1	1	2	3	2	3	3	2	3	2
11	1	2	1	1	2	2	3	3	2	2	3	3	1	1	2	2	2	2	3	1	3	1	1	3	1	3
12	1	2	1	1	2	2	3	3	3	3	1	1	2	2	3	3	3	3	1	2	1	2	2	1	2	1
13	1	2	2	2	3	3	1	1	1	1	2	2	3	3	2	3	2	3	3	2	3	2	1	1	1	1
14	1	2	2	2	3	3	1	1	2	2	3	3	1	1	3	1	3	1	1	3	1	3	2	2	2	2
15	1	2	2	2	3	3	1	1	3	3	1	1	2	2	1	2	1	2	2	1	2	1	3	3	3	3
16	1	2	3	3	1	1	2	2	1	1	2	2	3	3	3	2	3	2	1	1	1	1	2	3	2	3
17	1	2	3	3	1	1	2	2	2	2	3	3	1	1	1	3	1	3	2	2	2	2	3	1	3	1
18	1	2	3	3	1	1	2	2	3	3	1	1	2	2	2	1	2	1	3	3	3	3	1	2	1	2
19	1	3	1	2	1	3	2	3	1	2	1	3	2	3	1	1	2	3	1	1	3	2	2	3	3	2
20	1	3	1	2	1	3	2	3	2	3	2	1	3	1	2	2	3	1	2	2	1	3	3	1	1	3
21	1	3	1	2	1	3	2	3	3	1	3	2	1	2	3	3	1	2	3	3	2	1	1	2	2	1
22	1	3	2	3	2	1	3	1	1	2	1	3	2	3	2	3	3	2	2	3	1	1	3	2	1	1
23	1	3	2	3	2	1	3	1	2	3	2	1	3	1	3	1	1	3	3	1	2	2	1	3	2	2
24	1	3	2	3	2	1	3	1	3	1	3	2	1	2	1	2	2	1	1	2	3	3	2	1	3	3
25	1	3	3	1	3	2	1	2	1	2	1	3	2	3	3	2	1	1	3	2	2	3	1	1	2	3
26	1	3	3	1	3	2	1	2	2	3	2	1	3	1	1	3	2	2	1	3	3	1	2	2	3	1
27	1	3	3	1	3	2	1	2	3	1	3	2	1	2	2	1	3	3	2	1	1	2	3	3	1	2
28	2	1	1	3	3	2	2	1	1	3	3	2	2	1	1	1	3	2	3	2	2	3	2	3	1	1
29	2	1	1	3	3	2	2	1	2	1	1	3	3	2	2	2	1	3	1	3	3	1	3	1	2	2
30	2	1	1	3	3	2	2	1	3	2	2	1	1	3	3	3	2	1	2	1	1	2	1	2	3	3
31	2	1	2	1	1	3	3	2	1	3	3	2	2	1	2	3	1	1	1	1	3	2	3	2	2	3
32	2	1	2	1	1	3	3	2	2	1	1	3	3	2	3	1	2	2	2	2	1	3	1	3	3	1
33	2	1	2	1	1	3	3	2	3	2	2	1	1	3	1	2	3	3	3	3	2	1	2	1	1	2
34	2	1	3	2	2	1	1	3	1	3	3	2	2	1	3	2	2	3	2	3	1	1	1	1	3	2
35	2	1	3	2	2	1	1	3	2	1	1	3	3	2	1	3	3	1	3	1	2	2	2	2	1	3
36	2	1	3	2	2	1	1	3	3	2	2	1	1	3	2	1	1	2	1	2	3	3	3	3	2	1
37	2	2	1	2	3	1	3	2	1	2	3	1	3	2	1	1	2	3	3	2	1	1	3	2	2	3
38	2	2	1	2	3	1	3	2	2	3	1	2	1	3	2	2	3	1	1	3	2	2	1	3	3	1
39	2	2	1	2	3	1	3	2	3	1	2	3	2	1	3	3	1	2	2	1	3	3	2	1	1	2
40	2	2	2	3	1	2	1	3	1	2	3	1	3	2	2	3	3	2	1	1	2	3	1	1	3	2
41	2	2	2	3	1	2	1	3	2	3	1	2	1	3	3	1	1	3	2	2	3	1	2	2	1	3
42	2	2	2	3	1	2	1	3	3	1	2	3	2	1	1	2	2	1	3	3	1	2	3	3	2	1
43	2	2	3	1	2	3	2	1	1	2	3	1	3	2	3	2	1	1	2	3	3	2	2	3	1	1
44	2	2	3	1	2	3	2	1	2	3	1	2	1	3	1	3	2	2	3	1	1	3	3	1	2	2
45	2	2	3	1	2	3	2	1	3	1	2	3	2	1	2	1	3	3	1	2	2	1	1	2	3	3
46	2	3	1	3	2	3	1	2	1	3	2	3	1	2	1	1	3	2	2	3	3	2	1	1	2	3
47	2	3	1	3	2	3	1	2	2	1	3	1	2	3	2	2	1	3	3	1	1	3	2	2	3	1

48	2	3	1	3	2	3	1	2	3	2	1	2	3	1	3	3	2	1	1	2	2	1	3	3	1	2
49	2	3	2	1	3	1	2	3	1	3	2	3	1	2	2	3	1	1	3	2	1	1	2	3	3	2
50	2	3	2	1	3	1	2	3	2	1	3	1	2	3	3	1	2	2	1	3	2	2	3	1	1	3
51	2	3	2	1	3	1	2	3	3	2	1	2	3	1	1	2	3	3	2	1	3	3	1	2	2	1
52	2	3	3	2	1	2	3	1	1	3	2	3	1	2	3	2	2	3	1	1	2	3	3	2	1	1
53	2	3	3	2	1	2	3	1	2	1	3	1	2	3	1	3	3	1	2	2	3	1	1	3	2	2
54	2	3	3	2	1	2	3	1	3	2	1	2	3	1	2	1	1	2	3	3	1	2	2	1	3	3
roup	1	2			3											4										

$OA_{54}(2^1 \times 3^{25})$ (*continued*)
Interactions between two columns

(Col.) \ Col:	3	4	5	6	7	8
(9)	15 16	17 18	19 20	21 22	23 24	25 26

$OA_{16}(4^5)$

Tr.	Col: 1	2	3	4	5
1	1	1	1	1	1
2	1	2	2	2	2
3	1	3	3	3	3
4	1	4	4	4	4
5	2	1	2	3	4
6	2	2	1	4	3
7	2	3	4	1	2
8	2	4	3	2	1
9	3	1	3	4	2
10	3	2	4	3	1
11	3	3	1	2	4
12	3	4	2	1	3
13	4	1	4	2	3
14	4	2	3	1	4
15	4	3	2	4	1
16	4	4	1	3	2
Group	1		2		

$OA_{16}(4^5)$ (*continued*)
Interactions between two columns

(Col.) \ Col:	1	2	3	4	5
(1)		3 4 5	2 4 5	2 3 5	2 3 4
(2)			1 4 5	1 3 5	1 3 4
(3)				1 2 5	1 2 4
(4)					1 2 3

$OA_{64}(4^{21})$

Tr.	Col: 1	2	3	4	5	6	7	8	9	10	11	12	13	14	15	16	17	18	19	20	21
1	1	1	1	1	1	1	1	1	1	1	1	1	1	1	1	1	1	1	1	1	1
2	1	1	1	1	1	2	2	2	2	2	2	2	2	2	2	2	2	2	2	2	2
3	1	1	1	1	1	3	3	3	3	3	3	3	3	3	3	3	3	3	3	3	3
4	1	1	1	1	1	4	4	4	4	4	4	4	4	4	4	4	4	4	4	4	4
5	1	2	2	2	2	1	1	1	1	2	2	2	2	3	3	3	3	4	4	4	4
6	1	2	2	2	2	2	2	2	2	1	1	1	1	4	4	4	4	3	3	3	3
7	1	2	2	2	2	3	3	3	3	4	4	4	4	1	1	1	1	2	2	2	2
8	1	2	2	2	2	4	4	4	4	3	3	3	3	2	2	2	2	1	1	1	1
9	1	3	3	3	3	1	1	1	1	3	3	3	3	4	4	4	4	2	2	2	2
10	1	3	3	3	3	2	2	2	2	4	4	4	4	3	3	3	3	1	1	1	1
11	1	3	3	3	3	3	3	3	3	1	1	1	1	2	2	2	2	4	4	4	4
12	1	3	3	3	3	4	4	4	4	2	2	2	2	1	1	1	1	3	3	3	3
13	1	4	4	4	4	1	1	1	1	4	4	4	4	2	2	2	2	3	3	3	3
14	1	4	4	4	4	2	2	2	2	3	3	3	3	1	1	1	1	4	4	4	4
15	1	4	4	4	4	3	3	3	3	2	2	2	2	4	4	4	4	1	1	1	1
16	1	4	4	4	4	4	4	4	4	1	1	1	1	3	3	3	3	2	2	2	2
17	2	1	2	3	4	1	2	3	4	1	2	3	4	1	2	3	4	1	2	3	4
18	2	1	2	3	4	2	1	4	3	2	1	4	3	2	1	4	3	2	1	4	3
19	2	1	2	3	4	3	4	1	2	3	4	1	2	3	4	1	2	3	4	1	2
20	2	1	2	3	4	4	3	2	1	4	3	2	1	4	3	2	1	4	3	2	1
21	2	2	1	4	3	1	2	3	4	2	1	4	3	3	4	1	2	4	3	2	1
22	2	2	1	4	3	2	1	4	3	1	2	3	4	4	3	2	1	3	4	1	2
23	2	2	1	4	3	3	4	1	2	4	3	2	1	1	2	3	4	2	1	4	3
24	2	2	1	4	3	4	3	2	1	3	4	1	2	2	1	4	3	1	2	3	4
25	2	3	4	1	2	1	2	3	4	3	4	1	2	4	3	2	1	2	1	4	3
26	2	3	4	1	2	2	1	4	3	4	3	2	1	3	4	1	2	1	2	3	4
27	2	3	4	1	2	3	4	1	2	1	2	3	4	2	1	4	3	4	3	2	1
28	2	3	4	1	2	4	3	2	1	2	1	4	3	1	2	3	4	3	4	1	2

29	2	4	3	2	1	1	2	3	4	4	3	2	1	2	1	4	3	3	4	1	2
30	2	4	3	2	1	2	1	4	3	3	4	1	2	1	2	3	4	4	3	2	1
31	2	4	3	2	1	3	4	1	2	2	1	4	3	4	3	2	1	1	2	3	4
32	2	4	3	2	1	4	3	2	1	1	2	3	4	3	4	1	2	2	1	4	3
33	3	1	3	4	2	1	3	4	2	1	3	4	2	1	3	4	2	1	3	4	2
34	3	1	3	4	2	2	4	3	1	2	4	3	1	2	4	3	1	2	4	3	1
35	3	1	3	4	2	3	1	2	4	3	1	2	4	3	1	2	4	3	1	2	4
36	3	1	3	4	2	4	2	1	3	4	2	1	3	4	2	1	3	4	2	1	3
37	3	2	4	3	1	1	3	4	2	2	4	3	1	3	1	2	4	4	2	1	3
38	3	2	4	3	1	2	4	3	1	1	3	4	2	4	2	1	3	3	1	2	4
39	3	2	4	3	1	3	1	2	4	4	2	1	3	1	3	4	2	2	4	3	1
40	3	2	4	3	1	4	2	1	3	3	1	2	4	2	4	3	1	1	3	4	2
41	3	3	1	2	4	1	3	4	2	3	1	2	4	4	2	1	3	2	4	3	1
42	3	3	1	2	4	2	4	3	1	4	2	1	3	3	1	2	4	1	3	4	2
43	3	3	1	2	4	3	1	2	4	1	3	4	2	2	4	3	1	4	2	1	3
44	3	3	1	2	4	4	2	1	3	2	4	3	1	1	3	4	2	3	1	2	4
45	3	4	2	1	3	1	3	4	2	4	2	1	3	2	4	3	1	3	1	2	4
46	3	4	2	1	3	2	4	3	1	3	1	2	4	1	3	4	2	4	2	1	3
47	3	4	2	1	3	3	1	2	4	2	4	3	1	4	2	1	3	1	3	4	2
48	3	4	2	1	3	4	2	1	3	1	3	4	2	3	1	2	4	2	4	3	1
49	4	1	4	2	3	1	4	2	3	1	4	2	3	1	4	2	3	1	4	2	3
50	4	1	4	2	3	2	3	1	4	2	3	1	4	2	3	1	4	2	3	1	4
51	4	1	4	2	3	3	2	4	1	3	2	4	1	3	2	4	1	3	2	4	1
52	4	1	4	2	3	4	1	3	2	4	1	3	2	4	1	3	2	4	1	3	2
53	4	2	3	1	4	1	4	2	3	2	3	1	4	3	2	4	1	4	1	3	2
54	4	2	3	1	4	2	3	1	4	1	4	2	3	4	1	3	2	3	2	4	1
55	4	2	3	1	4	3	2	4	1	4	1	3	2	1	4	2	3	2	3	1	4
56	4	2	3	1	4	4	1	3	2	3	2	4	1	2	3	1	4	1	4	2	3
57	4	3	2	4	1	1	4	2	3	3	2	4	1	4	1	3	2	2	3	1	4
58	4	3	2	4	1	2	3	1	4	4	1	3	2	3	2	4	1	1	4	2	3
59	4	3	2	4	1	3	2	4	1	1	4	2	3	2	3	1	4	4	1	3	2
60	4	3	2	4	1	4	1	3	2	2	3	1	4	1	4	2	3	3	2	4	1
61	4	4	1	3	2	1	4	2	3	4	1	3	2	2	3	1	4	3	2	4	1
62	4	4	1	3	2	2	3	1	4	3	2	4	1	1	4	2	3	4	1	3	2
63	4	4	1	3	2	3	2	4	1	2	3	1	4	4	1	3	2	1	4	2	3
64	4	4	1	3	2	4	1	3	2	1	4	2	3	3	2	4	1	2	3	1	4
Group	1	2				3															

$OA_{64}(4^{21})$ (*continued*)

Interactions between two columns

(Col.)\Col:	1	2	3	4	5	6	7	8	9	10	11	12	13	14	15	16	17	18	19	20	21
		3	2	2	2	7	6	6	6	11	10	10	10	15	14	14	14	19	18	18	18
(1)		4	4	3	3	8	8	7	7	12	12	11	11	16	16	15	15	20	20	19	19
		5	5	5	4	9	9	9	8	13	13	13	12	17	17	17	16	21	21	21	20
			1	1	1	10	11	12	13	6	7	8	9	6	7	8	9	6	7	8	9
(2)			4	3	3	14	15	16	17	14	15	16	17	10	11	12	13	10	11	12	13
			5	5	4	18	19	20	21	18	19	20	21	18	19	20	21	14	15	16	17
				1	1	11	10	13	12	7	6	9	8	8	9	6	7	9	8	7	6
(3)				2	2	16	17	14	15	17	16	15	14	13	12	11	10	12	13	10	11
				5	4	21	20	19	18	20	21	18	19	19	18	21	20	15	14	17	16
					1	12	13	10	11	8	9	6	7	9	8	7	6	7	6	9	8
(4)					2	17	16	15	14	15	14	17	16	11	10	13	12	13	12	11	10
					3	19	18	21	20	21	20	19	18	20	21	18	19	16	17	14	15
						13	12	11	10	9	8	7	6	7	6	9	8	8	9	6	7
(5)						15	14	17	16	16	17	14	15	12	13	10	11	11	10	13	12
						20	21	18	19	19	18	21	20	21	20	19	18	17	16	15	14
							1	1	1	2	3	4	5	2	5	3	4	2	4	5	3
(6)							8	7	7	14	16	17	15	10	13	11	12	10	12	13	11
							9	9	8	18	21	19	20	18	20	21	19	14	17	15	16
								1	1	3	2	5	4	5	2	4	3	4	2	3	5
(7)								6	6	17	15	14	16	12	11	13	10	13	11	10	12
								9	8	20	19	21	18	21	19	18	20	16	15	17	14
									1	4	5	2	3	3	4	2	5	5	3	2	4
(8)									6	15	17	16	14	13	10	12	11	11	13	12	10
									7	21	18	20	19	19	21	20	18	17	14	16	15
										5	4	3	2	4	3	5	2	3	5	4	2
(9)										16	14	15	17	11	12	10	13	12	10	11	13
										19	20	18	21	20	18	19	21	15	16	14	17
											1	1	1	2	4	5	3	2	5	3	4
(10)											12	11	11	6	8	9	7	6	9	7	8
											13	13	12	18	21	19	20	14	16	17	15
												1	1	4	2	3	5	5	2	4	3
(11)												10	10	9	7	6	8	8	7	9	6
												13	12	20	19	21	18	17	15	14	16
													1	5	3	2	4	3	4	2	5
(12)													10	7	9	8	6	9	6	8	7
													11	21	18	20	19	15	17	16	14
														3	5	4	2	4	3	5	2
(13)														8	6	7	9	7	8	6	9
														19	20	18	21	16	14	15	17
															1	1	1	2	3	4	5
(14)															16	15	15	6	8	9	7
															17	17	16	10	13	11	12
																1	1	3	2	5	4
(15)																14	14	9	7	6	8
																17	16	12	11	13	10
																	1	4	5	2	3
(16)																	14	7	9	8	6
																	15	13	10	12	11
																		5	4	3	2
(17)																		8	6	7	9
																		11	12	10	13
																			1	1	1
(18)																			20	19	19
																			21	21	20
																				1	1
(19)																				18	18
																				21	20
																					1
(20)																					18
																					19

$OA_{32}(2^1 \times 4^9)$

Tr.	Col: 1	2	3	4	5	6	7	8	9	10
1	1	1	1	1	1	1	1	1	1	1
2	1	1	2	2	2	2	2	2	2	2
3	1	1	3	3	3	3	3	3	3	3
4	1	1	4	4	4	4	4	4	4	4
5	1	2	1	1	2	2	3	3	4	4
6	1	2	2	2	1	1	4	4	3	3
7	1	2	3	3	4	4	1	1	2	2
8	1	2	4	4	3	3	2	2	1	1
9	1	3	1	2	3	4	1	2	3	4
10	1	3	2	1	4	3	2	1	4	3
11	1	3	3	4	1	2	3	4	1	2
12	1	3	4	3	2	1	4	3	2	1
13	1	4	1	2	4	3	3	4	2	1
14	1	4	2	1	3	4	4	3	1	2
15	1	4	3	4	2	1	1	2	4	3
16	1	4	4	3	1	2	2	1	3	4
17	2	1	1	4	1	4	2	3	2	3
18	2	1	2	3	2	3	1	4	1	4
19	2	1	3	2	3	2	4	1	4	1
20	2	1	4	1	4	1	3	2	3	2
21	2	2	1	4	2	3	4	1	3	2
22	2	2	2	3	1	4	3	2	4	1
23	2	2	3	2	4	1	2	3	1	4
24	2	2	4	1	3	2	1	4	2	3
25	2	3	1	3	3	1	2	4	4	2
26	2	3	2	4	4	2	1	3	3	1
27	2	3	3	1	1	3	4	2	2	4
28	2	3	4	2	2	4	3	1	1	3
29	2	4	1	3	4	2	4	2	1	3
30	2	4	2	4	3	1	3	1	2	4
31	2	4	3	1	2	4	2	4	3	1
32	2	4	4	2	1	3	1	3	4	2
Group	1	2	3							

Note: The only interaction information that can be obtained from $OA_{32}(2^1 \times 4^9)$ is that between the first two columns.

$OA_{25}(5^6)$

Tr.	Col: 1	2	3	4	5	6
1	1	1	1	1	1	1
2	1	2	2	2	2	2
3	1	3	3	3	3	3
4	1	4	4	4	4	4
5	1	5	5	5	5	5
6	2	1	2	3	4	5
7	2	2	3	4	5	1
8	2	3	4	5	1	2
9	2	4	5	1	2	3
10	2	5	1	2	3	4
11	3	1	3	5	2	4
12	3	2	4	1	3	5
13	3	3	5	2	4	1
14	3	4	1	3	5	2
15	3	5	2	4	1	3
16	4	1	4	2	5	3
17	4	2	5	3	1	4
18	4	3	1	4	2	5
19	4	4	2	5	3	1
20	4	5	3	1	4	2
21	5	1	5	4	3	2
22	5	2	1	5	4	3
23	5	3	2	1	5	4
24	5	4	3	2	1	5
25	5	5	4	3	2	1
	a	b	ab	ab^2	ab^3	ab^4
Group	1	2				

$OA_{50}(2^1 \times 5^{11})$

Tr.	Col: 1	2	3	4	5	6	7	8	9	10	11	12
1	1	1	1	1	1	1	1	1	1	1	1	1
2	1	1	2	2	2	2	2	2	2	2	2	2
3	1	1	3	3	3	3	3	3	3	3	3	3
4	1	1	4	4	4	4	4	4	4	4	4	4
5	1	1	5	5	5	5	5	5	5	5	5	5
6	1	2	1	2	3	4	5	1	2	3	4	5
7	1	2	2	3	4	5	1	2	3	4	5	1
8	1	2	3	4	5	1	2	3	4	5	1	2
9	1	2	4	5	1	2	3	4	5	1	2	3
10	1	2	5	1	2	3	4	5	1	2	3	4
11	1	3	1	3	5	2	4	4	1	3	5	2
12	1	3	2	4	1	3	5	5	2	4	1	3
13	1	3	3	5	2	4	1	1	3	5	2	4
14	1	3	4	1	3	5	2	2	4	1	3	5
15	1	3	5	2	4	1	3	3	5	2	4	1
16	1	4	1	4	2	5	3	5	3	1	4	2
17	1	4	2	5	3	1	4	1	4	2	5	3
18	1	4	3	1	4	2	5	2	5	3	1	4
19	1	4	4	2	5	3	1	3	1	4	2	5
20	1	4	5	3	1	4	2	4	2	5	3	1
21	1	5	1	5	4	3	2	4	3	2	1	5
22	1	5	2	1	5	4	3	5	4	3	2	1
23	1	5	3	2	1	5	4	1	5	4	3	2
24	1	5	4	3	2	1	5	2	1	5	4	3
25	1	5	5	4	3	2	1	3	2	1	5	4
26	2	1	1	1	4	5	4	3	2	5	2	3
27	2	1	2	2	5	1	5	4	3	1	3	4
28	2	1	3	3	1	2	1	5	4	2	4	5
29	2	1	4	4	2	3	2	1	5	3	5	1
30	2	1	5	5	3	4	3	2	1	4	1	2
31	2	2	1	2	1	3	3	2	4	5	5	4
32	2	2	2	3	2	4	4	3	5	1	1	5
33	2	2	3	4	3	5	5	4	1	2	2	1
34	2	2	4	5	4	1	1	5	2	3	3	2
35	2	2	5	1	5	2	2	1	3	4	4	3
36	2	3	1	3	3	1	2	5	5	4	2	4
37	2	3	2	4	4	2	3	1	1	5	3	5
38	2	3	3	5	5	3	4	2	2	1	4	1
39	2	3	4	1	1	4	5	3	3	2	5	2
40	2	3	5	2	2	5	1	4	4	3	1	3

$OA_{50}(2^1 \times 5^{11})$ (*continued*)

Tr.	Col: 1	2	3	4	5	6	7	8	9	10	11	12
41	2	4	1	4	5	4	1	2	5	2	3	3
42	2	4	2	5	1	5	2	3	1	3	4	4
43	2	4	3	1	2	1	3	4	2	4	5	5
44	2	4	4	2	3	2	4	5	3	5	1	1
45	2	4	5	3	4	3	5	1	4	1	2	2
46	2	5	1	5	2	2	5	3	4	4	3	1
47	2	5	2	1	3	3	1	4	5	5	4	2
48	2	5	3	2	4	4	2	5	1	1	5	3
49	2	5	4	3	5	5	3	1	2	2	1	4
50	2	5	5	4	1	1	4	2	3	3	2	5
Group	1	2	3 (columns 3–12)									

Note: The only interaction information that can be obtained from $OA_{50}(2^1 \times 5^{11})$ is that between the first two columns.

$OA_{36}(2^{11} \times 3^{12})$ See note below for $OA_{36}(2^3 \times 3^{13})$

Tr.	Col: 1	2	3	4	5	6	7	8	9	10	11	12	13	14	15	16	17	18	19	20	21	22	23	1′	2′	3′	4′
1	1	1	1	1	1	1	1	1	1	1	1	1	1	1	1	1	1	1	1	1	1	1	1	1	1	1	1
2	1	1	1	1	1	1	1	1	1	1	1	2	2	2	2	2	2	2	2	2	2	2	2	1	1	1	1
3	1	1	1	1	1	1	1	1	1	1	1	3	3	3	3	3	3	3	3	3	3	3	3	1	1	1	1
4	1	1	1	1	1	2	2	2	2	2	2	1	1	1	1	2	2	2	2	3	3	3	3	1	2	2	1
5	1	1	1	1	1	2	2	2	2	2	2	2	2	2	2	3	3	3	3	1	1	1	1	1	2	2	1
6	1	1	1	1	1	2	2	2	2	2	2	3	3	3	3	1	1	1	1	2	2	2	2	1	2	2	1
7	1	1	2	2	2	1	1	1	2	2	2	1	1	2	3	1	2	3	3	1	2	2	3	2	1	2	1
8	1	1	2	2	2	1	1	1	2	2	2	2	2	3	1	2	3	1	1	2	3	3	1	2	1	2	1
9	1	1	2	2	2	1	1	1	2	2	2	3	3	1	2	3	1	2	2	3	1	1	2	2	1	2	1
10	1	2	1	2	2	1	2	2	1	1	2	1	1	3	2	1	3	2	3	2	1	3	2	2	2	1	1
11	1	2	1	2	2	1	2	2	1	1	2	2	2	1	3	2	1	3	1	3	2	1	3	2	2	1	1
12	1	2	1	2	2	1	2	2	1	1	2	3	3	2	1	3	2	1	2	1	3	2	1	2	2	1	1
13	1	2	2	1	2	2	1	2	1	2	1	1	2	3	1	3	2	1	3	3	2	1	2	1	1	1	2
14	1	2	2	1	2	2	1	2	1	2	1	2	3	1	2	1	3	2	1	1	3	2	3	1	1	1	2
15	1	2	2	1	2	2	1	2	1	2	1	3	1	2	3	2	1	3	2	2	1	3	1	1	1	1	2
16	1	2	2	2	1	2	2	1	2	1	1	1	2	3	2	1	1	3	2	3	3	2	1	1	2	2	2
17	1	2	2	2	1	2	2	1	2	1	1	2	3	1	3	2	2	1	3	1	1	3	2	1	2	2	2
18	1	2	2	2	1	2	2	1	2	1	1	3	1	2	1	3	3	2	1	2	2	1	3	1	2	2	2
19	2	1	2	2	1	1	2	2	1	2	1	1	2	1	3	3	3	1	2	2	1	2	3	2	1	2	2
20	2	1	2	2	1	1	2	2	1	2	1	2	3	2	1	1	1	2	3	3	2	3	1	2	1	2	2
21	2	1	2	2	1	1	2	2	1	2	1	3	1	3	2	2	2	3	1	1	3	1	2	2	1	2	2
22	2	1	2	1	2	2	2	1	1	1	2	1	2	2	3	3	1	2	1	1	3	3	2	2	2	1	2
23	2	1	2	1	2	2	2	1	1	1	2	2	3	3	1	1	2	3	2	2	1	1	3	2	2	1	2
24	2	1	2	1	2	2	2	1	1	1	2	3	1	1	2	2	3	1	3	3	2	2	1	2	2	1	2
25	2	1	1	2	2	2	1	2	2	1	1	1	3	2	1	2	3	3	1	3	1	2	2	1	1	1	3
26	2	1	1	2	2	2	1	2	2	1	1	2	1	3	2	3	1	1	2	1	2	3	3	1	1	1	3
27	2	1	1	2	2	2	1	2	2	1	1	3	2	1	3	1	2	2	3	2	3	1	1	1	1	1	3
28	2	2	2	1	1	1	1	2	2	1	2	1	3	2	2	2	1	1	3	2	3	1	3	1	2	2	3
29	2	2	2	1	1	1	1	2	2	1	2	2	1	3	3	3	2	2	1	3	1	2	1	1	2	2	3
30	2	2	2	1	1	1	1	2	2	1	2	3	2	1	1	1	3	3	2	1	2	3	2	1	2	2	3
31	2	2	1	2	1	2	1	1	1	2	2	1	3	3	3	2	3	2	2	1	2	1	1	2	1	2	3
32	2	2	1	2	1	2	1	1	1	2	2	2	1	1	1	3	1	3	3	2	3	2	2	2	1	2	3
33	2	2	1	2	1	2	1	1	1	2	2	3	2	2	2	1	2	1	1	3	1	3	3	2	1	2	3
34	2	2	1	1	2	1	2	1	2	2	1	1	3	1	2	3	2	3	1	2	2	3	1	2	2	1	3
35	2	2	1	1	2	1	2	1	2	2	1	2	1	2	3	1	3	1	2	3	3	1	2	2	2	1	3
36	2	2	1	1	2	1	2	1	2	2	1	3	2	3	1	2	1	2	3	1	1	2	3	2	2	1	3
Group	1					2											3										

Note: Replacing columns 1–11 with columns 1′–4′ gives $OA_{36}(2^3 \times 3^{13})$.

APPENDIX E

Statistical tables

TABLE T1. Critical values for the t-test

Degrees of freedom \ Significance level	Two-sided 10% (0.10)	5% (0.05)	1% (0.01)	One-sided 10% (0.10)	5% (0.05)	1% (0.01)
1	6.31	12.71	63.66	3.08	6.31	31.82
2	2.92	4.30	9.92	1.89	2.92	6.97
3	2.35	3.18	5.84	1.64	2.35	4.54
4	2.13	2.78	4.60	1.53	2.13	3.75
5	2.02	2.57	4.03	1.48	2.02	3.36
6	1.94	2.45	3.71	1.44	1.94	3.14
7	1.89	2.36	3.50	1.42	1.89	3.00
8	1.86	2.31	3.36	1.40	1.86	2.90
9	1.83	2.26	3.25	1.38	1.83	2.82
10	1.81	2.23	3.17	1.37	1.81	2.76
11	1.80	2.20	3.11	1.36	1.80	2.72
12	1.78	2.18	3.06	1.36	1.78	2.68
13	1.77	2.16	3.01	1.35	1.77	2.65
14	1.76	2.15	2.98	1.35	1.76	2.62
15	1.75	2.13	2.95	1.34	1.75	2.60
16	1.75	2.12	2.92	1.34	1.75	2.58
17	1.74	2.11	2.90	1.33	1.74	2.57
18	1.73	2.10	2.88	1.33	1.73	2.55
19	1.73	2.09	2.86	1.33	1.73	2.54
20	1.72	2.08	2.85	1.32	1.72	2.53
25	1.71	2.06	2.78	1.32	1.71	2.49
30	1.70	2.04	2.75	1.31	1.70	2.46
40	1.68	2.02	2.70	1.30	1.68	2.42
60	1.67	2.00	2.66	1.30	1.67	2.39
120	1.66	1.98	2.62	1.29	1.66	2.36
Infinite	1.64	1.96	2.58	1.28	1.64	2.33

TABLE T2.1. Critical values for the F-test

Two-sided at 5% significance level															
Degrees of freedom for denom.	Degrees of freedom for numerator														
	1	2	3	4	5	6	7	8	9	10	12	15	20	60	Infinity
1	647.8	799.5	864.2	899.6	921.8	937.1	948.2	956.7	963.3	968.6	976.7	984.9	993.1	1010	1018
2	38.51	39.00	39.17	39.25	39.30	39.33	39.36	39.37	39.39	39.40	39.41	39.43	39.45	39.48	39.50
3	17.44	16.04	15.44	15.10	14.88	14.73	14.62	14.54	14.47	14.42	14.34	14.25	14.17	13.99	13.90
4	12.22	10.65	9.98	9.60	9.36	9.20	9.07	8.98	8.90	8.84	8.75	8.66	8.56	8.36	8.26
5	10.01	8.43	7.76	7.39	7.15	6.98	6.85	6.76	6.68	6.62	6.52	6.43	6.33	6.12	6.02
6	8.81	7.26	6.60	6.23	5.99	5.82	5.70	5.60	5.52	5.46	5.37	5.27	5.17	4.96	4.85
7	8.07	6.54	5.89	5.52	5.29	5.12	4.99	4.90	4.82	4.76	4.67	4.57	4.47	4.25	4.14
8	7.57	6.06	5.42	5.05	4.82	4.65	4.53	4.43	4.36	4.30	4.20	4.10	4.00	3.78	3.67
9	7.21	5.71	5.08	4.72	4.48	4.32	4.20	4.10	4.03	3.96	3.87	3.77	3.67	3.45	3.33
10	6.94	5.46	4.83	4.47	4.24	4.07	3.95	3.85	3.78	3.72	3.62	3.52	3.42	3.20	3.08
12	6.55	5.10	4.47	4.12	3.89	3.73	3.61	3.51	3.44	3.37	3.28	3.18	3.07	2.85	2.72
15	6.20	4.77	4.15	3.80	3.58	3.41	3.29	3.20	3.12	3.06	2.96	2.86	2.76	2.52	2.40
20	5.87	4.46	3.86	3.51	3.29	3.13	3.01	2.91	2.84	2.77	2.68	2.57	2.46	2.22	2.09
60	5.29	3.93	3.34	3.01	2.79	2.63	2.51	2.41	2.33	2.27	2.17	2.06	1.94	1.67	1.48
Infinity	5.02	3.69	3.12	2.79	2.57	2.41	2.29	2.19	2.11	2.05	1.94	1.83	1.71	1.39	1.00

TABLE T2.1 (*continued*)
Two-sided at 1% significance level

Degrees of freedom for denom.	Degrees of freedom for numerator														
	1	2	3	4	5	6	7	8	9	10	12	15	20	60	Infinity
1	16211	20000	21615	22500	23056	23437	23715	23925	24091	24224	24426	24630	24836	25253	25465
2	198.5	199.0	199.2	199.2	199.3	199.3	199.4	199.4	199.4	199.4	199.4	199.4	199.4	199.5	199.5
3	55.55	49.80	47.47	46.19	45.39	44.84	44.43	44.13	43.88	43.69	43.29	43.08	42.78	42.15	41.83
4	31.33	26.28	24.26	23.15	22.46	21.97	21.62	21.35	21.14	20.97	20.70	20.04	20.17	19.61	19.32
5	22.78	18.31	16.53	15.56	14.94	14.51	14.20	13.96	13.77	13.62	13.38	13.15	12.90	12.40	12.14
6	18.63	14.54	12.92	12.03	11.46	11.07	10.79	10.57	10.39	10.25	10.03	9.81	9.59	9.12	8.88
7	16.24	12.40	10.88	10.05	9.52	9.16	8.89	8.68	8.51	8.38	8.18	7.97	7.75	7.31	7.08
8	14.69	11.04	9.60	8.81	8.30	7.95	7.69	7.50	7.34	7.21	7.01	6.81	6.61	6.18	5.95
9	13.61	10.11	8.72	7.96	7.47	7.13	6.88	6.69	6.54	6.42	6.23	6.03	5.83	5.41	5.19
10	12.83	9.43	8.08	7.34	6.87	6.54	6.30	6.12	5.97	5.85	5.66	5.47	5.27	4.86	4.64
12	11.75	8.51	7.23	6.52	6.07	5.76	5.52	5.35	5.20	5.09	4.91	4.72	4.53	4.12	3.90
15	10.80	7.70	6.48	5.80	5.37	5.07	4.85	4.67	4.54	4.42	4.25	4.07	3.88	3.48	3.26
20	9.94	6.99	5.82	5.17	4.76	4.47	4.26	4.09	3.96	3.85	3.68	3.50	3.32	2.92	2.69
60	8.49	5.79	4.73	4.14	3.76	3.49	3.29	3.13	3.01	2.90	2.74	2.57	2.39	1.96	1.69
Infinity	7.88	5.30	4.28	3.72	3.35	3.09	2.90	2.74	2.62	2.52	2.36	2.19	2.00	1.53	1.00

TABLE T2.2 Critical values for the F-test (one-sided).

df for denom.	α	df for numerator 1	2	3	4	5	6	7	8	9	10	11	12
1	0.10	39.9	49.5	53.6	55.8	57.2	58.2	58.9	59.4	59.9	60.2	60.5	60.7
	0.05	161	200	216	225	230	234	237	239	241	242	243	244
2	0.10	8.53	9.00	9.16	9.24	9.29	9.33	9.35	9.37	9.38	9.39	9.40	9.41
	0.05	18.5	19.0	19.2	19.2	19.3	19.3	19.4	19.4	19.4	19.4	19.4	19.4
	0.01	98.5	99.0	99.2	99.2	99.3	99.3	99.4	99.4	99.4	99.4	99.4	99.4
3	0.10	5.54	5.46	5.39	5.34	5.31	5.28	5.27	5.25	5.24	5.23	5.22	5.22
	0.05	10.1	9.55	9.28	9.12	9.10	8.94	8.89	8.85	8.81	8.79	8.76	8.74
	0.01	34.1	30.8	29.5	28.7	28.2	27.9	27.7	27.5	27.3	27.2	27.1	27.1
4	0.10	4.54	4.32	4.19	4.11	4.05	4.01	3.98	3.95	3.94	3.92	3.91	3.90
	0.05	7.71	6.94	6.59	6.39	6.26	6.16	6.09	6.04	6.00	5.96	5.94	5.91
	0.01	21.2	18.0	16.7	16.0	15.5	15.2	15.0	14.8	14.7	14.5	14.4	14.4
5	0.10	4.06	3.78	3.62	3.52	3.45	3.40	3.37	3.34	3.32	3.30	3.28	3.27
	0.05	6.61	5.79	5.41	5.19	5.05	4.95	4.88	4.82	4.77	4.74	4.71	4.68
	0.01	16.3	13.3	12.1	11.4	11.0	10.7	10.5	10.3	10.2	10.1	9.96	9.89
6	0.10	3.78	3.46	3.29	3.18	3.11	3.05	3.01	2.98	2.96	2.94	2.92	2.90
	0.05	5.99	5.14	4.76	4.53	4.39	4.28	4.21	4.15	4.10	4.06	4.03	4.00
	0.01	13.7	10.9	9.78	9.15	8.75	8.47	8.26	8.10	7.98	7.87	7.79	7.72
7	0.10	3.59	3.26	3.07	2.96	2.88	2.83	2.78	2.75	2.72	2.70	2.68	2.67
	0.05	5.59	4.74	4.35	4.12	3.97	3.87	3.79	3.73	3.68	3.64	3.60	3.57
	0.01	12.2	9.55	8.45	7.85	7.46	7.19	6.99	6.84	6.72	6.62	6.54	6.47
8	0.10	3.46	3.11	2.92	2.81	2.73	2.67	2.62	2.59	2.56	2.54	2.52	2.50
	0.05	5.32	4.46	4.07	3.84	3.69	3.58	3.50	3.44	3.39	3.35	3.31	3.28
	0.01	11.3	8.65	7.59	7.01	6.63	6.37	6.18	6.03	5.91	5.81	5.73	5.67
9	0.10	3.36	3.01	2.81	2.69	2.61	2.55	2.51	2.47	2.44	2.42	2.40	2.38
	0.05	5.12	4.26	3.86	3.63	3.48	3.37	3.29	3.23	3.18	3.14	3.10	3.07
	0.01	10.6	8.02	6.99	6.42	6.06	5.80	5.61	5.47	5.35	5.26	5.18	5.11
10	0.10	3.28	2.92	2.73	2.61	2.52	2.46	2.41	2.38	2.35	2.32	2.30	2.28
	0.05	4.96	4.10	3.71	3.48	3.33	3.22	3.14	3.07	3.02	2.98	2.94	2.91
	0.01	10.0	7.56	6.55	5.99	5.64	5.39	5.20	5.06	4.94	4.85	4.77	4.71
11	0.10	3.23	2.86	2.66	2.54	2.45	2.39	2.34	2.30	2.27	2.25	2.23	2.21
	0.05	4.84	3.98	3.59	3.36	3.20	3.09	3.01	2.95	2.90	2.85	2.82	2.79
	0.01	9.65	7.21	6.22	5.67	5.32	5.07	4.89	4.74	4.63	4.54	4.46	4.40
12	0.10	3.18	2.81	2.61	2.48	2.39	2.33	2.28	2.24	2.21	2.19	2.17	2.15
	0.05	4.75	3.89	3.49	3.26	3.11	3.00	2.91	2.85	2.80	2.75	2.72	2.69
	0.01	9.33	6.93	5.95	5.41	5.06	4.82	4.64	4.50	4.39	4.30	4.22	4.16

TABLE T2.2 (*continued*)

df for numerator													
15	20	24	30	40	50	60	100	120	200	500	∞	α	df for denom
61.2	61.7	62.0	62.3	62.5	62.7	62.8	63.0	63.1	63.2	63.3	63.3	0.10	1
246	248	249	250	251	252	252	253	253	254	254	254	0.05	
9.42	9.44	9.45	9.46	9.47	9.47	9.47	9.48	9.48	9.49	9.49	9.49	0.10	2
19.4	19.4	19.5	19.5	19.5	19.5	19.5	19.5	19.5	19.5	19.5	19.5	0.05	
99.4	99.4	99.5	99.5	99.5	99.5	99.5	99.5	99.5	99.5	99.5	99.5	0.01	
5.20	5.18	5.18	5.17	5.16	5.15	5.15	5.14	5.14	5.14	5.14	5.13	0.10	3
8.70	8.66	8.64	8.62	8.59	8.58	8.57	8.55	8.55	8.54	8.53	8.53	0.05	
26.9	26.7	26.6	26.5	26.4	26.4	26.3	26.2	26.2	26.2	26.1	26.1	0.01	
3.87	3.84	3.83	3.82	3.80	3.80	3.79	3.78	3.78	3.77	3.76	3.76	0.10	4
5.86	5.80	5.77	5.75	5.72	5.70	5.69	5.66	5.66	5.65	5.64	5.63	0.05	
14.2	14.0	13.9	13.8	13.7	13.7	13.7	13.6	13.6	13.5	13.5	13.5	0.01	
3.24	3.21	3.19	3.17	3.16	3.15	3.14	3.13	3.12	3.12	3.11	3.10	0.10	5
4.62	4.56	4.53	4.50	4.46	4.44	4.43	4.41	4.40	4.39	4.37	4.36	0.05	
9.72	9.55	9.47	9.38	9.29	9.24	9.20	9.13	9.11	9.08	9.04	9.02	0.01	
2.87	2.84	2.82	2.80	2.78	2.77	2.76	2.75	2.74	2.73	2.73	2.72	0.10	6
3.94	3.87	3.84	3.81	3.77	3.75	3.74	3.71	3.70	3.69	3.68	3.67	0.05	
7.56	7.40	7.31	7.23	7.14	7.09	7.06	6.99	6.97	6.93	6.90	6.88	0.01	
2.63	2.59	2.58	2.56	2.54	2.52	2.51	2.50	2.49	2.48	2.48	2.47	0.10	7
3.51	3.44	3.41	3.38	3.34	3.32	3.30	3.27	3.27	3.25	3.24	3.23	0.05	
6.31	6.16	6.07	5.99	5.91	5.86	5.82	5.75	5.74	5.70	5.67	5.65	0.01	
2.46	2.42	2.40	2.38	2.36	2.35	2.34	2.32	2.32	2.31	2.30	2.29	0.10	8
3.22	3.15	3.12	3.08	3.04	3.02	3.01	2.97	2.97	2.95	2.94	2.93	0.05	
5.52	5.36	5.28	5.20	5.12	5.07	5.03	4.96	4.95	4.91	4.88	4.86	0.01	
2.34	2.30	2.28	2.25	2.23	2.22	2.21	2.19	2.18	2.17	2.17	2.16	0.10	9
3.01	2.94	2.90	2.86	2.83	2.80	2.79	2.76	2.75	2.73	2.72	2.71	0.05	
4.96	4.81	4.73	4.65	4.57	4.52	4.48	4.42	4.40	4.36	4.33	4.31	0.01	
2.24	2.20	2.18	2.16	2.13	2.12	2.11	2.09	2.08	2.07	2.06	2.06	0.10	10
2.85	2.77	2.74	2.70	2.66	2.64	2.62	2.59	2.58	2.56	2.55	2.54	0.05	
4.56	4.41	4.33	4.25	4.17	4.12	4.08	4.01	4.00	3.96	3.93	3.91	0.01	
2.17	2.12	2.10	2.08	2.05	2.04	2.03	2.00	2.00	1.99	1.98	1.97	0.10	11
2.72	2.65	2.61	2.57	2.53	2.51	2.49	2.46	2.45	2.43	2.42	2.40	0.05	
4.25	4.10	4.02	3.94	3.86	3.81	3.78	3.71	3.69	3.66	3.62	3.60	0.01	
2.10	2.06	2.04	2.01	1.99	1.97	1.96	1.94	1.93	1.92	1.91	1.90	0.10	12
2.62	2.54	2.51	2.47	2.43	2.40	2.38	2.35	2.34	2.32	2.31	2.10	0.05	
4.01	3.86	3.78	3.70	3.62	3.57	3.54	3.47	3.45	3.41	3.38	3.36	0.01	

TABLE T2.2 (*continued*)

df for denom.	α	df for numerator											
		1	2	3	4	5	6	7	8	9	10	11	12
13	0.10	3.14	2.76	2.56	2.43	2.35	2.28	2.23	2.20	2.16	2.14	2.12	2.10
	0.05	4.67	3.81	3.41	3.18	3.03	2.92	2.83	2.77	2.71	2.67	2.63	2.60
	0.01	9.07	6.70	5.74	5.21	4.86	4.62	4.44	4.30	4.19	4.10	4.02	3.96
14	0.10	3.10	2.73	2.52	2.39	2.31	2.24	2.19	2.15	2.12	2.10	2.08	2.05
	0.05	4.60	3.74	3.34	3.11	2.96	2.85	2.76	2.70	2.65	2.60	2.57	2.53
	0.01	8.86	6.51	5.56	5.04	4.69	4.46	4.28	4.14	4.03	3.94	3.86	3.80
15	0.10	3.07	2.70	2.49	2.36	2.27	2.21	2.16	2.12	2.09	2.06	2.04	2.02
	0.05	4.54	3.68	3.29	3.06	2.90	2.79	2.71	2.64	2.59	2.54	2.51	2.48
	0.01	8.68	6.36	5.42	4.89	4.56	4.32	4.14	4.00	3.89	3.80	3.73	3.67
16	0.10	3.05	2.67	2.46	2.33	2.24	2.18	2.13	2.09	2.06	2.03	2.01	1.99
	0.05	4.49	3.63	3.24	3.01	2.85	2.74	2.66	2.59	2.54	2.49	2.46	2.42
	0.01	8.53	6.23	5.29	4.77	4.44	4.20	4.03	3.89	3.78	3.69	3.62	3.55
18	0.10	3.01	2.62	2.42	2.29	2.20	2.13	2.08	2.04	2.00	1.98	1.96	1.93
	0.05	4.41	3.55	3.16	2.93	2.77	2.66	2.58	2.51	2.46	2.41	2.37	2.34
	0.01	8.29	1.49	5.09	4.58	4.25	4.01	3.84	3.71	3.60	3.51	3.43	3.37
20	0.10	2.97	2.59	2.38	2.25	2.16	2.09	2.04	2.00	1.96	1.94	1.92	1.89
	0.05	4.35	3.49	3.10	2.87	2.71	2.60	2.51	2.45	2.39	2.35	2.31	2.28
	0.01	8.10	5.85	4.94	4.43	4.10	3.87	3.70	3.56	3.46	3.37	3.29	3.23
30	0.10	2.88	2.49	2.28	2.14	2.05	1.98	1.93	1.88	1.85	1.82	1.79	1.77
	0.05	4.17	3.32	2.92	2.69	2.53	2.42	2.33	2.27	2.21	2.16	2.13	2.09
	0.01	7.56	5.39	4.51	4.02	3.70	3.47	3.30	3.17	3.07	2.98	2.91	2.84
40	0.10	2.84	2.44	2.23	2.09	2.00	1.93	1.87	1.83	1.79	1.76	1.73	1.71
	0.05	4.08	3.23	2.84	2.61	2.45	2.34	2.25	2.18	2.12	2.08	2.04	2.00
	0.01	7.31	5.18	4.31	3.83	3.51	3.29	3.12	2.99	2.89	2.80	2.73	2.66
60	0.10	2.79	2.39	2.18	2.04	1.95	1.87	1.82	1.77	1.74	1.71	1.68	1.66
	0.05	4.00	3.15	2.76	2.53	2.37	2.25	2.17	2.10	2.04	1.99	1.95	1.92
	0.01	7.08	4.98	4.13	3.65	3.34	3.12	2.95	2.82	2.72	2.63	2.56	2.50
120	0.10	2.75	2.35	2.13	1.99	1.90	1.82	1.77	1.72	1.68	1.65	1.62	1.60
	0.05	3.92	3.07	2.68	2.45	2.29	2.17	2.09	2.02	1.96	1.91	1.87	1.83
	0.01	6.85	4.79	3.95	3.48	3.17	2.96	2.79	2.66	2.56	2.47	2.40	2.34
200	0.10	2.73	2.33	2.11	1.97	1.88	1.80	1.75	1.70	1.66	1.63	1.60	1.57
	0.05	3.89	3.04	2.65	2.42	2.26	2.14	2.06	1.98	1.93	1.88	1.84	1.80
	0.01	6.76	4.71	3.88	3.41	3.11	2.89	2.73	2.60	2.50	2.41	2.34	2.27
∞	0.10	2.71	2.30	2.08	1.94	1.85	1.77	1.72	1.67	1.63	1.60	1.57	1.55
	0.05	3.84	3.00	2.60	2.37	2.21	2.10	2.01	1.94	1.88	1.83	1.79	1.75
	0.01	6.63	4.61	3.78	3.32	3.02	2.80	2.64	2.51	2.41	2.32	2.25	2.18

TABLE T2.2 (*continued*)

df for numerator													
15	20	24	30	40	50	60	100	120	200	500	∞	α	df for denom.
2.05	2.01	1.98	1.96	1.93	1.92	1.90	1.88	1.88	1.86	1.85	1.85	0.10	13
2.53	2.46	2.42	2.38	2.34	2.31	2.30	2.26	2.25	2.23	2.22	2.21	0.05	
3.82	3.66	3.59	3.51	3.43	3.38	3.34	3.27	3.25	3.22	3.19	3.17	0.01	
2.01	1.96	1.94	1.91	1.89	1.87	1.86	1.83	1.83	1.82	1.80	1.80	0.10	14
2.46	2.39	2.35	2.31	2.27	2.24	2.22	2.19	2.18	2.16	2.14	2.13	0.05	
3.66	3.51	3.43	3.35	3.27	3.22	3.18	3.11	3.09	3.06	3.03	3.00	0.01	
1.97	1.92	1.90	1.87	1.85	1.83	1.82	1.79	1.79	1.77	1.76	1.76	0.10	15
2.40	2.33	2.29	2.25	2.20	2.18	2.16	2.12	2.11	2.10	2.08	2.07	0.05	
3.52	3.37	3.29	3.21	3.13	3.08	3.05	2.98	2.96	2.92	2.89	2.87	0.01	
1.94	1.89	1.87	1.84	1.81	1.79	1.78	1.76	1.75	1.74	1.73	1.72	0.10	16
2.35	2.28	2.24	2.19	2.15	2.12	2.11	2.07	2.06	2.04	2.02	2.01	0.05	
3.41	3.26	3.18	3.10	3.02	2.97	2.93	2.86	2.84	2.81	2.78	2.75	0.01	
1.89	1.84	1.81	1.78	1.75	1.74	1.72	1.70	1.69	1.68	1.67	1.66	0.10	18
2.27	2.19	2.15	2.11	2.06	2.04	2.02	1.98	1.97	1.95	1.93	1.92	0.05	
3.23	3.08	3.00	2.92	2.84	2.78	2.75	2.68	2.66	2.62	2.59	2.57	0.01	
1.84	1.79	1.77	1.74	1.71	1.69	1.68	1.65	1.64	1.63	1.62	1.61	0.10	20
2.20	2.12	2.08	2.04	1.99	1.97	1.95	1.91	1.90	1.88	1.86	1.84	0.05	
3.09	2.94	2.86	2.78	2.69	2.64	2.61	2.54	2.52	2.48	2.44	2.42	0.01	
1.72	1.67	1.64	1.61	1.57	1.55	1.54	1.51	1.50	1.48	1.47	1.46	0.10	30
2.01	1.93	1.89	1.84	1.79	1.76	1.74	1.70	1.68	1.66	1.64	1.62	0.05	
2.70	2.55	2.47	2.39	2.30	2.25	2.21	2.13	2.11	2.07	2.03	2.01	0.01	
1.66	1.61	1.57	1.54	1.51	1.48	1.47	1.43	1.42	1.41	1.39	1.38	0.10	40
1.92	1.84	1.79	1.74	1.69	1.66	1.64	1.59	1.58	1.55	1.53	1.51	0.05	
2.52	2.37	2.29	2.20	2.11	2.06	2.02	1.94	1.92	1.87	1.83	1.80	0.01	
1.60	1.54	1.51	1.48	1.44	1.41	1.40	1.36	1.35	1.33	1.31	1.29	0.10	60
1.84	1.75	1.70	1.65	1.59	1.56	1.53	1.48	1.47	1.44	1.41	1.39	0.05	
2.35	2.20	2.12	2.03	1.94	1.88	1.84	1.75	1.73	1.68	1.63	1.60	0.01	
1.55	1.48	1.45	1.41	1.37	1.34	1.32	1.27	1.26	1.24	1.21	1.19	0.10	120
1.75	1.66	1.61	1.55	1.50	1.46	1.43	1.37	1.35	1.32	1.28	1.25	0.05	
2.19	2.03	1.95	1.86	1.76	1.70	1.66	1.56	1.53	1.48	1.42	1.38	0.01	
1.52	1.46	1.42	1.38	1.34	1.31	1.28	1.24	1.22	1.20	1.17	1.14	0.10	200
1.72	1.62	1.57	1.52	1.46	1.41	1.39	1.32	1.29	1.26	1.22	1.19	0.05	
2.13	1.97	1.89	1.79	1.69	1.63	1.58	1.48	1.44	1.39	1.33	1.28	0.01	
1.49	1.42	1.38	1.34	1.30	1.26	1.24	1.18	1.17	1.13	1.08	1.00	0.10	∞
1.67	1.57	1.52	1.46	1.39	1.35	1.32	1.24	1.22	1.17	1.11	1.00	0.05	
2.04	1.88	1.79	1.70	1.59	1.52	1.47	1.36	1.32	1.25	1.15	1.00	0.01	

TABLE T3. Normal distribution probabilities

Z	Prob.	Z	Prob.	Z	Prob.	Z	Prob.	Z	Prob.	Z	Prob.	Z	Prob.	Z	Prob.
0.00	0.5000	0.31	0.3783	0.62	0.2676	0.93	0.1762	1.24	0.1075	1.55	0.0606	1.86	0.0314	2.85	0.00219
0.01	0.4960	0.32	0.3745	0.63	0.2643	0.94	0.1736	1.25	0.1056	1.56	0.0594	1.87	0.0307	2.90	0.00187
0.02	0.4920	0.33	0.3707	0.64	0.2611	0.95	0.1711	1.26	0.1038	1.57	0.0582	1.88	0.0301	2.95	0.00159
0.03	0.4880	0.34	0.3669	0.65	0.2578	0.96	0.1685	1.27	0.1020	1.58	0.0571	1.89	0.0294	3.00	0.00135
0.04	0.4840	0.35	0.3632	0.66	0.2546	0.97	0.1660	1.28	0.1003	1.59	0.0559	1.90	0.0287	3.05	0.00114
0.05	0.4801	0.36	0.3594	0.67	0.2515	0.98	0.1635	1.29	0.0985	1.60	0.0548	1.91	0.0281	3.10	0.00097
0.06	0.4761	0.37	0.3557	0.68	0.2483	0.99	0.1611	1.30	0.0968	1.61	0.0537	1.92	0.0274	3.15	0.00082
0.07	0.4721	0.38	0.3520	0.69	0.2451	1.00	0.1587	1.31	0.0951	1.62	0.0526	1.93	0.0268	3.20	0.00069
0.08	0.4681	0.39	0.3483	0.70	0.2420	1.01	0.1562	1.32	0.0934	1.63	0.0516	1.94	0.0262	3.25	0.00058
0.09	0.4641	0.40	0.3446	0.71	0.2389	1.02	0.1539	1.33	0.0918	1.64	0.0505	1.95	0.0256	3.30	0.00048
0.10	0.4602	0.41	0.3409	0.72	0.2358	1.03	0.1515	1.34	0.0901	1.65	0.0495	1.96	0.0250	3.35	0.00040
0.11	0.4562	0.42	0.3372	0.73	0.2327	1.04	0.1492	1.35	0.0885	1.66	0.0485	1.97	0.0244	3.40	0.00034
0.12	0.4522	0.43	0.3336	0.74	0.2296	1.05	0.1469	1.36	0.0869	1.67	0.0475	1.98	0.0239	3.45	0.00028
0.13	0.4483	0.44	0.3300	0.75	0.2266	1.06	0.1446	1.37	0.0853	1.68	0.0465	1.99	0.0233	3.50	0.00023
0.14	0.4443	0.45	0.3264	0.76	0.2236	1.07	0.1423	1.38	0.0838	1.69	0.0455	2.00	0.0228	3.55	0.00019
0.15	0.4404	0.46	0.3228	0.77	0.2206	1.08	0.1401	1.49	0.0823	1.70	0.0446	2.05	0.0202	3.60	0.00016
0.16	0.4364	0.47	0.3192	0.78	0.2177	1.09	0.1379	1.40	0.0808	1.71	0.0436	2.10	0.0179	3.65	0.00013
0.17	0.4325	0.48	0.3156	0.79	0.2148	1.10	0.1357	1.41	0.0793	1.72	0.0427	2.15	0.0158	3.70	0.00011
0.18	0.4286	0.49	0.3121	0.80	0.2119	1.11	0.1335	1.42	0.0778	1.73	0.0418	2.20	0.0139	3.75	0.00009

TABLE T3 (*continued*)

Z	Prob.	Z	Prob.	Z	Prob.	Z	Prob.	Z	Prob.	Z	Prob.	Z	Prob.	Z	Prob.
0.19	0.4247	0.50	0.3085	0.81	0.2090	1.12	0.1314	1.43	0.0764	1.74	0.0409	2.25	0.0122	3.80	0.00007
0.20	0.4207	0.51	0.3050	0.82	0.2061	1.13	0.1292	1.44	0.0749	1.75	0.0401	2.30	0.0107	3.85	0.00006
0.21	0.4168	0.52	0.3015	0.83	0.2033	1.14	0.1271	1.45	0.0735	1.76	0.0392	2.35	0.0094	3.90	0.00005
0.22	0.4129	0.53	0.2981	0.84	0.2005	1.15	0.1251	1.46	0.0721	1.77	0.0384	2.40	0.0082	3.95	0.00004
0.23	0.4090	0.54	0.2946	0.85	0.1977	1.16	0.1230	1.47	0.0708	1.78	0.0375	2.45	0.0071	4.00	0.00003
0.24	0.4052	0.55	0.2912	0.86	0.1949	1.17	0.1210	1.48	0.0694	1.79	0.0367	2.50	0.0062		
0.25	0.4013	0.56	0.2877	0.87	0.1922	1.18	0.1190	1.49	0.0680	1.80	0.0359	2.55	0.0054		
0.26	0.3974	0.57	0.2843	0.88	0.1894	1.19	0.1170	1.50	0.0668	1.81	0.0351	2.60	0.0047		
0.27	0.3936	0.58	0.2810	0.89	0.1867	1.20	0.1151	1.51	0.0665	1.82	0.0344	2.65	0.0040		
0.28	0.3897	0.59	0.2776	0.90	0.1841	1.21	0.1131	1.52	0.0643	1.83	0.0336	2.70	0.0035		
0.29	0.3859	0.60	0.2743	0.91	0.1814	1.22	0.1112	1.53	0.0630	1.84	0.0329	2.75	0.0030		
0.30	0.3821	0.61	0.2709	0.92	0.1788	1.23	0.1093	1.54	0.0618	1.85	0.0322	2.80	0.0026		

Percentage points

	Two-sided			One-sided		
Significance level	10% (0.10)	5% (0.05)	1% (0.01)	10% (0.10)	5% (0.05)	1% (0.01)
	1.64	1.96	2.58	1.28	1.64	2.33

TABLE T4. Coefficients of orthogonal polynomials

No. of levels	Coefficients	W_1	W_2	W_3	W_4	W_5	W_6	W_7	W_8	W_9	W_{10}	$\sum_i W_i^2 = S$
2	Linear	−1	1									2
3	Linear	−1	0	1								2
	Quadratic	1	−2	1								6
4	Linear	−3	−1	1	3							20
	Quadratic	1	−1	−1	1							4
	Cubic	−1	3	−3	1							20
5	Linear	−2	−1	0	1	2						10
	Quadratic	2	−1	−2	−1	2						14
	Cubic	−1	2	0	−2	1						10
	Quartic	1	−4	6	−4	1						70
6	Linear	−5	−3	−1	1	3	5					70
	Quadratic	5	−1	−4	−4	−1	5					84
	Cubic	−5	7	4	−4	−7	5					180
	Quartic	1	−3	2	2	−3	1					28
7	Linear	−3	−2	−1	0	1	2	3				28
	Quadratic	5	0	−3	−4	−3	0	5				84
	Cubic	−1	1	1	0	−1	−1	1				6
	Quartic	3	−7	1	6	1	−7	3				154
8	Linear	−7	−5	−3	−1	1	3	5	7			168
	Quadratic	7	1	−3	−5	−5	−3	1	7			168
	Cubic	−7	5	7	3	−3	−7	−5	7			264
	Quartic	7	−13	−3	9	9	−3	−13	7			616
	Quintic	−7	23	−17	−15	15	17	−23	7			2184
9	Linear	−4	−3	−2	−1	0	1	2	3	4		60
	Quadratic	28	7	−8	−17	−20	−17	−8	7	28		2772
	Cubic	−14	7	13	9	0	−9	−13	−7	14		990
	Quartic	14	−21	−11	9	18	9	−11	−21	14		2002
	Quintic	−4	11	−4	−9	0	9	4	−11	4		468
10	Linear	−9	−7	−5	−3	−1	1	3	5	7	9	330
	Quadratic	6	2	−1	−3	−4	−4	−3	−1	2	6	132
	Cubic	−42	14	35	31	12	−12	−31	−35	−14	42	8580
	Quartic	18	−22	−17	3	18	18	3	−17	−22	18	2860
	Quintic	−6	14	−1	−11	−6	6	11	1	−14	6	780

The sum of squares of the component effect (linear, quadratic, etc.) is given by

$$\frac{(W_1A_1 + \cdots + W_KA_K)^2}{r \times S}$$

where A_i is the sum total of the r observations in level i, $i = 1, \ldots, K$

Table T5. Duncan's new multiple range test coefficients for 1% significance level

Degrees of freedom \ No. of treatment means	2	3	4	5	6	7	8	9	10	12	14	16	18	20	50	100
1	90.0	90.0	90.0	90.0	90.0	90.0	90.0	90.0	90.0	90.0	90.0	90.0	90.0	90.0	90.0	90.0
2	14.0	14.0	14.0	14.0	14.0	14.0	14.0	14.0	14.0	14.0	14.0	14.0	14.0	14.0	14.0	14.0
3	8.26	8.5	8.6	8.7	8.8	8.9	8.9	9.0	9.0	9.0	9.1	9.2	9.3	9.3	9.3	9.3
4	6.51	6.8	6.9	7.0	7.1	7.1	7.2	7.2	7.3	7.3	7.4	7.4	7.5	7.5	7.5	7.5
5	5.70	5.96	6.11	6.18	6.26	6.33	6.40	6.44	6.5	6.6	6.6	6.7	6.7	6.8	6.8	6.8
6	5.24	5.51	5.65	5.73	5.81	5.88	5.95	6.00	6.0	6.1	6.2	6.2	6.3	6.3	6.3	6.3
7	4.95	5.22	5.37	5.45	5.53	5.61	5.69	5.73	5.8	5.8	5.9	5.9	6.0	6.0	6.0	6.0
8	4.74	5.00	5.14	5.23	5.32	5.40	5.47	5.51	5.5	5.6	5.7	5.7	5.8	5.8	5.8	5.8
9	4.60	4.86	4.99	5.08	5.17	5.25	5.32	5.36	5.4	5.5	5.5	5.6	5.7	5.7	5.7	5.7
10	4.48	4.73	4.88	4.96	5.06	5.13	5.20	5.24	5.28	5.36	5.42	5.48	5.54	5.55	5.55	5.55
11	4.39	4.63	4.77	4.86	4.94	5.01	5.06	5.12	5.15	5.24	5.20	5.34	5.38	5.39	5.39	5.39
12	4.32	4.55	4.68	4.76	4.84	4.92	4.96	5.02	5.07	5.13	5.17	5.22	5.24	5.26	5.26	5.26
13	4.26	4.48	4.62	4.69	4.74	4.84	4.88	4.94	4.98	5.04	5.08	5.13	5.14	5.15	5.15	5.15
14	4.21	4.42	4.55	4.63	4.70	4.78	4.83	4.87	4.91	4.96	5.00	5.04	5.06	5.07	5.07	5.07
15	4.17	4.37	4.50	4.58	4.64	4.72	4.77	4.81	4.84	4.90	4.94	4.97	4.99	5.00	5.00	5.00

16	4.13	4.34	4.45	4.54	4.60	4.67	4.72	4.76	4.79	4.84	4.88	4.91	4.93	4.94	4.94	4.94
17	4.10	4.30	4.41	4.50	4.56	4.63	4.68	4.73	4.75	4.80	4.83	4.86	4.88	4.89	4.89	4.89
18	4.07	4.27	4.38	4.46	4.53	4.59	4.64	4.68	4.71	4.76	4.79	4.82	4.84	4.85	4.85	4.85
19	4.05	4.24	4.35	4.43	4.50	4.56	4.61	4.64	4.67	4.72	4.76	4.79	4.81	4.82	4.82	4.82
20	4.02	4.22	4.33	4.40	4.47	4.53	4.58	4.61	4.65	4.69	4.73	4.76	4.78	4.79	4.79	4.79
22	3.99	4.17	4.28	4.36	4.42	4.48	4.53	4.57	4.60	4.65	4.68	4.71	4.74	4.75	4.75	4.75
24	3.96	4.14	4.24	4.33	4.39	4.44	4.49	4.53	4.57	4.62	4.64	4.67	4.70	4.72	4.74	4.74
26	3.93	4.11	4.21	4.30	4.36	4.41	4.46	4.50	4.53	4.58	4.62	4.65	4.67	4.69	4.73	4.73
28	3.91	4.08	4.18	4.28	4.34	4.39	4.43	4.48	4.51	4.56	4.60	4.62	4.65	4.67	4.72	4.72
30	3.89	4.06	4.16	4.22	4.32	4.36	4.41	4.45	4.48	4.54	4.58	4.61	4.63	4.65	4.71	4.71
40	3.82	3.99	4.10	4.17	4.24	4.30	4.34	4.37	4.41	4.46	4.51	4.54	4.57	4.59	4.69	4.69
60	3.76	3.92	4.03	4.12	4.17	4.23	4.27	4.31	4.34	4.39	4.44	4.47	4.50	4.53	4.66	4.66
100	3.71	3.86	3.98	4.06	4.11	4.17	4.21	4.25	4.29	4.35	4.38	4.42	4.45	4.48	4.64	4.65
∞	3.64	3.80	3.90	3.98	4.04	4.09	4.14	4.17	4.20	4.26	4.31	4.34	4.38	4.41	4.60	4.68

TABLE T5. Duncan's new multiple range test coefficients for 5% significance level

Degrees of freedom \ No. of treatment means	2	3	4	5	6	7	8	9	10	12	14	16	18	20	50	100
1	18.0	18.0	18.0	18.0	18.0	18.0	18.0	18.0	18.0	18.0	18.0	18.0	18.0	18.0	18.0	18.0
2	6.09	6.09	6.09	6.09	6.09	6.09	6.09	6.09	6.09	6.09	6.09	6.09	6.09	6.09	6.09	6.09
3	4.50	4.50	4.50	4.50	4.50	4.50	4.50	4.50	4.50	4.50	4.50	4.50	4.50	4.50	4.50	4.50
4	3.93	4.01	4.02	4.02	4.02	4.02	4.02	4.02	4.02	4.02	4.02	4.02	4.02	4.02	4.02	4.02
5	3.64	3.74	3.79	3.83	3.83	3.83	3.83	3.83	3.83	3.83	3.83	3.83	3.83	3.83	3.83	3.83
6	3.46	3.58	3.64	3.68	3.68	3.68	3.68	3.68	3.68	3.68	3.68	3.68	3.68	3.68	3.68	3.68
7	3.35	3.47	3.54	3.58	3.60	3.61	3.61	3.61	3.61	3.61	3.61	3.61	3.61	3.61	3.61	3.61
8	3.26	3.39	3.47	3.52	3.55	3.56	3.56	3.56	3.56	3.56	3.56	3.56	3.56	3.56	3.56	3.56
9	3.20	3.34	3.41	3.47	3.50	3.52	3.52	3.52	3.52	3.52	3.52	3.52	3.52	3.52	3.52	3.52
10	3.15	3.30	3.37	3.43	3.46	3.47	3.47	3.47	3.47	3.47	3.47	3.47	3.47	3.48	3.48	3.48
11	3.11	3.27	3.35	5.39	3.43	3.44	3.45	3.46	3.46	3.46	3.46	3.46	3.47	3.48	3.48	3.48
12	3.08	3.23	3.33	5.36	3.40	3.42	3.44	3.44	3.46	3.46	3.46	3.46	3.47	3.48	3.48	3.48
13	3.06	3.21	3.30	3.35	3.38	3.41	3.42	3.44	3.45	3.45	3.46	3.46	3.47	3.47	3.47	3.47
14	3.03	3.18	3.27	3.33	3.37	3.39	3.41	3.42	3.44	3.45	3.46	3.46	3.47	3.47	3.47	3.47

15	3.01	3.16	3.25	3.31	3.36	3.38	3.40	3.42	3.43	3.44	3.45	3.46	3.47	3.47	3.47	3.47
16	3.00	3.15	3.23	3.30	3.34	3.37	3.39	3.41	3.43	3.44	3.45	3.46	3.47	3.47	3.47	3.47
17	2.98	3.13	3.22	3.28	3.33	3.36	3.38	3.40	3.42	3.44	3.45	3.46	3.47	3.47	3.47	3.47
18	2.97	3.12	3.21	3.27	3.32	3.35	3.37	3.39	3.41	3.43	3.45	3.46	3.47	3.47	3.47	3.47
19	2.96	3.11	3.19	3.26	3.31	3.35	3.37	3.39	3.41	3.43	3.44	3.16	3.47	3.47	3.47	3.47
20	2.95	3.10	3.18	3.25	3.30	3.34	3.36	3.38	3.40	3.43	3.44	3.46	3.47	3.47	3.47	3.47
22	2.93	3.08	3.17	3.24	3.29	3.32	3.35	3.37	3.39	3.42	3.44	3.45	3.46	3.47	3.47	3.47
24	2.92	3.07	3.15	3.22	3.28	3.31	3.34	3.37	3.38	3.41	3.44	3.45	3.46	3.47	3.47	3.47
26	2.91	3.06	3.14	3.21	3.27	3.30	3.34	3.36	3.38	3.41	3.43	3.45	3.46	3.47	3.47	3.47
28	2.90	3.04	3.13	3.20	3.26	3.30	3.33	3.35	3.37	3.40	3.43	3.45	3.46	3.47	3.47	3.47
30	2.89	3.04	3.12	3.20	3.25	3.29	3.32	3.35	3.37	3.40	3.43	3.44	3.46	3.47	3.47	3.47
40	2.86	3.01	3.10	3.17	3.22	3.27	3.30	3.33	3.35	3.39	3.42	3.41	3.46	3.47	3.47	3.47
60	2.83	2.98	3.08	3.14	3.20	3.24	3.28	3.31	3.33	3.37	3.40	3.43	3.45	3.47	3.48	3.48
100	2.80	2.95	3.05	3.12	3.18	3.22	3.26	3.29	3.32	3.36	3.40	3.42	3.45	3.47	3.53	3.53
∞	2.77	2.92	3.02	3.09	3.15	3.19	3.23	3.26	3.29	3.34	3.38	3.41	3.44	3.47	3.61	3.67

APPENDIX F

Probability distributions

Some common probability distributions mentioned in this book are graphically depicted. Mathematical expressions for their density f, moments and moment-generating functions are given. (**E** denotes expectation, **V** variance.)

The contents of this Appendix (published here courtesy of the authors and publishers) have been adapted from Rothschild and Logothetis (1986) where more information about these and other probability distributions can be found.

Bernoulli distribution

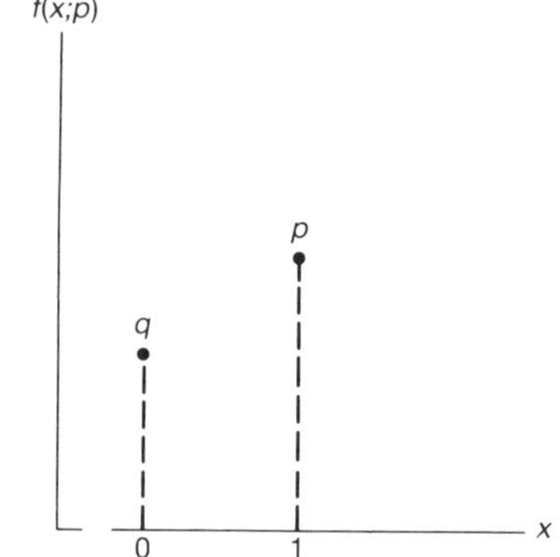

$$f(x;p) = \begin{cases} p^x q^{1-x}, & x = 0, 1; q = 1 - p \\ 0, & \text{otherwise} \end{cases}$$

$$\mathbf{E}(X) = p; \quad \mathbf{V}(X) = pq$$

$$\mu_r' = p, \qquad r \geq 1$$

$$\psi(t) = q + pe^t.$$

Binomial distributions

$$f(x;n,p) = \begin{cases} \binom{n}{x} p^x q^{n-x}, & x = 0, 1, 2, \ldots, n; 0 \leq q = 1 - p \leq 1 \\ 0, & \text{otherwise} \end{cases}$$

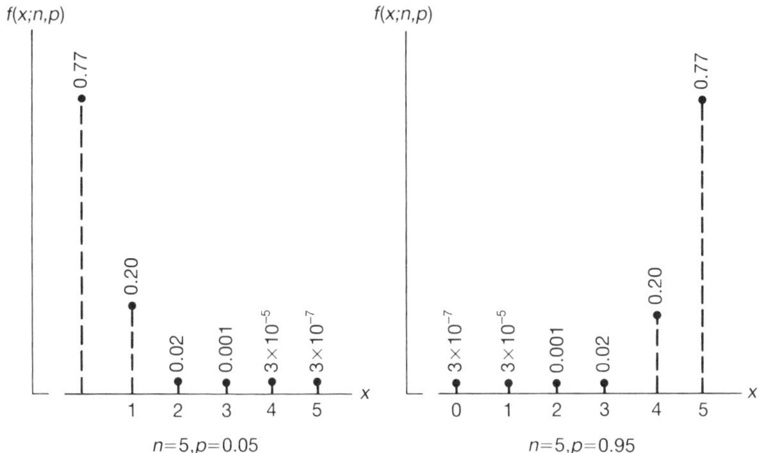

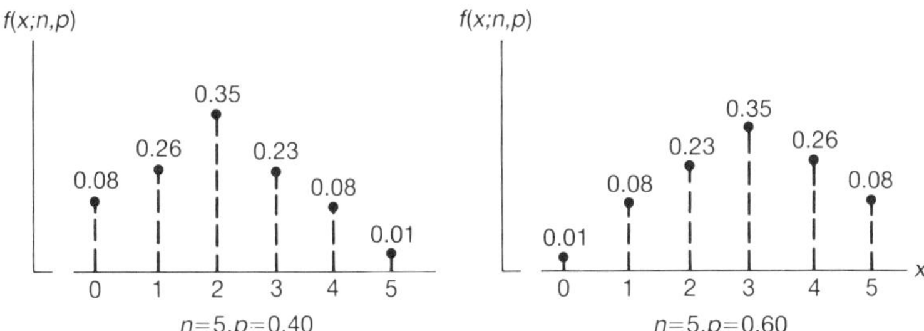

$$\mathbf{E}(X) = np; \qquad \mathbf{V}(X) = npq$$

$$\mu_3 = npq(q - p)$$

$$\mu_4 = 3n^2p^2q^2 + npq(1 - 6pq)$$

$$\psi(t) = (q + pe^t)^n.$$

Continuous Uniform or Rectangular distribution

$$f(x) = \begin{cases} \dfrac{1}{b-a}, & a < x < b \\ 0, & \text{otherwise} \end{cases}$$

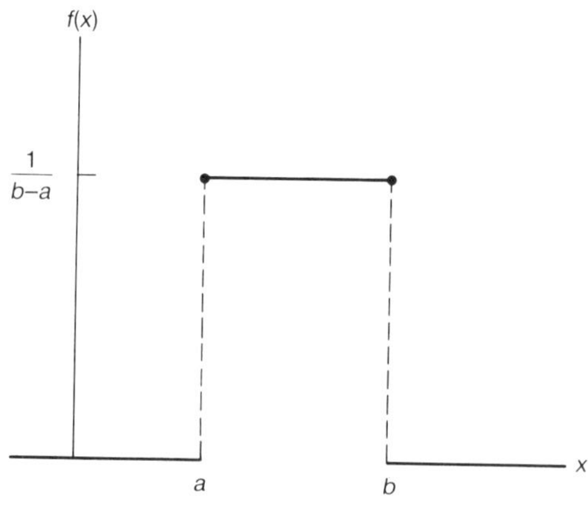

$$\mathbf{E}(X) = \frac{a+b}{2}; \qquad \mathbf{V}(X) = \frac{(b-a)^2}{12}$$

$$\mu_3 = 0$$

$$\mu_4 = \frac{(b-a)^4}{80}$$

$$\psi(t) = \frac{e^{tb} - e^{ta}}{t(b-a)}.$$

Normal (or Gaussian) distributions

$$f(x;\mu,\sigma) = \frac{1}{\sqrt{2\pi}\sigma}\exp[-(x-\mu)^2/2\sigma^2], \qquad -\infty < \mu < \infty;\ \sigma > 0$$

$$\mathbf{E}(X) = \mu; \qquad \mathbf{V}(X) = \sigma^2$$

$$\mu_r = \begin{cases} 0, & r \text{ odd} \\ \dfrac{r!\sigma^r}{(r/2)!2^{r/2}}, & r \text{ even} \end{cases}$$

$$\psi(t) = \exp[\mu t + (1/2)\sigma^2 t^2]$$

Note: If $\mu = 0$ and $\sigma = 1$, the Standard Normal distribution $\mathbf{N}(0,1) = \frac{1}{\sqrt{2\pi}}\exp[-(1/2)x^2]$ is obtained.

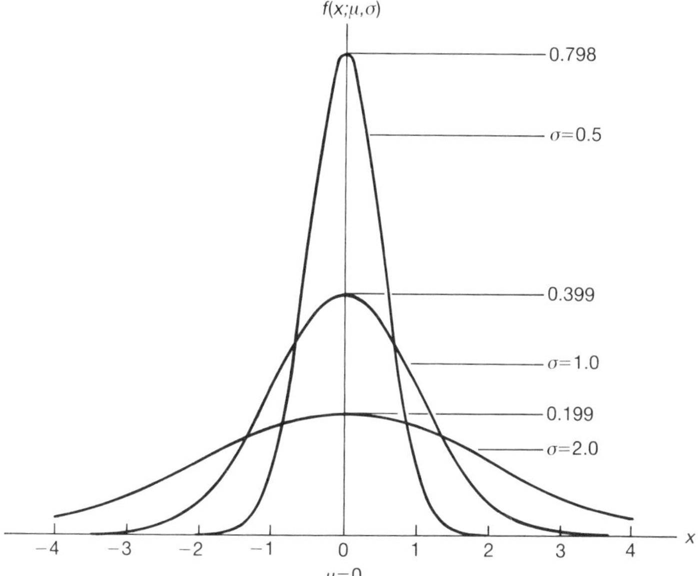

Student's t-distribution ($\nu = 4$ and 1) with the standard normal distribution N(0, 1) for comparison

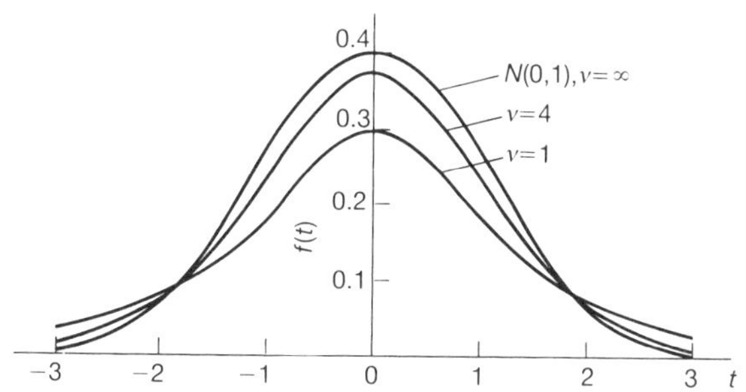

$$f(t) = \frac{\Gamma\left(\frac{\nu+1}{2}\right)}{\sqrt{\nu\pi}\,\Gamma(\nu/2)}\left(1+\frac{t^2}{\nu}\right)^{-(\nu+1)/2}, \qquad -\infty < t < \infty$$

$$\mathbf{E}(T) = 0,\ \nu > 1; \qquad \mathbf{V}(T) = \frac{\nu}{\nu-2},\ \nu > 2$$

$$\mu_r = \begin{cases} 0, & \nu > r,\ r \text{ odd} \\ \nu^{r/2} \dfrac{B[(r+1)/2, (\nu - r)/2]^*}{B(1/2, \nu/2)}, & \nu > r,\ r \text{ even} \end{cases}$$

$\psi(t)$ does not exist.

See Appendix F1 for properties of B[ ] (p. 438).

F distributions

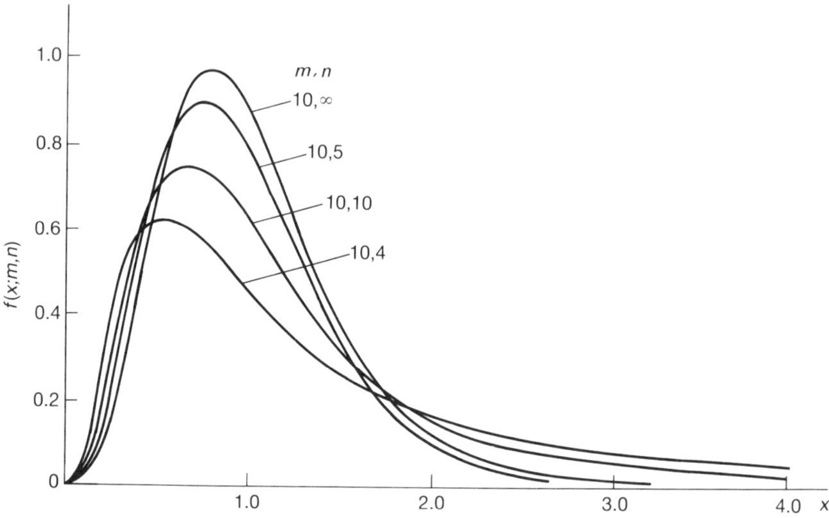

$$f(x; m, n) = \frac{\Gamma[(m+n)/2]}{\Gamma(m/2)\Gamma(n/2)} \left(\frac{m}{2}\right)^{m/2} \frac{x^{(m-2)/2}}{[1 + (m/n)x]^{(m+n)/2}}, \qquad m, n = 1, 2, \ldots$$

$$\mathbf{E}(X) = \frac{n}{n-2}, \quad n > 2; \quad \mathbf{V}(X) = \frac{2n^2(m+n-2)}{m(n-2)^2(n-4)}, \qquad n > 4$$

$$\mu_r' = \left(\frac{n}{m}\right)^r \frac{\Gamma(m/2 + r)\Gamma(n/2 - r)}{\Gamma(m/2)\Gamma(n/2)}, \qquad n/2 > r$$

$\psi(t)$ does not exist.

χ^2 Distributions for various degrees of freedom ν

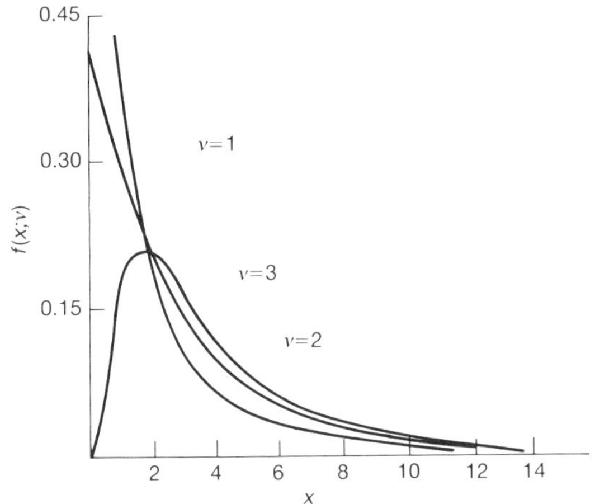

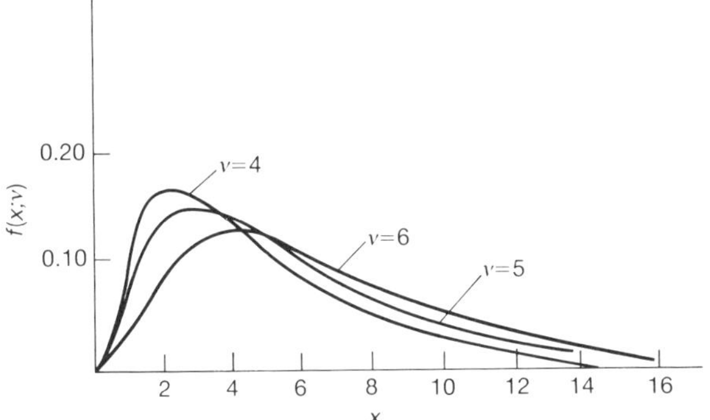

$$f(x;\nu) = \begin{cases} \dfrac{1}{2^{\nu/2}\Gamma^*(\nu/2)} x^{(\nu/2)-1} e^{-x/2}, & x > 0 \\ 0, & x \leq 0 \end{cases}$$

$$\mathbf{E}(X) = \nu; \qquad \mathbf{V}(X) = 2\nu$$

$$\mu_r' = \frac{2^r \Gamma[(\nu/2) + r]}{\Gamma(\nu/2)}, \qquad r \geq 1$$

$$\psi(t) = (1 - 2t)^{-\nu/2}.$$

See Appendix F1 for properties of Γ[ ] (p. 438).

Weibull distributions

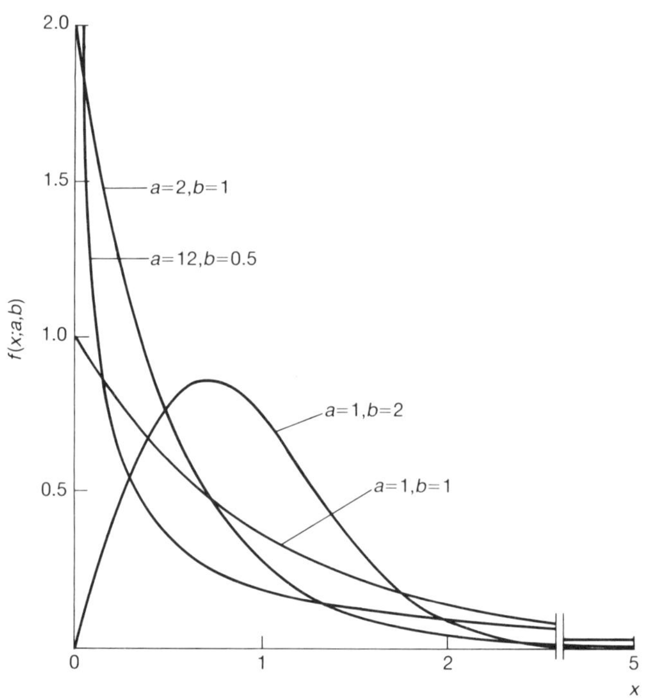

$$f(x;a,b)=\begin{cases} abx^{b-1}e^{-ax^b}, & x>0; a>0; b>0 \\ 0, & x\le 0 \end{cases}$$

$$\mathbf{E}(X)=(1/a)^{1/b}\Gamma(1+b^{-1}); \qquad \mathbf{V}(X)=a^{-2/b}[\Gamma(1+2b^{-1})-\Gamma^2(1+b^{-1})]$$

$$\mu_r'=a^{-r/b}\Gamma(1+r/b),\ r\ge 1$$

$$\psi(t)=a^{-t/b}\Gamma\left(1+\frac{t}{b}\right).$$

Lognormal distributions

$$f(x;\mu,\sigma)=\frac{1}{\sqrt{2\pi\sigma^2x^2}}\exp[-(\log x-\mu)^2/2\sigma^2], \qquad 0<x<\infty; -\infty<\mu<\infty; \sigma>0$$

$$\mathbf{E}(X)=\exp[\mu+1/2\sigma^2]; \qquad \mathbf{V}(X)=e^{2\mu}(e^{2\sigma^2}-e^{\sigma^2})$$

$$\mu_r'=\exp[1/2r^2\sigma^2+r\mu], \quad r\geqslant 1$$

$\psi(t)$, of no use.

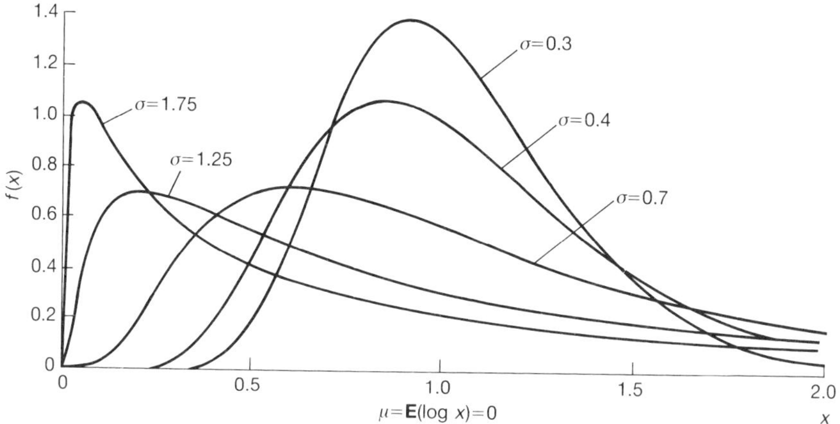
1.4
1.2
1.0
0.8
0.6
0.4
0.2
0
f(x)
σ=1.75
σ=1.25
σ=0.3
σ=0.4
σ=0.7
0
0.5
1.0
1.5
2.0
μ=E(log x)=0
x

APPENDIX F1

Some mathematical relationships

$$\mathbf{E}[g(x)] = \begin{cases} \sum_x g(x)f(x) & \text{discrete} \\ \int_x g(x)f(x)\,\mathrm{d}x & \text{continuous} \end{cases}$$

$$\mathbf{E}[g(x)] = \begin{cases} \mu'_r \text{ when } g(x) = X^r, \ (\mu'_1 = \mathbf{E}(X) = \mu), & r\text{th moment} \\ \mu_r \text{ when } g(x) = [X - \mathbf{E}(X)]^r, \ (\mu_2 = \mathbf{V}(X) = \sigma^2), & r\text{th central moment} \\ \psi(t) \text{ when } g(x) = \mathrm{e}^{tX}, & \text{moment generating function} \end{cases}$$

$$\frac{\mathrm{d}^r}{\mathrm{d}t^r}\psi(t)|_{t=0} = \mu'_r$$

$$\mu_r = \sum_{i=0}^{r} (-1)^i \binom{r}{i} (\mu'_1)^i \mu'_{r-i}$$

$$\mu'_r = \sum_{i=0}^{r} \binom{r}{i} (\mu)^i \mu_{r-i}$$

so that

$$\begin{aligned} &\mu'_0 = 1 && \mu_0 = 1 \\ &\mu'_1 = \mu && \mu_1 = 0 \\ &\mu'_2 = \mu_2 + \mu^2 && \mu_2 = \mu'_2 - \mu_1'^2 \\ &\mu'_3 = \mu_3 + 3\mu\mu_2 + \mu^3 && \mu_3 = \mu'_3 - 3\mu'_1\mu'_2 + 2\mu_1'^3 \\ &\mu'_4 = \mu_4 + 4\mu\mu_3 + 6\mu^2\mu_2 + \mu^4 && \mu_4 = \mu'_4 - 4\mu'_1\mu'_3 + 6\mu_1'^2\mu'_2 - 3\mu_1'^4 \end{aligned}$$

$$\mathbf{V}(X) = \mathbf{E}(X)^2 - [\mathbf{E}(X)]^2$$

$$\mathrm{cov}(X, Y) = \mathbf{E}[(X - \mu_x)(Y - \mu_Y)]$$

$$\sum_{i=1}^{n} x_i = x_1 + x_2 + \cdots + x_n; \ \prod_{i=1}^{n} x_i = x_1 \cdot x_2 \cdot \cdots \cdot x_n$$

$$\Gamma(n+1) = n\Gamma(n) = n!; \ \Gamma(1/2) = \sqrt{\pi} \qquad \Gamma(x+1) = \int_0^\infty \mathrm{e}^{-t} t^x \,\mathrm{d}t$$

$$B(p, q) = \int_0^1 x^{p-1}(1-x)^{q-1}\,\mathrm{d}x = \frac{\Gamma(p)\Gamma(q)}{\Gamma(p+q)}.$$

APPENDIX G

Glossary of terms

Alias effect An effect in a fractional-factorial design which 'looks like' or cannot be distinguished from another effect. An alternative terminology for a confounded effect (see Section 2.4).
Alpha (α) In decimal form it is the probability of rejecting a true hypothesis (see **Type I error**); in percentage form it is usually referred to as the 'size of the test' or 'level of significance' (Section 2.1).
Alpha error *see* **Type I error.**
Accumulating analysis An analysis technique favoured by Taguchi for analysing ranked data (Section 4.7).
Additive model A model in which the factors influencing the dependent variable have an additive effect, i.e. there are no curvature or interaction effects among these factors.
Allowance design The stage when expensive and difficult to change factors are studied (Section 6.2.3).
Alternative hypothesis In the theory of testing hypothesis, any admissible hypothesis alternative to the 'null hypothesis', i.e. to the one under test (Section 2.1.2).
Analysis of variance (ANOVA) The separation of the total variation displayed by a set of observations (as measured by the sums of squares of deviations from the mean) into components associated with defined sources of variations (such as certain controllable factors) (Section 2.6).
Arithmetic mean For a set of values $x_1, \ldots, x_n$, it is their sum divided by their size number n. It is often denoted by $\bar{x}$. For a continuous distribution with probability density function $f(x)$, it is defined as the integral

$$\int_{-\infty}^{\infty} xf(x)\,\mathrm{d}x$$

and is usually denoted by μ $[=E(x)]$. The word 'arithmetic' is usually omitted (see Section 2.1.1).

Balanced factorial design A factorial experimental design in which a level of a particular factor is replicated the same number of times as any other level of this factor, and this holds for all factors under consideration.
β-technique A simple technique for choosing a data-transformation with the objective of making the trial mean independent of the trial variance (Section 6.4.2).
Beta (β) In decimal form it is the probability of accepting a wrong hypothesis (see **Type II error**).
Beta-coefficients See **Discount coefficients**.
Between-groups variance In ANOVA it is the sum of squares of deviations of

the group means from the total (grand) mean of the observations, divided by the number of degrees of freedom (usually one less than the number of groups).

Bias Generally, an effect which deprives a statistical result of representativeness by systematically distorting it, as distinct from a random error which may distort on any one occasion but balances out on the average. Technically, the deviation of a parameter from the mean value of the estimator.

Biased sample A sample obtained by a biased sampling process.

Binary data A series of numerals, usually 0 and 1, representing two categories—live or die, effective or non defective, etc. (Section 4.6).

Block The name given to a group of items under treatment or observation. For example, a block may comprise a group of continuous plots of land, all the animals in a litter, the experimental results obtained by a single observer, etc. The general purpose of 'blocking' is to isolate sources of heterogeneity, i.e. to remove the 'nuisance' variation from the random (residual) variation. (See also **Randomized block design, Incomplete block design**.) Another reason for dividing into blocks is to construct fractional designs (see Section 2.4.1).

'Boot-strap' technique A resampling technique for estimating population parameters through successive sampling with replacement (Section 7.3.1).

Box–Cox technique A technique developed for choosing a data-transformation (see Section 6.4.1).

Central composite designs Experimental factorial designs which are supplemented by additional experimental points such as centre points and star points (Section 5.3).

Coefficient of determination (R^2) See **Determination.**

Coefficient of variation The ratio of the standard deviation to the mean (Section 6.2.2.3).

Completely randomized design A simple design in which all levels of a factor (treatments) are assigned to the experimental units in a completely random manner (Section 3.2.1).

Combining technique A technique for assigning two two-level factors on a single three-level column (Section 3.3.6.1b).

Compounding technique A technique for studying numerous responses of interest using a relatively small experimental design (Section 4.1.2).

Confidence interval An interval within which a parameter of interest is said to lie with a specified level of confidence. The extreme values of a confidence interval are called confidence limits (Section 2.1.1).

Confidence level The probability of the truth of a statement that a parameter value will lie within two confidence limits (Section 2.1.1).

Confounding An experimental arrangement in which certain effects cannot be distinguished from others (see Section 2.4).

Confounding technique A technique for studying effects totalling 'n' degrees of freedom with designs allowing for 'm' degrees of freedom where $m < n$ (Section 4.1.1).

Contribution ratio An estimate of the variability contribution of a source to the total variability of the experimental results (Section 4.3).

Contrast A linear combination of level totals or averages where the sum of all coefficients is zero (see Section 2.3).

Correction factor In ANOVA the square of the total sum of the values divided by the sample size (Section 2.6.3.2).

Correlation A measure of the interdependence between quantitative or qualitative data (Section B.4).

Correlation matrix For a set of variates $x_1, \ldots, x_n$ with correlations between x_i and x_j denoted by r_{ij}, the correlation matrix is the square matrix of values (r_{ij}).

Covariance The first product moment of two variates about their mean value (Section B.4).

Covariance matrix The square matrix (C_{ij}) where C_{ij} is the covariance of x_i and x_j, $i, j = 1, \ldots, n$, with the diagonal elements C_{ij} being the variances $\operatorname{var} x_i$, $i = 1, \ldots, n$. Alternative names are the dispersion matrix and the variance–covariance matrix.

Critical value The lowest of a set of values of a test statistic where the hypothesis under test is rejected (Section 2.1.2).

Defining contrast An expression that indicates which effects are to be confounded with blocks in a factorial design that is confounded. It is used in fractional designs for finding alias sets of effects (Section 2.4).

Degrees of freedom (df) The number of observations that can be varied independently of each other (Section 2.1.1).

Dependent variable Another word for the response variable or 'predictand' or 'regressand'. Usually the factor of interest whose variation we want to explain and control.

Design matrix The exhibition in matrix form of an experimental design (Section 3.3.1).

Determination, coefficient of For the case of two variables the square of the correlation between the two variables r^2. More generally, if a dependent variable has multiple correlation R with a set of independent variables, $\mathbf{R}^2$ is known as the coefficient of determination.

Discount (beta) coefficients Coefficients favoured by Taguchi for improving the long-term prediction (or estimation) of a process performance (Section 4.4.3).

Distribution $F(x)$ The proportion of members with values $\leq x$.

Dummy-level technique A technique for assigning a p-level factor to a q-level column when $p < q$, by repeating some of the actual factor levels (Section 3.3.6.1a).

Duncan's test A multiple comparison test (Section 2.6.3.5).

Effective number of replications See Section 4.4.2.

Effect (of a factor) The change in response produced by a change in factor levels.

Error, mean-square (or residual MS) The error sum of squares (the sum of squares of the contributions from the stochastic (random) components) divided by the number of degrees of freedom.

Error of first kind See **Type I error.**

Error of second kind See **Type II error.**

Error sum of squares (or residual SS) In ANOVA, when estimates are made of the source effects and subtracted from the observations, the residuals are estimates of the contribution (in the variation) of the random components. The sum of squares of these residuals are known as error or residual sum of squares (Section 2.6).
Estimation An inference about the numerical value of an unknown population parameter from incomplete data such as a sample.
Evolutionary operation A sequential experimental procedure for collecting information during production, to improve a process without disturbing production (Section 5.5.1).
'Expanding-shrinking' method A method for dealing with mixing experiments (Section 4.2.3.1).
Expected value of a statistic The average value of a statistic over repeated sampling from a given population. [Usual symbol $E(\ )$.]
'Experimental regression' analysis A technique favoured by Taguchi for fitting a model to a limited number of actual data values through a 'multivariate successive approximation' using orthogonal arrays (Section 4.8).

F-test A test based on the ratio of two independent statistics which are quadratic estimators of the parent variance. Widely employed in ANOVA to test the homogeneity of a set of means (Section 2.6).
Factor A variable selected for experimentation (Section 2.2.3).
Factorial experiment An experiment designed to examine the effect of one or more factors, each factor being applied at at least two levels so that different effects can be observed (Section 2.3). If the experimental design considers all possible level combinations of the factors under study, then the experiment (and the design) is called a full-factorial (Section 3.3.2).
Fractional (replication) design An experiment design in which usually, due to cost restrictions, only a fraction of a complete factorial is run. This is likely to be useful only when certain high-order interactions can be regarded as negligible (Section 2.4).
Frequency distribution A representation of a large set of data in compact form achieved by dividing the data-range into intervals and counting the frequency of the data values in each interval (Section 2.1.3.1).

Geometric mean The nth root of the product of the n observations (Appendix B1).
Gaussian distribution Another name for the normal distribution (Section 2.1.3.2).
Graeco-Latin square A design in which, for example, four factors are so arranged that each level of each factor is combined only once with each level of the other three factors (Section 3.2.4).

Half normal plot A graphical method for interpreting the factor effects in a factorial experiment (Section 2.6.4.2).
Harmonic mean The reciprocal of the average of the reciprocals of the observations (Appendix B1).

Histogram A graphic representation of a frequency distribution (Section 2.1.3.1).

Idle-column technique A technique, favoured by Taguchi, for assigning multi-level factors to two- or three-level arrays (Section 3.3.6.2b).
Incomplete block design A randomized block design in which not all treatment combinations can be included in one block. Such a design is called a 'balanced' design if each pair of treatments occurs together the same number of times (Section 3.2.2).
Independent variable A variable (or factor) of interest which can be controlled and is suspected of affecting a response (dependent) variable. (See also **Regression.)**
Interaction A measure of the extent to which dependent variable effects associated with changes in the level of one factor, depend on the levels of one or more factors.
Inter-experimental error The error between experimental trials. Otherwise called the 'first-order variance' (Section 2.6.3.3).

Jack-knife technique A resampling technique which operates by re-estimating the required parameters (mean or variance) several times, each time omitting a subset of the original data values.

Latin square An experimental design in which the allocation of k treatments in the cells of a k by k square is such that each treatment occurs exactly once in each row or column.
Least squares estimates Estimates determined through the application of the least squares method (Section 2.2.2).
Least squares method A technique of estimation by which the estimated quantities are determined by minimizing the squares of the deviations of observations from the model (Section 2.1.4).
Linear model A model in which the equations connecting the variables are in a linear form (Section 2.1.4).
Level of a factor One of the different values of a factor which can be used in an experiment. It may be measured qualitatively or quantitatively (Section 2.2.3).
Level of significance See **Alpha.**
Logit transformation A transformation of proportions p into decibels through

$$-10\log_{10}\left\{\frac{1}{p}-1\right\}$$

Otherwise known as 'omega method' (Section 4.6.3).

Main effect An estimate of the effect of an experimental variable or treatment measured independently of other treatments which may form part of the experiment (Section 2.3.3).

Mean Of a sample $x_1, \ldots, x_n$:

$$\bar{x} = \sum_{i=1}^{n} x_i/n.$$

Of a population $\mu = E(x)$ (Section 2.1.1 and Appendix B).

Mean (average) deviation A measure of dispersion derived from the average deviation of observations from some central value, such deviations being taken absolutely (Appendix B1).

Mean sums of squares (MS) An unbiased estimate of a population variability of a factor determined by dividing a sum of squares due to this factor by its degrees of freedom (Section 2.6.2).

Median That variate value which divides the total frequency into two halves (Appendix B1).

Median rank A crude estimate of the cumulative percentage of items under consideration up to a specific value (Section 2.1.3.4).

Mid-range For a set of values $x_1, \ldots, x_n$ arranged in order of magnitude, it is defined as $\frac{1}{2}(x_n + x_1)$ (Appendix B1).

Minute analysis A method favoured by Taguchi for analysing life-test data (Section 4.6.4).

Missing values Data values that have been lost as a result of carelessness or accident (Section 4.5).

Mixture experiments Experiments involving the mixing of various substances in percentages which have to add up to 100 (Section 4.2.3).

Mode The variate value with the highest frequency (Appendix B1).

Model A formalized expression of a causal situation which is regarded as having generated observed data.

Multiple correlation The correlation between the dependent variable and the independent variables in a multiple regression.

Multiple regression The regression of a dependent variable on more than one independent variable (Section 2.1.4).

'Multivariate successive approximation' See **'Experimental regression' analysis**.

Nested design (or 'hierarchical' design) A design in which the levels of one factor are chosen within the levels of another factor. Factors which are not nested are said to be crossed (Section 3.3.5.3).

Net variation An estimate of the net effect of a source (Section 4.3).

Noise A convenient term for a series of random disturbances borrowed from communication engineering. It represents the effects of uncontrollable factors (Section 6.2).

Noise performance measure (NPM) A statistical measure reflecting the variability in a response (Section 6.2.2.3).

Normal (or Gaussian) distribution The continuous frequency distribution of infinite range represented by the probability density function

$$f(x) = \frac{1}{\sqrt{2\pi}\,\sigma} \exp\{-\tfrac{1}{2}[(x-\mu)/\sigma]^2\} \quad -\infty < x < \infty$$

where μ is the mean and σ is the standard deviation (Section 2.1.3).

Normal probability paper A specially ruled graph paper with a variate x as abscissa and an ordinate y scaled in such a way that the graph of the cumulative distribution function, y, of the normal distribution, is a straight line (Section 2.1.3.4).
Null hypothesis The particular hypothesis under test (Section 2.1.2).

Off-line quality control Quality control applied before production, at the product-development stage, or during installation or recommissioning of a process (Section 1.1.3).
Omega factor A type of factor used in Taguchi's minute analysis (Section 4.6.4).
Omega method See **Logit transformation.**
One-factor-at-a-time method An experimentation method when a single factor is varied while all other factors are kept fixed at a specific set of conditions (Section 3.1.2).
One-sided test A test of a hypothesis for which the region of rejection is wholly located at one end of the distribution of the test statistic, e.g. to test for $\mu = \theta$ against the alternative that $\mu > \theta$ (Section 2.1.2).
On-line quality control Quality control applied during full production (see Section 1.4).
Optimality criteria Criteria for choosing an experimental design (Section 5.4).
Orthogonal contrasts Two contrasts for which the products of their corresponding coefficients add to zero (Section 2.2.3).
Orthogonal design A design in which certain variables or linear combinations of them, can be regarded as statistically independent. One of the reasons for introducing orthogonality into an experimental design is to enable main effects to be separately estimated. In an orthogonal array (of 'order 2'), for every pair of columns, every combination of levels appears the same number of times (Section 3.1.4).
Orthogonal Latin squares Latin squares which when they are superimposed on each other, every entry of one square (letter or numeral) occurs once and only once with every entry of the other (Section 3.2.4).
Outliers Observations which are so far separated in value from the rest of the observations in a sample, that they lead one to question whether they are from a different population, or whether the sampling technique is at fault.

Parameter In statistical theory usually this is the name used for a characteristic of a population, such as the population mean or variance. In industry, it is used to mean variables or factors affecting a process, e.g. process parameter.
Parameter design The stage when the best nominal settings of easy-to-control parameters are decided upon (Section 6.2.2).
Percent fit See **Determination, coefficient of**.
Partially supplemented design A design which is partially supplemented by additional trials so that more levels of a factor can be studied (Section 3.3.6.3).
Plackett and Burman designs These are special fractional factorial designs introduced by Plackett and Burman (1946) and are most useful for 'screening' experiments (Section 3.3.4 and Appendix C).

Planning out method A method for dealing with mixture experiments (Section 4.2.3.2).
Population A finite or infinite collection of individuals (not necessarily living organisms) (Section 2.1.1).
Probability The 'degree of belief' or the limiting frequency in an infinite random series.
Probability density function An alternative term for the frequency function when the distribution concerned is considered as one of probability (Section 2.1.3.2).
Probability distribution A distribution giving the probability of a value x as a function of x (Section 2.1.3.2).
Pseudo-factor A factor whose 'levels' or whose 'Type' changes according to the levels of another factor (Sections 3.3.5.3 and 3.3.6.2b).

Quality by design Designing quality into the product and process prior to the manufacturing stage (Section 6.1).

Random (selection) (sample) A process of selection applied to a set of objects giving to each one an equal chance of being chosen (Section 2.1.1).
Randomization When a set of objects or set of treatments applied to any given unit is chosen at random (Section 2.1.1).
Randomized block design An experimental design in which the treatment combinations are randomized within a block and several blocks are run (Section 3.2.2).
Range The largest minus the smallest of a set of variate values (Appendix B1).
Regression The process of establishing a relationship (linear or curvilinear) between a dependent variable Y and certain independent variables $x_1, \ldots, x_n$ (Section 2.1.4).
Regression coefficient The coefficient of an independent variable in a regression equation (Section 2.1.4).
Regression line/curve A diagrammatic exposition of a regression equation (Section 2.1.4).
Regression sum of squares The amount of variability in a model. The difference between the total sum of squares and the residual sum of squares (Section 2.1.4).
Replication A repeated observation within a trial (Section 2.2.3).
Replication error The experimental error among replications within a trial.
Resolution designs These are two-level fractional factorial designs, classified (by Box and Hunter 1961) according to their degree of fractionation into Resolution III, IV, and V designs (see Section 3.3.3).
Residual The difference between a theoretical response value (from a fitted response model) and the real (observed) value.
Residual sum of squares See **Error sum of squares**.
Response The reaction of an individual unit to some form of stimulus, due to the effect of controllable or uncontrollable factors (Section 2.1.4).
Response surface experimentation A study of the response surface $Y = f(x_1, x_2, \ldots, x_n)$ where Y is the response variable and $x_1, \ldots, x_n$ independent or controlled variables (Chapter 5).

Rotatable design A design that has equal predictiveness in all directions from a centre point (Section 5.3).
R-square See **Determination, coefficient of**.

Sample A part of a population (or a subset from a set of units) used to investigate properties of the parent population (Section 2.1.1).
Sample size The total number of observations in a sample (Section 2.1.1).
Screening variables Secondary variables included in a design only to investigate their main effects (Section 3.3.4).
Signal-factor See **Target-control factors**.
Signal-to-noise ratio (SNR) The inverse of the coefficient of variation (see Section 6.2.2.3).
Significance test A test using a test statistic which provides a test of the hypothesis that the effect is absent (Section 2.1.2).
Simple regression A regression with only one independent variable (Section 2.1.4).
Size of test *See* **Alpha**.
Split-plot design A design in which additional or subsidiary treatments are introduced by dividing each plot into two or more portions. For example, the division of experimental plots into halves enables an additional factor or treatment to be included at two levels.
Split-unit design A terminology introduced by Taguchi as equivalent to split-plot designs used in situations when there are certain restrictions on randomization in the order of experimentation, e.g. when there are factors with levels which are difficult to change (Section 3.3.5.4).
Standard deviation A measure of dispersion, the square root of the variance (Section 2.1.1).
Standard error of a statistic The standard deviation of the sampling distribution of the statistic (Section 2.1.4).
Standardization If x is a variable with mean μ and standard deviation σ then

$$y = \frac{x - \mu}{\sigma}$$

is the standardization of x. (The standardized x, y, has zero mean and unit standard deviation) (Section 2.1.3.3).
Statistic A measure computed from a sample, such as the sample mean or sample variance.
Statistics Decision making in the light of uncertainty!
Statistical inference Inferring something about a population from a sample of that population (Section 2.1).
Steepest ascent method A method that directs the experimenter towards the maximum of a response surface (Section 5.1) by always moving upwards.
Stepwise regression A regression method which selects the most significant variables first.
System design The stage when the prototype for a product or process is developed (Section 6.2.1).

Target-control factors (TCF) The controllable factors that affect the mean response (Section 6.2.2.1). Also known as 'signal-factors'.

Target performance-measure (TPM) A statistical measure reflecting the average response (Section 6.2.2.3).

Tolerance design A procedure for rational specification of components tolerances (Section 6.2.3 and Chapter 7).

***t*-test** A test based on the distribution known as t-distribution (or 'Student's t'). Can be used to test for significance of variables (Section 2.1.2).

Test of hypothesis A rule by which a hypothesis is accepted or rejected (Section 2.1.2).

Test statistic A statistic used to test a hypothesis (Section 2.1.2).

Transfactor technique A technique for constructing nested factorial designs (Section 3.3.5.3).

Treatment combination A given combination showing the levels of all factors to be run for that set of experimental conditions.

Trial A particular setting of levels of the different factors under study (Section 2.2.3).

Two-sample *t*-test A significance test for comparing two population means (Section 2.1.2.1).

Two-sided test A test for which the rejection region comprises areas of both extremes of the sampling distribution of the test function, e.g. to test for $\mu = \theta$ against the alternative that $\mu \neq \theta$ (i.e. either $\mu > \theta$ or $\mu < \theta$) (Section 2.1.2).

Type I error The error of rejecting a true hypothesis. Alternative to α-error or error of the first kind or producer's risk (Section 2.1.2).

Type II error The error of accepting a false hypothesis. Alternative to β-error or error of the second kind or consumer's risk (Section 2.1.2).

Unbiased estimator A statistic whose expected value equals the parameter it is estimating. (Appendix B).

Variability control factor (VCF) A controllable factor that affects the variability in the response (Section 6.2.2.1). Also known simply as 'control factor'.

Variable Any quantity which varies and may take any one of a specified set of values.

Variance The mean of the squares of deviations from the arithmetic mean (see Section 2.1.1).

Youden square A special incomplete Latin square (Section 3.2.5).

APPENDIX H

References

Books on 'Design of experiments'

Anderson, V. L. and McLean, R. A. (1974). *Design of experiments: a realistic approach.* Dekker, New York.

Bannerjee, K. S. (1975). *Weighing designs: For chemistry, medicine, operations research, statistics.* Dekker, New York.

Barker, T. B. (1985). *Quality experimental design.* Dekker, New York; ASQC Quality Press, Milwaukee.

Bose, R. C. and Manvel, B. (1984). *Introduction to combinatorial theory.* Wiley, New York.

Box, G. E. P. and Draper, N. R. (1969). *Evolutionary operation. A statistical method for process improvement.* Wiley, New York.

Box, G. E. P. and Draper, N. R. (1987). *Empirical model building and response surfaces.* Wiley, New York.

Box, G. E. P., Hunter, W. G., and Hunter, J. S. (1978). Statistics for Experimenters. *An introduction to design, data analysis, and model building.* Wiley, New York.

Caulcutt, R. (1982). *Statistics for research and development.* Chapman & Hall, London.

Cochran, W. G. and Cox, G. M. (1957). *Experimental designs* (2nd edn). Wiley, New York.

Cornell, J. A. (1981). *Experiments with mixtures. Designs, models, and the analysis of mixture data.* Wiley, New York.

Cox, D. R. (1958). *Planning of experiments.* Wiley, New York.

Daniel, C. (1976). *Applications of statistics to industrial experimentation,* Wiley, New York.

Davies, O. L. (ed.) (1956). *The design and analysis of industrial experiments* (2nd edn). Oliver & Boyd, Edinburgh. (Now Longman, Harlow).

Davies, O. L. and Goldsmith, P. L. (1972) *Statistical methods for research and production.* Oliver & Boyd, Edinburgh (Now Longman, Harlow).

Dey, A. (1985). *Orthogonal fractional factorial designs.* Wiley Eastern, New Delhi.

Dey, A. (1986). *Theory of block designs.* Wiley Eastern, New Delhi.

Federer, W. T. (1955). *Experimental design. Theory and applications.* Macmillan, New York.

Federov, V. V. (1972) *Theory of optimal designs* (translated and edited by W. J. Studden and E. M. Klimko). Academic Press, New York.

Finney, D. J. (1955, 1974). *Experimental design and its statistical basis*. University of Chicago Press, Chicago, and Cambridge University Press, Cambridge.

Finney, D. J. (1960). *The theory of experimental design*. University of Chicago Press, Chicago.

Fisher, R. A. (1925). *Statistical methods for research workers*. Oliver & Boyd, London.

Fisher, R. A. (1966). *Design of experiments*. (8th edn). Oliver & Boyd, Edinburgh. (now Longman, Harlow.)

Fisher, R. A. and Yates, F. (1982). *Statistical tables for biological, agricultural and medical research* (6th reprint of 6th edn, revised and enlarged). Longman, Harlow.

Hicks, C. R. (1982). *Fundamental concepts of design of experiments*. (3rd edn). Holt, Rinehart and Winston, New York.

John, J. A. (1987). *Cyclic designs*. Chapman & Hall, London.

John, J. A. and Quenouille, M. H. (1977). *Experiments: Design and analysis* (2nd edn). Griffin, London.

John, P. W. M. (1971). *Statistical design and analysis of experiments*. Macmillan, New York.

John, P. W. M. (1980). *Incomplete block designs*. Dekker, New York.

Kempthrone, O. (1952). *The design and analysis of experiments*. Wiley, New York.

Kushner, H. J. and Clark, D. S. (1978). *Stochastic approximation for constrained and unconstrained systems*. Springer-Verlag, New York.

Pazman, A. C. (1980). *Foundation of optimum experimental design*. Reidel, Berlin.

Plackett, R. L. (1960). *Principles of regression analysis*. Clarendon Press, Oxford.

Quenouille, M. H. (1953). *Design and analysis of experiment*. Griffin, London.

Raghavarao, D. (1971). *Constructions and combinatorial problems in designs of experiments*. Wiley, New York.

Raktoe, B. L., Hedayat, A., and Federer, W. T. (1981). *Factorial designs*. Wiley, New York.

Scheffe, H. (1959). *Analysis of variance*. Wiley, New York.

Silvey, S. D. (1980). *Optimal design. An Introduction to the theory for parameter estimation*. Chapman & Hall, London.

Snee, R. D., Hare, L. B., and Trout, J. R. (1985). *Experiments in industry. Design, analysis, and interpretation of results*. American Society for Quality Control, Milwaukee, Wisconsin.

Street, A. P. and Street, D. J. (1987). *Combinatorics of experimental design*. Oxford University Press, Oxford.

Vajda, S. (1967). *The mathematics of experimental design. Incomplete block designs and Latin squares*. Griffin, London.

Yates, F. (1937). *Design and analysis of factorial experiments,* Technical Communication No. 35. Commonwealth Bureau of Soils, Harpenden, England.

Yates, F. (1970). *Experimental design. Selected papers of Frank Yates*. Griffin, London.

Papers on 'Design of experiments'

Addelman, S. (1961). Irregular fractions of the 2^n-factorial experiments. *Technometrics* **3,** 479–96.

Addelman, S. (1962). Orthogonal main-effect plans for asymmetrical factorial experiments. *Technometrics* **4,** 21–4.

Atkinson, A. C. (1982). Developments in the design of experiments. *International Statistical Review* **50,** 161–77.

Atwood, G. L. (1969). Optimal and efficient design of experiments. *Annals of Mathematical Statistics* **40,** 1570–602.

Barnett, E. H. (1960). Introduction to evolutionary operation. *Industrial and Engineering Chemistry* **52,** 500.

Baumert, L. D., Golumb, S. W., and Hall, M. Jun. (1962). Discovery of an Hadamard matrix of order 92. *Bulletin of the American Mathematical Society* **68,** 237–38.

Box, G. E. P. (1957). Evolutionary operation: A method for increasing industrial productivity: *Applied Statistics* VI, No. 2

Box, G. E. P. and Behnken, D. W. (1960). Some new three-level designs for the study of quantitative variables. *Technometrics* **2,** 455–76.

Box, G. E. P. and Draper, N. R. (1959). A basis for the selection of a response surface design. *Journal of the American Statistical Association* **53,** 622–54.

Box, G. E. P. and Draper, N. R. (1975). Robust designs. *Biometrica* **62,** 347–52.

Box, G. E. P. and Hunter, J. S. (1957). Multifactor experimental design for exploring response surfaces. *Annals of Mathematical Statistics* **28,** 195–241.

Box, G. E. P. and Hunter, J. S. (1961). The 2^{K-P} fractional factorial designs. *Technometrics* **3,** 311–52.

Box, G. E. P. and Wilson, K. B. (1951). On the experimental attainment of optimum conditions. *Journal of the Royal Statistical Society Ser. B* **13,** 1–45.

Cook, R. D. and Nachtsheim, C. J. (1980). A comparison of algorithms for constructing D-optimal designs. *Technometrics* **22,** 315–24.

Daniel, C. (1959). Use of half-normal plots in interpreting factorial two-level experiments. *Technometrics* **1,** 311–41.

Dean, A. M. and John, J. A. (1975). Single replicate factorial experiments in generalized cycle designs: (I) Symmetrical arrangements, *Journal of the Royal Statistical Society Ser. B* **37,** 72–76. (II) Asymmetrical arrangements. *Journal of the Royal Statistical Society Ser. B* **37,** 63–71.

Finney, D. J. (1945). Fractional replication of factorial arrangements. *Annals of Eugenics* **12,** 291–301.

Ford, I., Titterington, D. M., and Wu, C. J. F. (1985). Inference and sequential design. *Biometrika* **72,** 545–53.

Hedayat, A. and Wallis, W. D. (1978). Hadamard matrices and their applications. *Annals of Statistics* **6,** 1184–238.

Hunter, J. S. (1985). Statistical design applied to product design. *Journal of Quality Technology* **17,** 210–21.

Kiefer, J. (1984). *Collected papers,* Vol. III, *Design of experiments.* Springer-Verlag, New York.

Kiefer, J. and Wolfowitz, J. (1952). Stochastic estimation of the maximum of a regression function. *Annals of Mathematical Statistics* **23,** 402–66.

Lai, T. L. (1985). Stochastic approximation and sequential search for optimum. *Proceedings of the Berkeley Conference,* Vol. II, pp. 557–77. Wadsworth.

Ljung, L. (1977). Analysis of recursive stochastic algorithms. *IEEE Trans. Aut. Contr.* **AC-22,** 551–75.

Mead, R. and Pike, D. J. (1975). A review of response surface methodology from a biometric view-point. *Biometrics* **31,** 803–51.

Mitchell, T. J. (1974). An algorithm for the construction of D-optimal experimental designs. *Technometrics* **16,** 203–10.

Mitchell, T. J., Sacks, J., Welch, W. J., and Wynn, H. P. (1989). Design and analysis of computer experiments (DACE). *Statistical Science* (to appear).

Nelson, L. S. (1982). Analysis of two-level factorial experiments. *Journal of Quality Technology* **14,** 95–8.

Plackett, R. L. and Burman, J. P. (1946). The design of optimum multifactorial experiments. *Biometrika* **33,** 305–25.

Rao, C. R. (1947). Factorial experiments derivable from combinatorial arrangements of arrays. *Journal of the Royal Statistical Society* (*Suppl.*) **9,** 128–39.

Robbins, H. and Munro, S. (1951). A stochastic approximation method. *Annals of Mathematical Statistics* **22,** 400–7.

Sacks, J., Schiller, S., and Welch, W. J. (1988). Design for computational experiments. *Technometrics* **30.**

Singhal, K. and Pinel, J. F. (1981). Statistical design centering and tolerancing using parametric sampling. *IEEE Transactions of Circuits and Systems* **CS-28,** 692.

Snee, R. D. (1971). Design and analysis of mixture experiments. *Journal of Quality Technology* **3,** 159–69.

Snee, R. D. (1985). Computer aided design of experiments—some practical experiences. *Journal of Quality Technology* **17,** 222–36.

Tippett, L. H. C. (1935). Some applications of statistical methods to the study of variation of quality in the production of cotton yarns. *Journal of the Royal Statistical Society* (*Suppl.*) **II,** 27–55.

Turyn, R. J. (1973). The computation of certain Hadamard matrices. *Notices of the American Mathematical Society* **20,** A-1.

Vuchkov, I. N. and Boyadjieva, L. N. (1988). The robustness against tolerances of performance characteristics described by second order polynomials. In *Optimal design and analysis of experiments.* North Holland, Amsterdam.

Welch, W. J. (1982). Branch-and-bound search for experimental designs based on D-optimality and other criteria. *Technometrics* **24,** 41–8.

Welch, W. J. (1983). A mean squared error criterion for the design of experiments. *Biometrika* **70,** 205–13.

Welch, W. J. (1984). Computer-aided design of experiments for response estimation. *Technometrics* **26,** 217–24.

Wu, C. J. F. (1985). Asymptotic inference from sequential design in a nonlinear situation. *Biometrika* **72,** 553–58.

Wu, C. J. F. and Wynn, H. P. (1978). The convergence of general step-length algorithms for regular optimum design criteria. *Annals of Statistics* **6,** 1273–85.

Youden, W. J. and Hunter, J. S. (1955). Partially replicated latin squares. *Biometrics* **11,** 399–405.

Zelen, M. (1959). Factorial experiments in life-testing. *Technometrics* **1,** 269–88.

Sources of 'Taguchi information'

Barker, T. B. (1986*a*). Simulation by experimental design: A Taguchi Concept. *ASQC Quality Congress Transactions* pp. 452–9.

Barker, T. B. (1986*b*). Quality engineering by design; Taguchi's philosophy. *Quality Progress,* December 1986, pp. 32–42.

Barker, T. B. and Clausing, D. P. (1984). Quality by design: The Taguchi method. Presented at the 40th annual ASQC conference.

Basso, L., Winterbottom, A., and Wynn, H. P. (1986). Review of the Taguchi methods for off-line quality control. *Quality and Reliability Engineering International* **2,** 71–9.

Bisgaard, S. and Box, G. E. P. (1988). *A practical aid for experiments.* Centre for Quality and Productivity Improvement. University of Wisconsin, Madison, Wisconsin.

Box, G. E. P. (1986). *Studies in quality improvement: Signal to noise ratios, Performance criteria and statistical analysis: Part* 1. Report No. 11, Center for Quality and Productivity Improvement, University of Wisconsin, Madison, Wisconsin.

Box, G. E. P. (1988). Signal-to-noise ratios, performance criteria and transformations (with discussion). *Technometrics* **30** (No. 1), 1–40.

Box, G. E. P. and Fung, C. A. (1986). *Studies in quality improvement minimizing transmitted variation by parameter design.* Report No. 8, Center for Quality and Productivity Improvement, University of Wisconsin, Madison, Wisconsin.

Box, G. E. P. and Jones, S. (1986). *An investigation of the method of accumulation analysis.* Report No. 19, Center for Quality and Productivity Improvement, University of Wisconsin, Madison, Wisconsin.

Box, G. E. P. and Meyer, R. D. (1986*a*). An analysis for unreplicated fractional factorials. *Technometrics* **28** (No. 1), 11–18.

Box, G. E. P. and Meyer, R. D. (1986*b*). Dispersion effects from fractional designs. *Technometrics* **29** (No. 1), 19–27.

Box, G. E. P. and Ramirez, J. (1986). *Studies in quality improvement: signal to noise ratios, performance criteria and statistical analysis; Part II.* Report No. 12, Center for Quality and Productivity Improvement, University of Wisconsin, Madison, Wisconsin.

Box, G. E. P., Bisgaard, S., and Fung, C. (1988) An explanation and critique of Taguchi's contributions to quality engineering. *Quality and Reliability Engineering International* **4** (No. 2), 123–31.

Burgam, P. M. (1985). Design of experiments—The Taguchi way. *Manufacturing Engineering* May 1985, 44–7.

Byrne, D. M. and Taguchi, Shin (1986). The Taguchi approach to parameter design. *ASQC Quality Congress Transactions* 168–77.

Dunn, J. (1987). Cure the effect, not the cause. *The Engineer* **14,** 22–3.

Fuchs, E. (1986). Quality—theory and practice. *A.T.&T. Technical Journal* **65** (No. 2), 4–8.

Ganter, W. A. (1988). Quality in design. *Quality and Reliability. Engineering International* **4** (No. 1) 4–6.

Gunter, B. (1987). A perspective on the Taguchi methods. *Quality Assurance* **13** (No. 3), 81–7.

Hamada, M. and Wu, C. F. J. (1986). *A critical look at accumulation analysis and related methods.* Report No. 20, Center for Quality and Productivity Improvement, University of Wisconsin, Madison, Wisconsin.

Illumoka, A. and Spence, D. (1988). Parameter tolerance design for electrical circuits. *Quality and Reliability Engineering International* **4** (No. 2), 87–94.

Kackar, R. N. (1985). Off-line quality control, parameter design and the Taguchi method. *Journal of Quality Technology* **17** (No. 4), 176–88. Discussion, pp. 189–209.

Kackar, R. N. and Shoemaker, A. C. (1986). Robust design: A cost effective method for improving manufacturing processes. *A.T.&.T. Technical Journal* **65** (No. 2), 39–50.

Leon, R. V., Shoemaker, A. C. & Kackar, R. N. (1987). Performance measures independent of adjustment: An explanation and extension of Taguchi's signal to noise ratio (with discussion). *Technometrics* **29** (No. 3), 253–85.

Logothetis, N. (1987*a*). Off-line quality control with prior exploration of data. *GEC Journal of Research* **5** (No. 1), 40–8.

Logothetis, N. (1987*b*). Off-line quality control and ill-designed data. *Quality and Reliability Engineering International* **3** (No. 4), 227–38.

Logothetis, N. (1988). The role of data-transformation in the Taguchi analysis. *Quality and Reliability Engineering International* **4** (No. 1), 49–61.

Logothetis, N. and Haigh, A. (1987). The statistical flexibility of the Taguchi method in the optimization of multi-response processes. *Professional Statistician* **6** (7), 10–16.

Logothetis, N. and Haigh, A. (1988). Characterizing and optimizing multi-response processes by the Taguchi method. *Quality and Reliability Engineering International* **4** (No. 2), 159–69.

Meyer, R. D. (1987). *Further details of an analysis of unreplicated fractional factorials.* Report No. 22, Center for Quality and Productivity Improvement, University of Wisconsin, Madison, Wisconsin.

Meyer, R. D. and Box, G. E. P. (1987). *Identification of active factors in unreplicated fractional factorial experiments.* Report No 23, Center for Quality and Productivity, Improvement, University of Wisconsin, Madison, Wisconsin.

Nair, V. N. (1986). Testing in industrial experiments with ordered categorical data. *Technometrics* **28,** 283–91.

Nair, V. N. and Pregibon, D. (1986). A data-analysis strategy for quality engineering experiments. *A.T.&T. Technical Journal* **65** (No. 3), 73–84.

Pao, T. W., Phadke, M. S. and Sherrerd, C. S. (1985). *Computer response time optimization using orthogonal array experiments.* Proceedings of ICC, IEEE Communications Society (28.2), pp. 890–5.

Phadke, M. S. (1982). *Quality engineering using design of experiments.*

Proceedings of the American Statistical Association, Section on Statistical Education, pp. 11–20.

Phadke, M. S. (1986). Design optimization case studies. *A.T.&T. Technical Journal* **65** (No. 2), 51–68.

Phadke, M. S. and Dehnad, K. (1988). Optimization of product and process design for quality and cost. *Quality and Reliability Engineering International* **4** (No. 2), 105–12.

Phadke, M. S., Kackar, R. N., Speeney, D. V., and Grieco, M. J. (1983). Off-line quality control in integrated circuit fabrication using experimental design. *Bell System Technical Journal* **62** (No. 5), 1273–309.

Pignatiello, J. J. and Ramberg, J. S. (1985). Discussion on Kackar's paper. *Journal of Quality Technology* **17** (No. 4), 198–209.

Quinlan, J. (1985). *Product improvement by application of Taguchi methods.* Third Supplier Symposium on Taguchi Methods, ASI Inc., Dearborn, Michigan.

Shoemaker, A. C. and Kackar, R. N. (1988). A methodology for planning experiments in robust product and process design. *Quality and Reliability Engineering International* **4** (No. 2), 95–103.

Song, J. and Lawson, J. (1988). Use of 2^{k-p} designs in parameter design. *Quality and Reliability Engineering International* **4** (No. 2), 151–58.

Spence, R. (1984). The tolerance analysis and design of electronic circuits. *Computer Aided Engineering Journal* **1,** Part 2, 91–9.

Sullivan, L. P. (1984). Reducing variability: A new approach to quality. *Quality Progress* July 1984, pp. 15–21.

Taguchi, G. (1978*a*). *Off-line and on-line quality control systems,* pp. B4-1 to B4-5 ICQC 1978, Tokyo.

Taguchi, G. (1978*b*) Performance Analysis Design. *International Journal of Production Research* **16** (No. 6), 521–30.

Taguchi, G. (1981). *On-line quality control during production.* Japanese Standards Association.

Taguchi, G. (1985). Quality engineering in Japan. *Communications in Statistics Theory and Methods,* Japanese Standards Association, **14** (Part II), 22785–85–801.

Taguchi, G. (1986). *Introduction of quality engineering.* ASI Center for Taguchi Methods 1987. Asian Productivity Organization, and UNIPUB/Krams International Publications.

Taguchi, G. (1987). *Systems of experimental design* (Vols 1 and 2, 1976, 1977, with 1987 translation published by UNIPUB).

Taguchi, G. and Konishi, S. (1987). *Taguchi methods: Orthogonal arrays and linear graphs; tools for quality engineering.* ASI Inc., Dearborn, Michigan.

Taguchi, G. and Phadke, M. S. (1984). *Quality engineering through design optimization.* (IEEE Globecom 1984 Conference, Atlanta, Georgia, Vol. 3, pp. 1106–13.)

Taguchi, G. and Wu, Y.-I. (1980). *Introduction of off-line quality control.* Central Japan Quality Control Association, ASI Inc., 32100 Detroit Industrial Expressway, Romulus, Michigan 48174.

Wu, C. F. J., Mao, S. S. and Ma, F. S. (1987). *An investigation of OA-based methods for parameter design optimization*, Report No. 24. Center for Quality and Productivity Improvement, University of Wisconsin, Madison, Wisconsin.

Sources of general information

Abramowitz, M. and Stegun, A. (1968) *Handbook of mathematical functions*. Dover, New York.

Atkinson, A. C. (1986). *Plots, transformations and regression: Introduction to graphical methods of diagnostic regression analysis*. Oxford University Press, Oxford.

Barlow, R. E. and Proschan, F. (1965). *Mathematical theory of reliability*. Wiley, New York.

Bartlett, M. S. (1947). The use of transformations. *Biometrics* **3,** 39–52.

Bartlett, M. S. and Kendall, D. G. (1949). The statistical analysis of variance—heterogeneity and the logarithmic transformation. *Journal of the Royal Statistical Society* (*Suppl.*) **8,** 128–38.

Box, G. E. P. and Cox, D. R. (1964). An analysis of transformations (with discussion). *Journal of the Royal Statistical Society, Ser. B* **26,** 211–46.

Box, G. E. P. and Tidwell, P. W. (1962). Transformations of the independent variables. *Technometrics* **4,** 531–50.

Carroll, R. J. and Ruppert, D. (1982). Robust estimation in heterscedastic linear models. *Annals of Statistics* **10,** 429–41.

Casiati, F. and Faravelli, L. (1985). Methods on non-linear stochastic dynamics for assessment of structure fragility. *Nuclear Engineering Design* **90,** 341–66.

Chatfield, C. (1983). *Statistics for technology*. Chapman & Hall, London.

Cook, R. D. and Weisberg, S. (1982). *Residuals and influence in regression*. Chapman & Hall, London.

Curtiss, J. H. (1943). On transformations used in the analysis of variance. *Annals of Statistics* **14,** 107–22.

Deming, W. E. (1982). *Quality, productivity and competitive position*. Massachusetts Institute of Technology Center for Advanced Engineering Study, Cambridge, Massachusetts.

Deming, W. E. (1986). *Out of the crisis*. MIT Press, Cambridge, Massachusetts.

Draper, N. R. and Smith, H. (1981). *Applied regression analysis*. Wiley, New York.

Duncan, D. B. (1956). Multiple range and multiple F-tests. *Biometrics* No. 11.

Eder, W. E. (ed.) (1987). *Proceedings of the* 1987 *International Conference on Engineering Design* (2 vols). ICED 87, American Society of Mechanical Engineers, New York.

Gitlow, S. and Gitlow, H. (1987). *The Deming guide to quality and competitive position*. Prentice-Hall, Englewood Cliffs, New Jersey.

Guttman, I. (1981). *Statistical tolerance regions, classical and Bayesian*. Griffin, London.

Hammersley, J. M. and Handscomb, D. C. (1964). *Monte-Carlo methods.* Methuen, London.

Ishikawa, K. (1976). *Guide to quality control.* Asian Productivity Organization, Tokyo.

Ishikawa, K. (1978). *Quality control in Japan.* International Academy for Quality, Kyota, Japan.

Jebb, A. and Wynn, H. P. (1989). Robust engineering design, post-Taguchi. *Philosophical Transactions of the Royal Society,* **327,** 605–16.

Joiner, B. L. (1985). The key role of statisticians in the transformation of North American industry. *American Statistician* **39** (No. 3).

Joiner, B. L. and Scholtes, P. R. (1986). The quality manager's new job. *Quality Progress* **19,** 10.

Juran, J. M. (1964). *Managerial breakthrough,* McGraw-Hill, New York.

Mann, N. R. (1985). *The keys to excellence—The story of Deming philosophy.* Prestwick Books, Los Angeles, California.

Moheiman, F. and Wynn, H. P. (1988). *Statistical methods in tolerance design I: the use of cumulant expansion.* (Submitted for publication.)

Nassif, S. R., Strojwas, A. J., and Director, S. W. (1984). Fabrics II: A statistically based IC fabrication process simulator. *IEEE Transactions on Computer-Aided Design* **CAD-3,** 40–6.

Neave, H. (1987). Deming's 14 Points for management: Framework for success. *Statistician* **36** (No. 5).

Niederreiter, H. (1978). Quasi-Monte Carlo methods and pseudo-random numbers. *Bulletin of the American Mathematical Society* **84,** 957–1041.

Pahl, G. and Beitz, W. (1984). *Engineering design.* The Design Council, London.

Perry, J. N. (1987). Iterative improvement of a power transformation to stabilize variance. *Applied Statistics, Journal of the Royal Statistical Society* C **36,** 15–21.

Ripley, B. D. (1987). *Stochastic simulation.* Wiley, New York.

Rabins, M. J. (1986). *Goals and priorities for research in engineering design, A report.* American Society of Mechanical Engineers.

Rothschild, V. and Logothetis, N. (1986). *Probability distributions.* Wiley, New York.

Scherkenbach, W. W. (1986). *The Deming route to quality and productivity—Road maps and road blocks.* CEEP Press, Washington DC.

Shewhart, W. A. (1925). The application of statistics as an aid in maintaining quality of a manufactured product. *Journal of the American Statistical Association* **20,** 546–48.

Shewhart, W. A. (1931). *Economic control of quality of manufactured products.* Van Nostrand, New York; Republished by the American Society of Quality Control (1980).

Snedecor, G. W. and Cochran, W. G. (1980). *Statistical methods* (7th edn). The Iowa State University Press, Ames, Iowa.

Spotts, M. F. (1983). *Dimensioning and tolerancing for quantity production.* Prentice-Hall, Englewood Cliffs, New Jersey.

Thoft-Christensen, P. and Baker, M. J. (1982). *Structural reliability theory and its applications.* Springer-Verlag, New York.

Tippett, L. H. C. (1934). Application of statistical methods to the control of quality in industrial production. *Journal of Manchester Statistical Society*.

Walton, M. (1986). *The Deming management method*. Dodd, Mead, New York.

Wheeler, D. and Chambers, D. (1986). *Understanding statistical process control*. Statistical Process Controls Inc., Knoxville, Tennessee.

INDEX